The ARRL 1986- Extra Class License Manual for the Radio Amateur

Edited By

Larry D. Wolfgang, WA3VIL
Bruce S. Hale, KB1MW

Contributors

Mark J. Wilson, AA2Z
Bruce O. Williams, WA6IVC
Paul K. Pagel, N1FB
Charles L. Hutchinson, K8CH

Production Staff

Deborah Strzeszkowski
David Pingree
Steffie Nelson, KA1IFB
Joel Kleinman, N1BKE
Sue Fagan, Cover Design
Michelle Chrisjohn, WB1ENT
Leslie Bartoloth, KA1MJP

American Radio Relay League
Newington, CT 06111 USA

This work is publication No. 60 of the Radio Amateur's Library, published by the League.

Printed in USA

ISBN: 0-87259-017-8

$5.00 in USA
$6.00 elsewhere

Second edition
Third printing

When To Use This Book

In mid-1987, a committee of Volunteer Examiner Coordinators (VECs) agreed on a three-year revision cycle for question pools from which Amateur Radio examinations are designed. Under the current schedule, the question pool for the Element 4B (Extra class) exam will be revised during 1987 and 1988.

The revised Element 4B pool is scheduled to be released to VECs and publishers of Amateur Radio study material on March 1, 1988 for use in Extra class examinations administered beginning November 1, 1988. Therefore, this **Second Edition Extra Class License Manual may be used to prepare for Extra class exams administered before November 1, 1988.**

Do not use this book to study for Extra class exams administered after November 1, 1988.

Foreword

Welcome to the second edition of *The ARRL Extra Class License Manual for the Radio Amateur.* The material in this manual has been carefully organized to provide a logical progression of topics, and to follow the FCC study guide for Element 4B as closely as possible. The text has been written in a manner specifically designed to teach the electronics theory and Amateur Radio techniques required to pass the Extra Class exam. Each drawing has been carefully selected and rendered to illustrate clearly the more vexing concepts you will come across as you study. Your understanding of the concepts is tested along the way by directing you to appropriate sections of the Element 4B question pool. By the time you have studied the entire *Manual*, you will be familiar with the topics covered by all of those questions.

The examinations for Amateur Radio licenses are given by Volunteer Examiners, using questions taken from the Element 4B question pool, published in advance. This procedure protects the integrity of the examination process and ensures that every applicant will receive a fair exam: You need not be afraid that the questions will take you by surprise, if you have prepared by studying and being familiar with all of the material in this book.

Chapter 1 includes the complete FCC Syllabus for the Element 4B examination, which is the one you must pass to upgrade from an Advanced class license. If you don't have an Advanced class ticket, it also explains what other examination elements you will need to pass. Chapter 10 lists all of the more than 400 questions in the question pool released in April 1986, along with the multiple-choice answers and distractors that will be used by many Volunteer Examiners until November 1988. Between those two chapters you will find complete explanations of every electronics topic covered on the Extra Class exam. Reference material for the questions on FCC regulations is found in the ARRL publication, *The FCC Rule Book*, which contains a complete copy of Part 97, the FCC Rules for Amateur Radio, and full explanations to help you decipher the legalese. You should have a copy of that book as a companion to this one; together they provide all the information you need to pass the written test.

League publications represent a real team effort. Writing, editing, typesetting, preparing technical illustrations and final layout all require special talents. The people performing those jobs put a lot of themselves into our publications to bring you the best possible books. Within our Publications Office the Technical Department editors and secretarial staff; the Production Department copy editors, technical illustrators, typesetters, layout staff and proofreaders; and the sales and shipping staffs of our Circulation Department have all been involved in helping you obtain the proper material to study for your Extra Class license exam. We couldn't possibly list all the names on the title page, but every one of us wishes you the very best on your exam. We think you will find the new privileges to be well worth the effort. We are proud of our work, and we hope you are proud to say you are a League member.

David Sumner, K1ZZ

Executive Vice President, ARRL
Newington, Connecticut

Table of Contents

HOW TO USE THIS BOOK

To earn an Extra Class Amateur Radio license, you will have to know some rather advanced electronics theory, and the rules and regulations governing the Amateur Radio Service, as contained in Part 97 of the FCC Rules. You'll also have to be able to send and receive the international Morse code at a rate of 20 WPM. This book provides a brief description of the Amateur Radio Service, and the Amateur Extra Class license in particular. The major portion of the book is designed to guide you, step by step, through the theory that you must know to pass your amateur license exam. The material is presented in a manner that closely follows the FCC study guide, or syllabus, printed at the end of Chapter 1. Chapter 10 contains the complete set of 400 questions used on Element 4B exams. It also includes the multiple-choice answers and distractors that will be used by many VECs, including the ARRL VEC, on these exams.

At the beginning of each chapter, you will find a list of key words that appear in that chapter, along with a simple definition for each word or phrase. As you read the text, you will find these words printed in *italic* type the first time they appear. You may want to refer back to the beginning of the chapter at that point, to learn the definition. At the end of the chapter, you may also want to go back and review those definitions.

As you study the material, you will be instructed to turn to sections of the questions in Chapter 10. Be sure to use these questions to review your understanding of the material at the suggested times. This will break the material into bite-sized pieces, and make it easier for you to learn. Do not try to memorize all 400 questions. That will be impossible! Instead, by using the questions for review, you will be familiar with them when you take the test, but you will also understand the electronics theory behind the questions.

In addition to this book, you will want to purchase a copy of the ARRL publication, *The FCC Rule Book*, which covers all the rules and regulations you'll need to know. You will probably need to increase your code speed, and ARRL also offers a complete set of cassette tapes to help you with that. Even with the tapes, you'll want to tune in to the code practice sessions transmitted by W1AW, the ARRL Headquarters station. When you are almost able to copy the code at 20 WPM, you may find it helpful to listen to code at 30 or 35 WPM at the beginning of your practice session, and then decrease the speed to 25 and 20 WPM. The W1AW fast code-practice sessions are transmitted in this order to aid your learning. For more information about W1AW or how to order any ARRL publication or code tape, write to: ARRL Hq., 225 Main St., Newington, CT 06111.

Chapter 1

The Extra Class License

Every Amateur Radio operator probably thinks about working toward his or her Amateur Extra Class license at one time or another. It certainly is a worthy goal to work toward! Many amateurs hesitate to actually take the exam, however, because they think they can't reach the 20 WPM code speed, or they think the theory is much too hard for them to understand. Once you make the commitment to study and learn what it takes to pass the exam, however, you will be able to do it. It often takes more than one try to pass the exam, but many amateurs do succeed the first time. The key is that you must make the commitment, and be willing to study.

There are many good Morse code training techniques, including the ARRL code tapes, W1AW code practice and even some computer programs. So with this book, carefully designed to teach the required electronics theory, *The FCC Rule Book* published by the ARRL, and plenty of code practice, you will soon have an Amateur Extra license. Whether you now hold a Novice, Technician, General or Advanced license, or even if you don't have any license yet, you will find the exclusive operating privileges available to an Amateur Extra Class licensee to be worth the time spent learning about your hobby. After passing the FCC Element 4B exam, you will be able to operate on every frequency band assigned to the Amateur Radio Service. There are segments of the 80, 40, 20 and 15-meter CW and phone bands reserved exclusively for Amateur Extra Class operators. So if you find the other portions of the bands getting too crowded, just move down to these less-used segments.

IF YOU'RE A NEWCOMER TO AMATEUR RADIO

Earning an Amateur Radio license, at whatever level, is a special achievement. The half a million or so people in the U.S. who call themselves Amateur Radio operators, or hams, are part of a global fraternity. Radio amateurs serve the public as a voluntary, noncommercial, communication service, especially during natural disasters or other emergencies. Hams continue to make important contributions to the field of electronics. Amateur Radio experimentation is yet another reason many people become part of this self-disciplined group of trained operators, technicians and electronics experts—an asset to any country. Hams pursue their hobby purely for personal enrichment in technical and operating skills, without consideration of any type of payment.

Because radio signals do not know territorial boundaries, hams have a unique ability to enhance international goodwill. A ham becomes an ambassador of his country every time he puts his station on the air.

Amateur Radio has been around since before World War I, and hams have always been at the forefront of technology. Today hams relay signals through their own satellites in the OSCAR (Orbiting Satellite Carrying Amateur Radio) series, bounce signals off the moon, and use any number of other exotic communications techniques. Amateurs talk from hand-held transceivers through mountaintop repeater stations that can relay their signals to transceivers in other hams' cars or homes. Hams send their own pictures by television, chat with other hams around the world by voice or, keeping alive a distinctive traditional skill, tap out messages in Morse code. When emergencies arise, radio amateurs are on the spot to relay information to and from disaster-stricken areas that have lost normal lines of communication.

The U.S. government, through the Federal Communications Commission (FCC), grants all U.S. Amateur Radio licenses. This licensing procedure ensures operating skill and electronics know how. Without this skill, radio operators might unknowingly cause interference to other services using the radio spectrum because of improperly adjusted equipment or neglected regulations.

Who Can Be a Ham?

The FCC doesn't care how old you are or whether you're a U.S. citizen: If you pass the examination, the Commission will issue you an amateur license. Any person (except the agent of a foreign government) may take the exam, and, if successful, receive an amateur license. It's important to understand that if a citizen of a foreign country receives an amateur license in this manner, he or she is a U.S. Amateur Radio operator. (This should not be confused with reciprocal licensing, which allows visitors from certain countries who hold valid amateur licenses in their homelands to operate their own stations in the U.S. without having to take an FCC exam.)

Licensing Structure

By examining Table 1-1, you'll see that there are five amateur license classes.

Table 1-1
Amateur Operator Licenses†

Class	*Code Test*	*Written Examination*	*Privileges*
Novice	5 WPM (Element 1A)	Elementary theory and regulations (Element 2)	Telegraphy in 3700-3750, 7100-7150 and 21,100-21,200 kHz with 200 watts PEP output maximum; telegraphy and RTTY on 28,100-28,300 kHz and telegraphy and SSB voice on 28,300-28,500 kHz with 200 W PEP max; all amateur modes authorized on 222.1-223.91 MHz, 25 W PEP max; all amateur modes authorized on 1270-1295 MHz, 5 W PEP max.
Technician	5 WPM (Element 1A)	Elementary theory and regulations; general-level theory and regulations. (Elements 2 and 3A)	All amateur privileges above 50.0 MHz plus Novice privileges.
General	13 WPM (Element 1B)	Elementary theory and regulations; general theory and regulations. (Elements 2, 3A and 3B)	All amateur privileges except those reserved for Advanced and Amateur Extra Class; see Table 1-2.
Advanced	13 WPM (Element 1B)	General theory and regulations, plus intermediate theory. (Elements 2, 3A, 3B and 4A)	All amateur privileges except those reserved for Amateur Extra Class; see Table 1-2.
Amateur Extra	20 WPM (Element 1C)	General theory and regulations, intermediate theory, plus special exam on advanced techniques. (Elements 2, 3A, 3B, 4A and 4B)	All amateur privileges

†A licensed radio amateur will be required to pass only those elements that are not included in the examination for the amateur license currently held.

Table 1-2
Amateur Operating Privileges

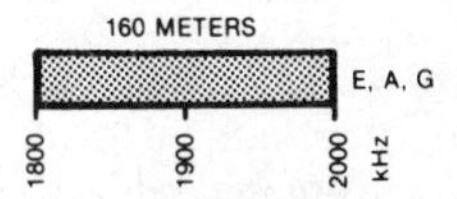

Amateur stations operating at 1900-2000 kHz must not cause harmful interference to the radiolocation service and are afforded no protection from radiolocation operations; see January 1986 Happenings for details.

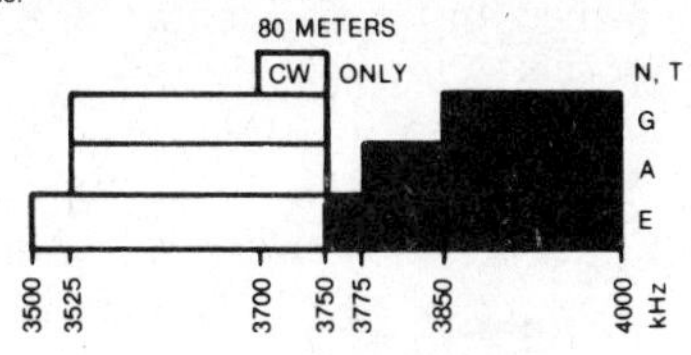

5167.5 kHz Alaska emergency use only. (SSB only)

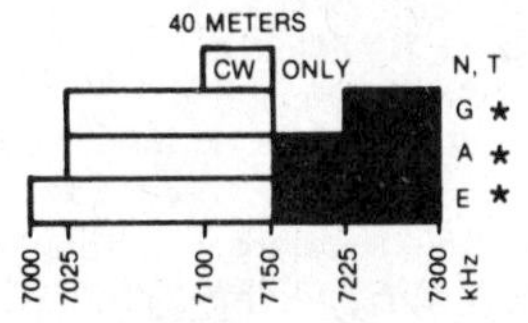

★ Phone operation is allowed on 7075-7100 kHz in Puerto Rico, US Virgin Islands and areas of the Caribbean south of 20 degrees north latitude; and in Hawaii and areas near ITU Region 3, including Alaska.

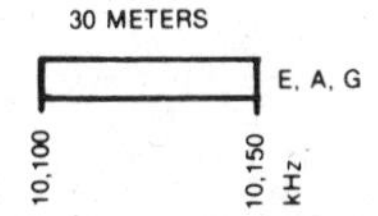

Maximum power limit on 30 meters is 200 watts PEP output. Amateurs must avoid interference to the fixed service outside the US.

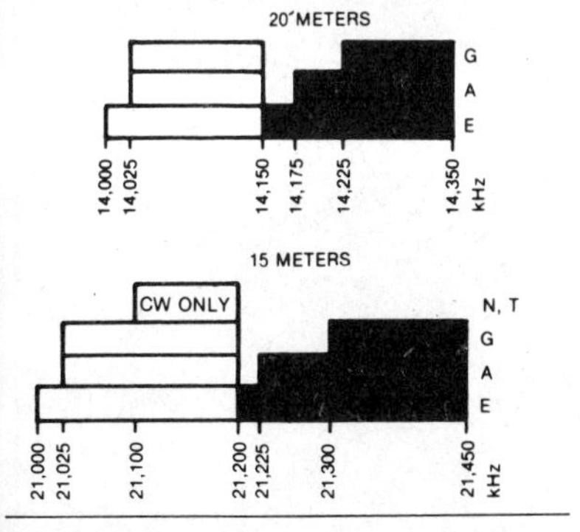

Consult *The ARRL Repeater Directory* for information on recommended operating frequencies and band plans for the bands above 50 MHz.

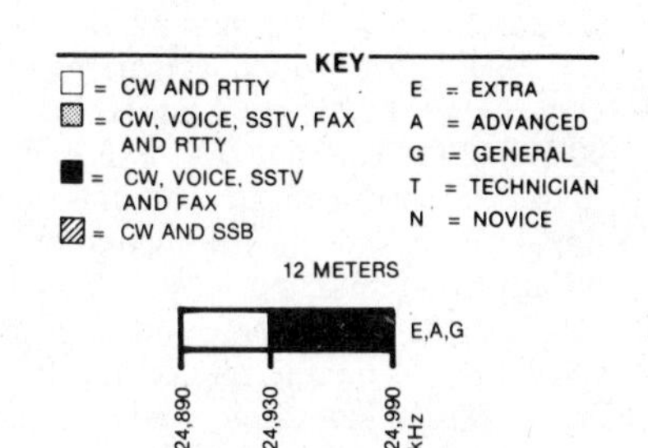

Amateurs must avoid interference to the fixed service outside the US.

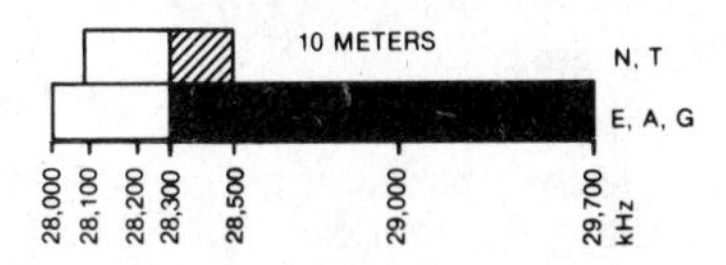

Novices and Technicians are limited to 200 watts on 10 meters.

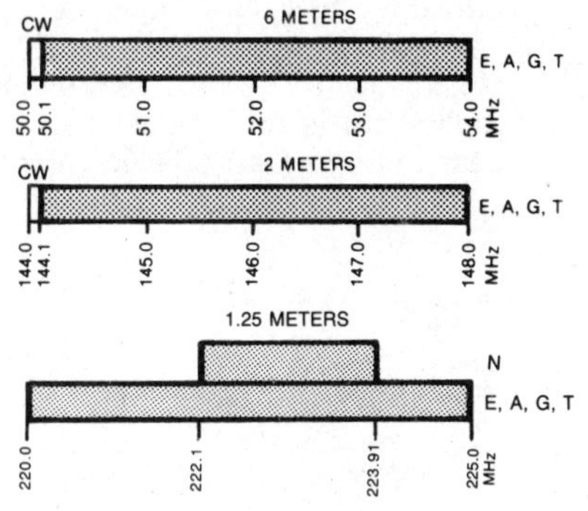

Novices are limited to 25 watts PEP on 222.1 to 223.91 MHz

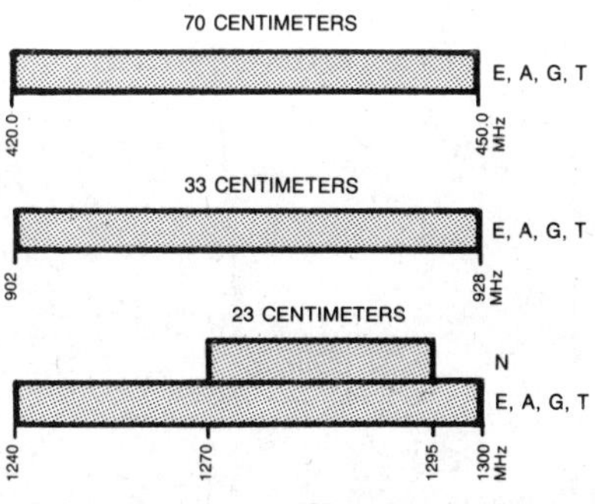

Novices are limited to 5 watts PEP on 1270 to 1295 MHz.

Each class has its own requirements and privileges. The FCC requires proof of your ability to operate an amateur station properly. The required knowledge is in line with the privileges of the license you hold. Higher license classes require more knowledge—and offer greater operating privileges. The specific operating privileges for Amateur Extra Class licensees are shown on the charts in Table 1-2. As you upgrade your license class, you must pass more-challenging written examinations.

In addition, you must demonstrate an ability to receive international Morse code at 5 WPM for Novice and Technician, 13 WPM for General and Advanced and 20 WPM for Amateur Extra. It's important to stress that although you may intend to use voice rather than code, this doesn't excuse you from the code test. By international treaty, knowing the international Morse code is a basic requirement for operating on any amateur band below 30 MHz.

Learning the Morse code is a matter of practice. Instructions on learning the code, how to handle a telegraph key, and so on can be found in the ARRL *Tune in the World with Ham Radio* package. This package includes two code-teaching cassettes for beginners. Additional cassettes for code practice at speeds of 5, 7½, 10, 13, 15 and 20 WPM are available from the American Radio Relay League, Newington, CT 06111.

Station Call Signs

Many years ago, by international agreement, the nations of the world decided to allocate certain call-sign prefixes to each country. This means that if you hear a radio station call sign beginning with W or K, for example, you know the station is licensed by the United States. A call sign beginning with the letter U is licensed by the USSR, and so on.

International Telecommunication Union (ITU) radio regulations outline the basic principles used in forming amateur call signs. According to these regulations, an amateur call sign must be made of one or two characters (the first one may be a

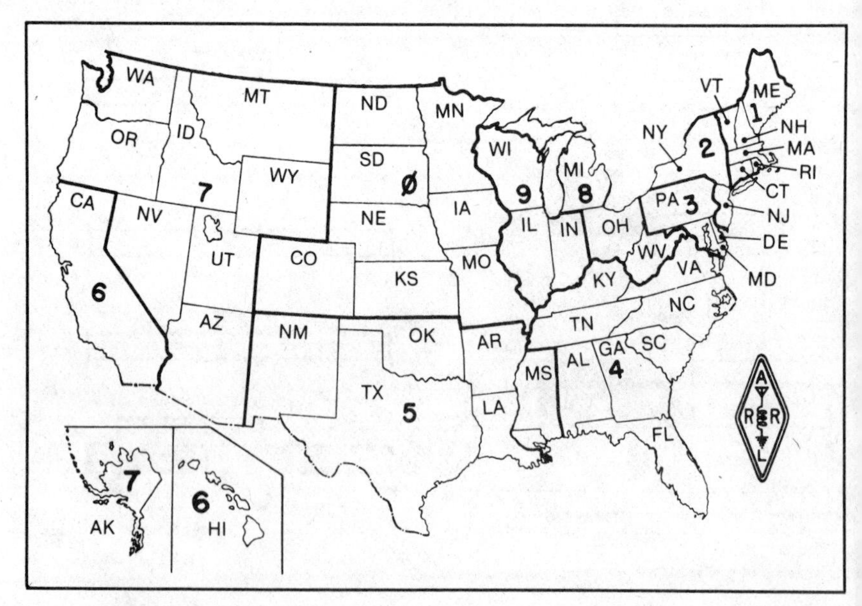

Fig. 1-1—There are 10 U.S. call areas. Hawaii is part of the sixth call area, and Alaska is part of the seventh.

numeral) as a prefix, followed by a numeral, and then a suffix of not more than three letters. The prefixes W, K, N and A are used in the United States. The continental U.S. is divided into 10 Amateur Radio call districts (sometimes called areas), numbered Ø through 9). Fig. 1-1 is a map showing the U.S. call districts.

All U.S. Amateur Radio call signs assigned by the FCC after March 1978 can be categorized into one of five groups, each corresponding to a class, or classes, of license. Call signs are issued systematically by the FCC; requests for special call signs are not granted. For further information on the FCC's call-sign assignment system, and a table listing the blocks of call signs for each license class, see the ARRL publication, *The FCC Rule Book*. If you already have an amateur call sign, you may keep the same one when you change license class, if you wish. You must indicate your preference to receive a new call sign when you fill out an FCC Form 610 to apply for the exam or change your address.

EARNING A LICENSE

Applying for an Exam: FCC Form 610

Before you can take an FCC exam, you'll have to fill out a Form 610. This form is used as an application for a new license, an upgraded license, a renewal of a license or a modification to a license. In addition, hams who have held a valid license that has expired within the past two years may apply for reinstatement with a Form 610. Each form comes with detailed instructions. (These instructions also appear in *The FCC Rule Book*.) Form 610s showing a revision date before June 1984 are no longer valid, and will be returned if you attempt to use one. If you have any old forms, throw them away. To obtain a new Form 610, send a business-size, self-addressed, stamped envelope to: Form 610, ARRL, 225 Main St., Newington, CT 06111.

Volunteer Examiner Program

Since January 1, 1985, U.S. amateur exams above the Novice level have been administered under the Volunteer-Examiner Program. *The FCC Rule Book* contains details on this program. Novice exams have always been given by volunteer examiners, and that is still true. Those exams do not come under the regulations involving Volunteer Examiner Coordinators (VECs), however.

To qualify for an Amateur Extra Class license, you must pass all elements through Element 1C and Element 4B. If you already hold a valid license, then you have credit for passing at least some of those elements, and will not have to retake them when you go for your Amateur Extra exam. See Table 1-1 for details.

The Element 4B exam consists of 40 questions taken from a pool of more than 400. The question pools are maintained by Volunteer-Examiner Coordinators, and the FCC now allows Volunteer Examiners to select the questions for an amateur exam. The Volunteer Examiners must use the questions exactly as they are released by the VEC coordinating the test session. If you attend a test session coordinated by the ARRL/VEC, your test will be designed by the ARRL/VEC, and the questions and answers will be exactly as they are printed in Chapter 10.

The question pool printed in Chapter 10 was released by the FCC in April 1986. VECs may now maintain their own question pools, but all VECs have declared a moratorium on changes to the pools until January 1988, except for changes required by FCC rules changes and to fix typographical errors. The pool printed in Chapter 10 will be in use until November 1988. More information about VEC maintenance of question pools and the release dates for new pools will appear in *QST*.

Finding an Exam Opportunity

To determine where and when an exam will be given, contact the ARRL/VEC Office, or watch for announcements in the Hamfest Calendar and Coming Conventions columns in *QST*. Many local clubs sponsor exams, so they are another good source of information on exam opportunities. ARRL officials such as Directors, Vice Directors and Section Managers receive notices about test sessions in their area. See page 8 in the latest issue of *QST* for names and addresses.

To register for an exam, send a completed Form 610 to the Volunteer Examining team responsible for the exam session if preregistration is required. Otherwise, bring the form to the session. Registration deadlines, and the time and location of the exams, are mentioned prominently in publicity releases about upcoming sessions.

Taking The Exam

By the time examination day rolls around, you should have already prepared yourself. This means getting your schedule, supplies and mental attitude ready. Plan your schedule so you'll get to the examination site with plenty of time to spare. There's no harm in being early. In fact, you might have time to discuss hamming with another applicant, which is a great way to calm pre-test nerves. Try not to discuss the material that will be on the examination, as this may make you even more nervous. By this time, it's too late to study anyway!

What supplies will you need? First, be sure you bring your current *original* Amateur Radio license, if you have one. Bring along several sharpened no. 2 pencils and two pens (blue or black ink). Be sure to have a good eraser. A pocket calculator will also come in handy. You may use a programmable calculator if that is the kind you have, but take it into your exam "empty." Don't program a lot of equations ahead of time, because you may be asked to demonstrate that there is nothing in the calculator's memory. The Volunteer Examining Team is required to check two forms of identification before you enter the test room. A photo ID of some type is best, but not required by FCC. Other acceptable forms of identification include a driver's license, a piece of mail addressed to you, a birth certificate, or some other such document.

Before taking the code test, you'll be handed a piece of paper to copy the code as it's sent. The test will begin with about a minute of practice copy. Then comes the actual test: five minutes of Morse code. You are responsible for knowing the 26 letters of the alphabet, the numerals Ø through 9, the period, comma, question mark, and procedural signals $\overline{\text{AR}}$, $\overline{\text{SK}}$, $\overline{\text{BT}}$ and $\overline{\text{DN}}$. You may copy the entire text word for word, or just take notes on the content. At the end of the transmission, the examiner will hand you 10 questions about the text. Simply fill in the blanks with your answers. If you get at least 7 correct, you pass! Alternatively, the exam team has the option to look at your copy sheet if you fail the 10-question exam. If you have one minute of solid copy, they can certify that you passed the test on that basis. The format of the test transmission is similar to one side of a normal on-the-air amateur conversation.

A sending test may not be required. The Commission has decided that if applicants can demonstrate receiving ability, they most likely can also send at that speed. But be prepared, just in case!

If all has gone well with the code test, you'll then take the written examination. The examiner will give all applicants a test booklet, an answer sheet and scratch paper. After that, you're on your own. The first thing to do is read the instructions. Be sure to sign your name every place it's called for. Do all of this at the beginning to get it out of the way.

Next, check the examination to see that all pages and questions are there. If not, report this to the examiner immediately. When filling in your answer sheet, make sure your answers are marked next to the numbers that correspond to each question.

Go through the entire exam, and answer the easy questions first. Next, go back to the beginning and try the harder questions. The really tough questions should be left for last. Guessing can only help, as there is no additional penalty for answering incorrectly.

If you have to guess, do it intelligently: At first glance, you may find that you can eliminate one or more distractors. Of the remaining responses, more than one may seem correct; only one is the *best* answer, however. To the applicant who is fully prepared, incorrect distractors to each question are obvious. Nothing beats preparation!

After you've finished, check the examination thoroughly. You may have read a question wrong or goofed in your arithmetic. Don't be overconfident. There's no rush, so take your time. Think, and check your answer sheet. When you feel you've done your best and can do no more, return the test booklet, answer sheet and scratch pad to the examiner.

The Volunteer Examiner team will grade the exam right away. 74% is the passing mark. (That means no more than 10 incorrect answers.) If you are already licensed, and you pass the exam elements required to earn a higher class of license, you will receive a certificate allowing you to operate with your new privileges. The certificate has a special identifier code that must be used on the air when you use your new privileges, until your permanent license arrives from the FCC.

AND NOW, LET'S BEGIN

The section of the questions in Chapter 10 that begin with numbers 4BA- cover the rules and regulations for the Amateur Extra exam. You should use the *FCC Rule Book* to find the material covered by those questions. Then go over that section of the question pool to check your understanding of the rules. Perhaps you will want to study the rules a few at a time, using that as a break from your study in the rest of this book.

There you have it. The remainder of this book will provide the background in electronics theory that you will need to pass the Element 4B Extra Class written exam.

Taken from FCC PR Bulletin 1035, May 1986

STUDY GUIDE FOR AMATEUR RADIO OPERATOR LICENSE EXAMINATIONS

This bulletin contains the syllabi for amateur operator examinations. License candidates should become knowledgeable about all topics in the syllabus for the examination element they will be taking. Volunteer Examiners grade candidates' answers to written questions based upon the following topics to determine their qualifications for one of the five classes of amateur operator license.

Each higher license class conveys additional privileges to the holder. Those privileges are many and they are diverse. As an amateur operator, one is allowed to build, repair, and modify amateur station transmitters. In general, the FCC equipment authorization requirements do not apply to amateur stations. The operator alone is responsible for the technical quality of the station's transmissions. An amateur operator is allowed to communicate with amateur operators in other countries and, in some cases, send messages for friends. All license classes are allowed to communicate using telegraphy. Higher classes may also use voice, teleprinting, facsimile, and television.

For such a flexible radio service to be practical, all amateur operators must thoroughly understand their responsibilities and have the skills necessary to properly operate an amateur station.

Element 4(B) syllabus.
(Required for Amateur Extra license.)

Subelement 4BA.
Commission's Rules.

4BA-1 Examination elements. 97.21.
4BA-2 Examination requirements. 97.23.
4BA-3 Examination credit. 97.25.
4BA-4 Examination procedure. 97.26.
4BA-5 Telegraphy examination preparation. 97.27(b).
4BA-6 Written examination preparation. 97.27 (d).
4BA-7 Examination administration. 97.28(a), (e), (h), (i), (j).
4BA-8 Volunteer examiner compensation. 97.31(c).
4BA-9 Volunteer examiner requirements. 97.31(d).
4BA-10 Volunteer examiner accreditation. 97.31(e).
4BA-11 Temporary operating authority. 97.35.
4BA-12 Reimbursement for expenses. 97.36.
4BA-13 Emission purity. 97.73.
4BA-14 Point of communications. 97.89.
4BA-15 Portable and mobile operation. 97.95.
4BA-16 Operation aboard ship or aircraft. 97.101.
4BA-17 RACES purpose. 97.161.
4BA-18 RACES definition. 97.163(a).
4BA-19 RACES registration. 97.175.
4BA-20 RACES operator requirements. 97.177.
4BA-21 RACES operator privileges. 97.179.
4BA-22 Frequencies available to RACES. 97.185.
4BA-23 RACES point of communications. 97.189.
4BA-24 RACES permissible communications. 97.191.
4BA-25 Limitations on use of RACES frequencies. 97.185(b), (c).
4BA-26 Purpose of reciprocal agreement rules. 97.301(b).
4BA-27 Reciprocal permit. 97.303.
4BA-28 Reciprocal operating privileges. 97.311.
4BA-29 Reciprocal operating station identification. 97.313.
4BA-30 Purpose of the Amateur-Satellite Service. 97.401.
4BA-31 Space speration definition. 97.403(a).
4BA-32 Earth operation definition. 97.403(b).
4BA-33 Telecommand operation definition. 97.403(c).
4BA-34 Telemetry definition. 97.403(d).
4BA-35 Eligibility for space operation. 97.407.
4BA-36 Eligibility for earth operation. 97.409.
4BA-37 Eligibility for telecommand operation. 97.411.
4BA-38 Frequencies available to Amateur-Satellite Service. 97.415.
4BA-39 Special provisions for space operation. 97.417.
4BA-40 Special provisions for telemetry. 97.419.
4BA-41 Special provisions for telecommand. 97.421.
4BA-42 Special provisions for earth operation. 97.422.
4BA-43 Space operation notifications. 97.423.
4BA-44 Volunteer-Examiner Coordinator definition. 97.503(a).
4BA-45 Volunteer Examiner definition. 97.503(b).
4BA-46 VEC qualifications. 97.507.
4BA-47 VEC agreement. 97.511.
4BA-48 Examination scheduling. 97.513.
4BA-49 Coordinating VE's. 97.515.
4BA-50 Conflict of Interest. 97.31(b); 97.509.

Subelement 4BB.
Operating Procedures.

4BB-1 Use of amateur satellite.
4BB-2 Emission C3F and A3F operation.

Subelement 4BC.
Radio-wave propagation.

4BC-1 EME moonbounce.
4BC-2 Meteor burst.
4BC-3 Trans-equatorial.

4BC-4 Long path.
4BC-5 Crooked path.
4BC-6 Gray line.

Subelement 4BD.
Amateur practices.

4BD-1 Spectrum analyzer.
4BD-2 Logic probe.
4BD-3 Vehicle noise suppression.
4BD-4 Direction finding techniques.

Subelement 4BE.
Electrical principles.

4BE-1 Photoconductive effect.
4BE-2 Exponential charge/discharge.
4BE-3 Time constant for R-C and R-L curcuits.
4BE-4 Impedance diagrams.
4BE-5 Impedance of R-L-C networks at a specified frequency.
4BE-6 Algebraic operations using complex numbers: real, imaginary, magnitude, angle.

Subelement 4BF.
Circuit components.

4BF-1 Field-effect transistor.
4BF-2 Operational amplifier.
4BF-3 Phase-locked loop.
4BF-4 7400 series TTL digital integrated circuits.
4BF-5 4000 series CMOS digital integrated circuits.
4BF-6 Vidicon.
4BF-7 Cathode ray tube.

Subelement 4BG.
Practical circuits.

4BG-1 Digital logic circuits.
4BG-2 Digital frequency divider circuits.
4BG-3 Active audio filters.
4BG-4 Receiver noise figure, sensitivity.
4BG-5 Receiver selectivity.
4BG-6 Receiver dynamic range.
4BG-7 Integrated operational amplifier.
4BG-8 FET common-source amplifier.
4BG-9 Preselector.
4BG-10 Single stage amplifier frequency response.

Subelement 4BH.
Signals and emissions.

4BH-1 Pulse modulation.
4BH-2 Digital signals.
4BH-3 Amplitude compandored single-sideband.
4BH-4 Information rate vs bandwidth.
4BH-5 Peak amplitude.
4BH-6 Peak-to-peak values.

Subelement 4BI.
Antennas and feed lines.

4BI-1 Space communications antennas.
4BI-2 Isotropic radiator.
4BI-3 Phased vertical antennas.
4BI-4 Rhombic antennas.
4BI-5 Matching antenna to feed line.
4BI-6 Characteristics of 1/8 wavelength feed line.
4BI-7 Characteristics of 1/4 wavelength feed line.
4BI-8 Characteristics of 3/8 wavelength feed line.
4BI-9 Characteristics of 1/2 wavelength feed line.

Key Words

Acquisition of signal (AOS)—The time you can first hear a satellite, usually just after it rises above the horizon.

Apogee—That point in a satellite's orbit at which it is farthest from the earth.

ATV (amateur television)—A wideband TV system that can use commercial transmission standards. Only permitted on the 70-cm band and higher frequencies.

Blanking—Portion of a video signal that is "blacker than black," used to be certain that the return trace is invisible.

Composite video signal—A complete video signal, consisting of the output from the vidicon tube, blanking pulses and sync pulses.

Doppler shift—A change in the observed frequency of a signal, as compared with the transmitted frequency, caused by satellite movement toward or away from you.

Eccentricity—The orbital parameter used to describe how much an elliptical orbit deviates from a circle; eccentricity values vary between e = 0 for a circle and e = 1 for a straight line.

Faraday rotation—A rotation of the polarization plane of radio waves when the waves travel through the ionized magnetic field of the ionosphere.

Fast-scan TV—Another name for ATV, used because a new frame is transmitted every 1/30 of a second, as compared to every 8 seconds for slow-scan TV.

Horizontal sync pulse—Part of a TV signal used by the receiver to keep the CRT electron-beam scan in step with the camera scanning beam. This pulse is transmitted at the beginning of each horizontal scan line.

Inclination—The angle at which a satellite crosses the equator at an ascending node. Inclination is also equal to the highest latitude reached in an orbit.

Kepler's Laws—Three laws of planetary motion, used to mathematically describe satellite-orbit parameters.

Loss of signal (LOS)—The time when a satellite passes out of range.

Major axis—An axis passing through the foci of an ellipse. The longest straight line passing through an ellipse.

Minor axis—An axis passing through the center of an ellipse, perpendicular to the major axis. The shortest straight line passing through the center of an ellipse.

Node—A point where a satellite crosses the plane passing through the earth's equator. It is an ascending node if the satellite is moving from south to north, and a descending node if the satellite is moving from north to south.

Perigee—That point in a satellite's orbit when it is closest to the earth.

Polarization—A property of an electromagnetic wave that describes the orientation of the electric field of the wave with respect to earth.

Scanning—The process of analyzing or synthesizing, in a predetermined manner, the light values or equivalent characteristics of elements constituting a picture area. Also the process of recreating those values to produce a picture on a CRT screen.

Spin modulation—Periodic amplitude fade-and-peak variations resulting from the Phase III satellite's 60 r/min spin.

Telecommand operation—earth-to-space Amateur Radio communication to initiate, modify or terminate functions of a station in space operation (an Amateur Radio satellite)

Television raster—A predetermined pattern of scanning lines that provides substantially uniform coverage of an area.

Transponder—A repeater aboard a satellite that retransmits, on another frequency band, any type of signals it receives. Signals within a certain receiver bandwidth are translated to the new frequency, so many signals can share a transponder simultaneously.

Vertical sync pulse—Part of a TV signal used by the receiver to keep the CRT electron-beam scan in step with the camera scanning beam. This pulse returns the beam to the top edge of the screen at the proper time.

Vestigial sideband (VSB)—A signal-transmission method in which one sideband, the carrier and part of the second sideband are transmitted. The bandwidth is not as wide as for a double-sideband AM signal, but not as narrow as a single-sideband signal.

Vidicon tube—A TV-camera tube in which a charge-density pattern is formed by photoconduction, and is stored on the surface of the photoconductor that is scanned by an electron beam. This converts the light and dark areas of a picture into a varying electric signal.

A more complete glossary of satellite terminology is given in Table 2-4.

Chapter 2

Operating Procedures

There are two major sections to this chapter. The first deals with amateur satellites; the second with fast-scan television. When you have studied the information in each section, you will be directed to the examination questions in Chapter 10 to check your understanding of the material. If you are unable to answer a question correctly, go back and review the appropriate part of this chapter.

AMATEUR RADIO SATELLITES

Terrestrial communication is limited by the spherical shape of the earth. There are numerous propagation mechanisms which can be used to transmit a signal around the earth at HF. Long-haul communication at VHF and UHF, however, may require the use of higher effective radiated power and may not be possible at all over some paths. The communication range of amateur stations is increased greatly by using relay equipment mounted in orbiting satellites.

As an Extra Class licensee, you will be eligible to use your call as the station call sign on an OSCAR satellite, or other type of space station. Part 97, Subpart H of the FCC rules covers the Amateur Satellite Service. These rules outline space operation, earth operation and telecommand operation. They also list frequencies available for space operation and describe the requirements for notifying FCC about your space operation. You should be especially familiar with this section of the rules.

Before you run out and start building a satellite, you should realize that there are other hurdles to clear before you can launch a rocket or hitch a ride for your satellite on another rocket. If you find yourself in a position to ride along on a space shuttle flight, and can get all the other necessary permission, the Amateur Radio Satellite Service rules will govern your "space mobile" operation as well!

Understanding Satellite Orbits

Two factors affect a body in orbit about the earth. These factors are forward motion and the force of gravitational attraction. Forward motion tends to move the body away from the earth in a straight line in the direction it is moving at that instant. Gravity tends to pull the body toward the earth. Motion and gravity must balance to maintain an orbit.

Kepler described the planetary orbits of our solar system. His three laws of planetary motion also describe the lunar orbit and the orbits of artificial earth satellites. Those laws can be restated for artificial earth satellites as follows:

Kepler's Laws

First law: The orbit of a satellite with respect to the earth is an ellipse, the earth being at one of the foci.

Second law: A line drawn from the earth to a satellite sweeps across equal areas in equal times.

Third law: The square of the time a satellite takes to complete one orbit is proportional to the cube of its mean (average) distance from the earth.

At first, these laws may seem difficult to understand, but they are not. Let's look at them one at a time.

Kepler's First Law

The first law says that orbits are ellipses (Fig. 2-1). The distance through the thickest part of an ellipse is called the *major axis*—through the thinnest it is called the *minor axis*. The *semimajor axis* is half the length of the major axis. The *semiminor axis* is also one half the minor-axis length. *Eccentricity* is the distance from the center to one of the focal points divided by the semimajor axis. Eccentricity ranges between 0 (a circle) to 1 (a straight line)—the larger the number the "thinner" the ellipse. AMSAT-OSCAR 10 orbit eccentricity is approximately 0.6.

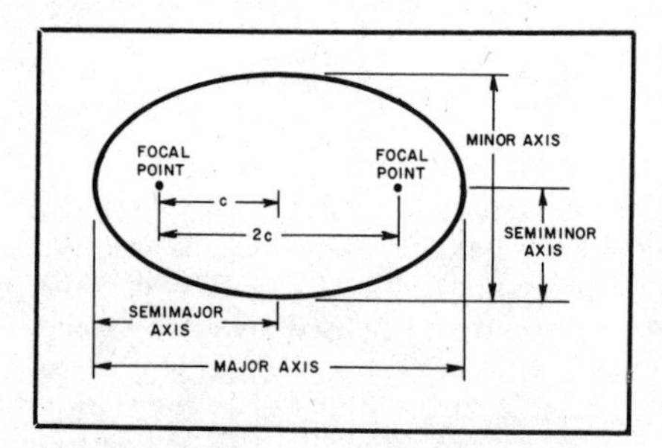

Fig. 2-1—Geometry of an ellipse. Labeled "c," the distance from the center to either focal point is called linear eccentricity. This distance should not be confused with numerical eccentricity, which is the distance c divided by the semimajor axis. Being a ratio, numerical eccentricity is a unitless number—unless otherwise stated this is simply referred to as eccentricity.

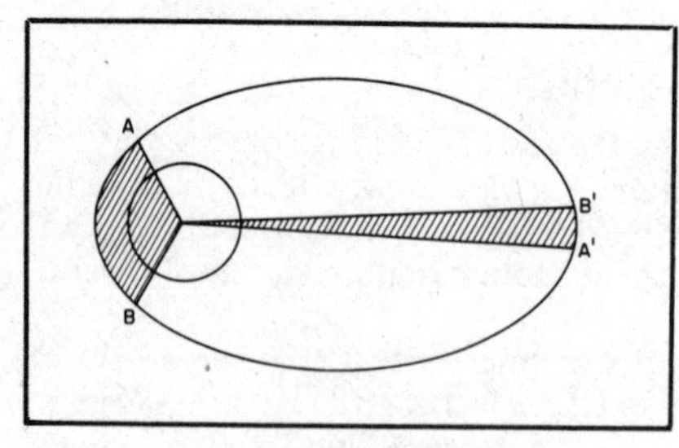

Fig. 2-2—A graphical representation of Kepler's second and third laws. The two shaded areas are equal. The time for the satellite to move along its orbit from A to B or from A′ to B′ is the same.

Kepler's Second Law

Kepler's second law is illustrated in Fig. 2-2. The time required for a satellite to move in its orbit from point A to point B is the same as required to move from A′ to B′. Another way of saying this is that a satellite moves faster in its elliptical orbit when it is closer to the earth, and slower when it is farther away.

Kepler's Third Law

The third law shows that the greater the average distance from the earth, the longer it takes for a satellite to complete each orbit. The time required for a satellite to make a complete orbit is called the orbital period. Low-flying amateur satellites typically had periods of approximately 90 minutes. OSCAR 10, in its high elliptical orbit, has a period of almost 700 minutes (11 hours, 40 minutes).

An easy way to remember the subjects of Kepler's laws is to think of them as follows: The first law deals with the shape of a satellite orbit. The second law covers the satellite speed at various points along the orbit. The third law has to do with the orbital period.

Orbital Terminology

Inclination is the angle of a satellite orbit with respect to earth. Inclination is measured between the plane of the orbit and the plane of the equator (Fig. 2-3).

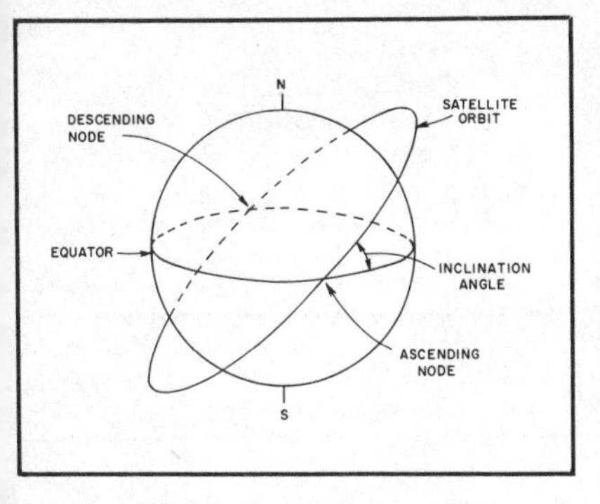

Fig. 2-3—Graphical representation of satellite-orbit terminology. Definitions can be found in Table 2-4.

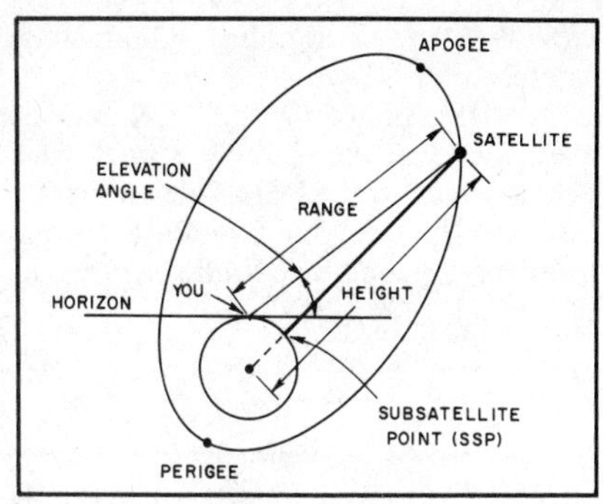

Fig. 2-4—Graphical representation of satellite-orbit terminology. Definitions can be found in Table 2-4.

If a satellite is always over the equator as it travels through its orbit, the orbit has an inclination of zero degrees. Should the orbit take the satellite over the poles (if it goes over one it will go over the other), the inclination is 90 degrees.

A *node* is the point where a satellite crosses the equator. The *ascending node* is where it crosses the equator when it is traveling from south to north. Inclination is measured at the ascending node, as shown in Fig. 2-3. Equator crossing (EQX) is specified at the ascending node. EQX is usually given in time (UTC) of crossing, and in degrees west longitude. The *descending node* is the point where the satellite crosses the equator traveling from north to south. When the satellite is within range of your location, it is common to describe the pass as either an ascending pass or a descending pass. This depends on whether the satellite is traveling from south to north or north to south over your area.

The point of greatest height in a satellite orbit is called the *apogee* (Fig. 2-4). *Perigee* is the point of least height. Half the distance between the apogee and perigee is equal to the semimajor axis of the satellite orbit. Orbital parameters for a few amateur satellites can be found in Table 2-1.

More information on satellite orbits and methods of tracking satellites can be found in *The Satellite Experimenter's Handbook* published by ARRL.

Frequencies and Propagation

Satellites used for two-way communication generally use one amateur band to receive signals from earth (*uplink*) and another to transmit back to earth (*downlink*). A variety of uplink and downlink band combinations, or modes, are used in the amateur satellites. Table 2-2 is a list of modes. Table 2-3 shows specific frequen-

Table 2-1
Amateur Radio Spacecraft Orbital Parameters

Satellite	*Inclination (Deg.)*	*Apogee (km)*	*Perigee (km)*	*Period (Min.)*	*Increment (Deg. W)*
UoSAT-OSCAR 9	97.48*	544*	536*	94.4*	23.6*
AMSAT-OSCAR 10	26	35,406	4049	699.52	175.36
UoSAT-OSCAR 11	98.25	705	697	98.56	24.64

*UoSAT-OSCAR 9 is in a low orbit that is affected by atmospheric drag. The apogee, perigee, period and increment, all interrelated parameters, are approximate values that will change significantly with time.

cies used by a number of amateur satellites.

A complete list of frequencies available to the Amateur Satellite Service can be found in the FCC Rules, Part 97, Subpart H. You can find Part 97, the rules for the Amateur Radio Service, in the ARRL's *The FCC Rule Book*.

Table 2-2
Bands used for Satellite Communications

Mode	*Uplink*	*Downlink*
A	2 m	10 m
B	70 cm	2 m
J	2 m	70 cm
L	23 cm	70 cm

Table 2-3
Spacecraft Frequencies

	Uplink	*Downlink*	*Beacon*
OSCAR 9			
HF Beacons—7,050, 14,002, 21,002 and 29,510 kHz. On-off keying with Morse telemetry. Interspersed with a carrier or continuous carrier.			
VHF Beacon—145.825 MHz NBFM ±5 kHz. ASCII, Baudot, voice, AFSK and Morse.			
UHF Beacon—435.025 MHz NBFM ±5 kHz. ASCII, Baudot, voice, AFSK and Morse.			
S-Band Beacon—2401.0-MHz NBFM ±10 kHz. ASCII, Baudot, voice, AFSK and Morse.			
X-Band Beacon—10.470-GHz steady carrier. S- and X-band beacons use LHCP.			
OSCAR 10			
Mode B	435.025-435.175 MHz	145.975-145.825 MHz	General 145.812 MHz Engin. 145.990 MHz
Mode L	1269.050-1269.850 MHz	436.950-436.150 MHz	General 436.040 MHz Engin. 436.020 MHz
OSCAR 11			
VHF Beacon—145.825 MHz NBFM FSK			
UHF Beacon—435.025 MHz NBFM FSK, PSK			
S-Band Beacon—2401.5 MHz AFSK, PSK			

Doppler Shift

Doppler shift is caused by the relative motion between you and the satellite. If there were no relative motion, you could predict precise downlink frequencies coming from the satellite. In operation, as the satellite is moving toward you, the frequency of a downlink signal appears increased by a small amount. When the satellite passes overhead and begins to move away from you, there will be a sudden frequency drop of a few kilohertz, in much the same way as the tone of a car horn or train whistle drops as the vehicle moves past you. The result is that signals passing through the satellite move around the expected or calculated downlink frequency, depending on whether the satellite is moving toward you or away from you. Those signals appear higher when the satellite is moving toward you, lower when it is moving away. Locating

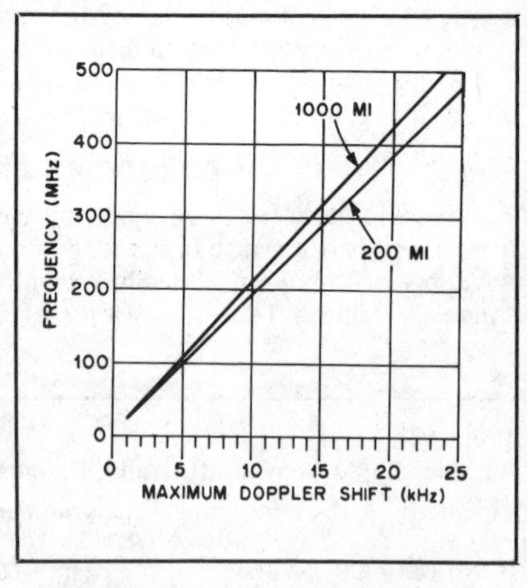

Fig. 2-5—Satellite operating frequency versus Doppler shift for satellites in 200- or 1000-statute-mile orbits. For noninverting transponders, use the difference between uplink and downlink frequencies as the "frequency."

your own signal is, therefore, a little more difficult than simply computing the relation between input and output frequency; the effects of Doppler must be taken into account.

Since the speed of the satellite, relative to earth, is greater for a satellite that is close to earth, Doppler shift is generally more noticeable with low-orbit satellites than with high-orbit ones like OSCAR 10. Doppler shift is also frequency dependent. The higher the operating frequency, the greater the shift. See Fig. 2-5.

Faraday Rotation

The *polarization* of a radio signal passing through the ionosphere does not remain constant. A "horizontally polarized" signal leaving a satellite will not be horizontally polarized when it reaches Earth. That signal will in fact seem to be changing polarization at a receiving station. This effect is called *Faraday rotation*. The best way to deal with Faraday rotation is to use circularly polarized antennas for transmitting and receiving.

Spin Modulation

Spin modulation is a phenomenon that has emerged with the introduction of the AMSAT Phase III satellites. As the satellite orbits overhead, the on-board computer pulses an electromagnet that works against the earth's magnetic field. This spins the spacecraft at approximately 1 revolution per second, thereby stabilizing it. A side effect, however, is the relatively rapid (3 Hz) periodic fade of the transmitted signal amplitude, called spin modulation. The 3-Hz spin modulation is caused by having a transmitting antenna on each satellite "arm," so three antennas spin past you each second. It is important to note that the passband is not electronically modulated in the sense to which amateurs are accustomed; rather, the apparent modulation is a residual effect of physical rotation.

Use of linear antennas will deepen the spin-modulation fades to a point where they may become annoying. Circularly polarized antennas of the proper sense will minimize the effect.

Satellite Systems and Hardware

Present communications satellites are functionally integrated systems. Rechargeable batteries, solar cell arrays, voltage regulators, command decoders, antenna-deployment mechanisms, stabilization systems, sensors, telemetry encoders and even on-board computers and kick motors each serve a unique and indispensable purpose. But to the radio amateur interested in communicating through a satellite, the transponder is of primary importance.

Transponders

By convention, *transponder* is the name given to any linear translator that is installed in a satellite. The translator or transponder is similar to a repeater in many ways. Each is a combination of a receiver and a transmitter that is used to extend the range of mobile, portable and fixed stations. A typical FM voice repeater receives on a single frequency or channel and retransmits what it receives on another channel. By contrast, a transponder receive passband includes enough spectrum for many channels. An amateur satellite transponder does not use channels in the way that voice repeaters do. Received signals from a band segment are amplified, shifted to a new frequency range and retransmitted by the transponder. Fig. 2-6 shows block diagrams for a simple voice repeater and a simple transponder.

By comparing block diagrams, you can see that the major hardware difference between a repeater and a transponder is signal detection. In a repeater, the signal is reduced to baseband (audio) before it is retransmitted. In a transponder, signals in the passband are moved to an IF for amplification and retransmission.

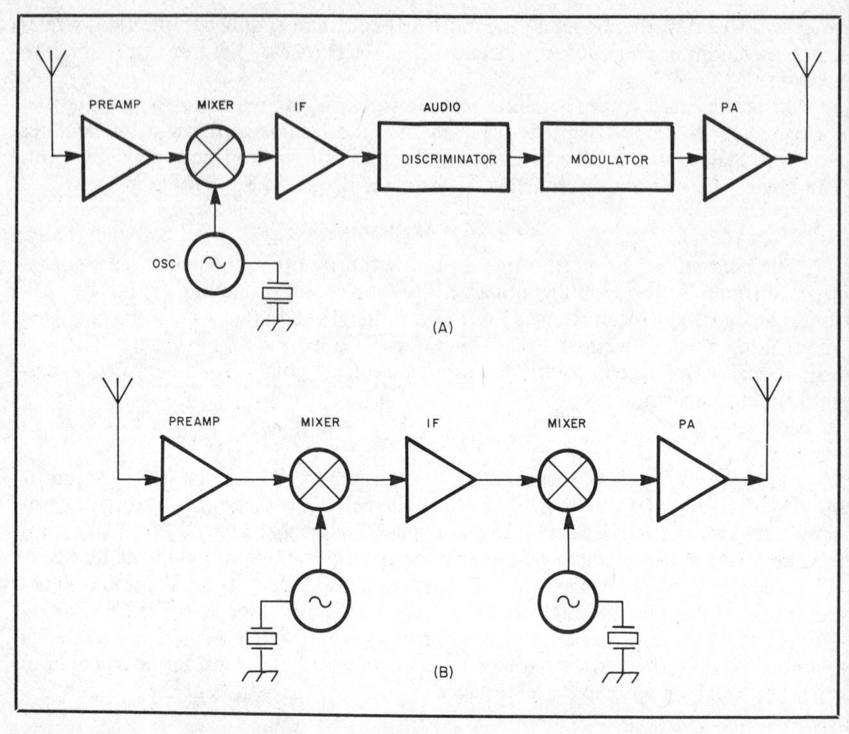

Fig. 2-6—Block diagram of an FM voice repeater at A; of a linear transponder at B.

Operationally, the contrast is much greater. An FM voice repeater is a one-signal, one-mode-input and one-signal, one-mode-output device. A transponder can receive several signals at once and convert them to a new range. Further, a transponder can be thought of as a multimode repeater. Whatever goes in is what comes out. The same transponder can simultaneously handle SSB, ACSSB and CW signals. The use of a transponder rather than a channelized repeater allows more stations to use the satellite at one time. In fact, the number of different stations using a transponder at any one time is limited only by mutual interference, and the fact that the output power of the satellite (a couple of watts on the low-orbit satellites and about 50 watts on Phase III) is divided among the users. Because all users must share the power output, continuous-carrier modes such as FM and RTTY are not used through the amateur satellites.

Repeaters are often referred to by their operating frequencies. For example, a 34/94 machine receives on 146.34 MHz and transmits on 146.94 MHz. A different convention is used with transponders: The input band is given, followed by the corresponding output band. For example, a 2-m/10-m transponder would have an input passband centered near 146 MHz and an output passband centered near 29 MHz. Transponders usually are identified by mode—not mode of transmission such as SSB or CW. Mode has an entirely different meaning in this case. As mentioned earlier, the various frequency band combinations used for uplink and downlink are referred to as modes. See Table 2-2. For example, OSCAR 10 has Mode B and Mode L transponders.

For several reasons, transponders used aboard satellites are more complex than the one shown in Fig. 2-6. Just as a receiver design provides band-pass filtering, image

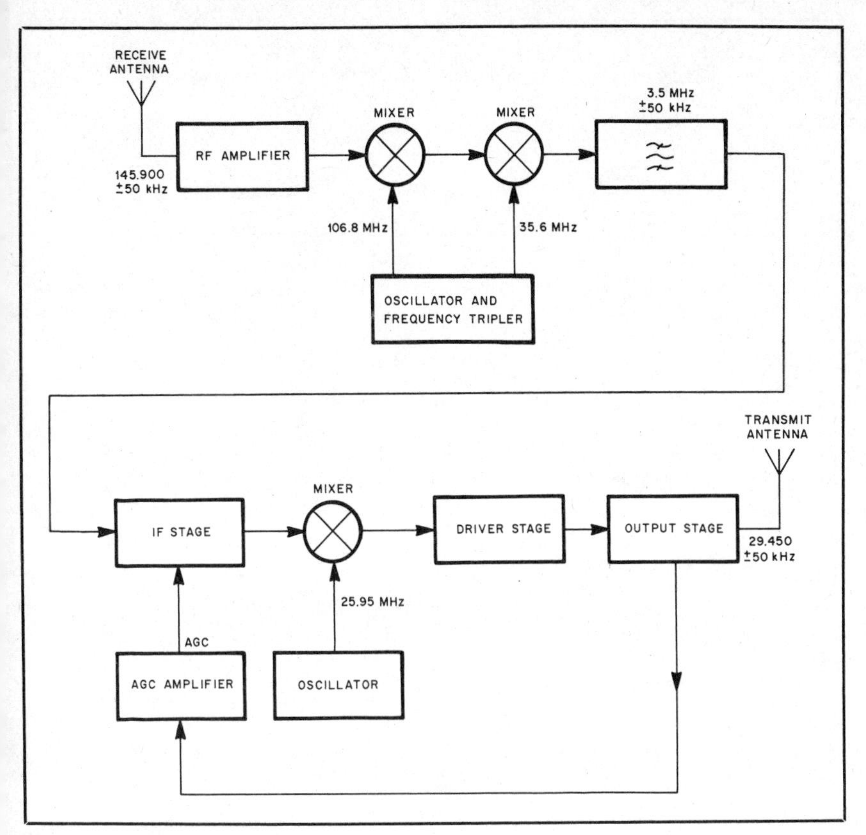

Fig. 2-7—Block diagram of AMSAT-OSCAR 8 Mode A transponder.

rejection, AGC and overall gain, a transponder design must meet similar requirements. For that reason, transponder designers use multiple-frequency conversions. A block diagram of the basic Mode A transponder used on OSCAR 8 is shown in Fig. 2-7.

Inverting and Noninverting Transponders

Any single- or multiple-conversion transponder may be designed so that entering signals are inverted. That type design changes upper-sideband signals into lower-sideband (and vice versa), transposes relative mark/space placement in RTTY, and so on. When this scheme is used, the result is called an inverting transponder. Both transponder types are illustrated in Fig. 2-8.

The major benefit of an inverting transponder is that the Doppler shift on the uplink is translated in the opposite direction and will partially cancel the Doppler shift on the downlink. With the 146/29-MHz (Mode A) link combination, Doppler is not serious; transponders using this frequency combination are typically non-inverting. Transponders using higher frequency combinations are usually inverting.

Power and Bandwidth

The power output, bandwidth and operating frequencies of a transponder must be compatible. When a transponder is fully loaded with equal-strength signals, each signal should provide an adequate signal-to-noise ratio at the ground.

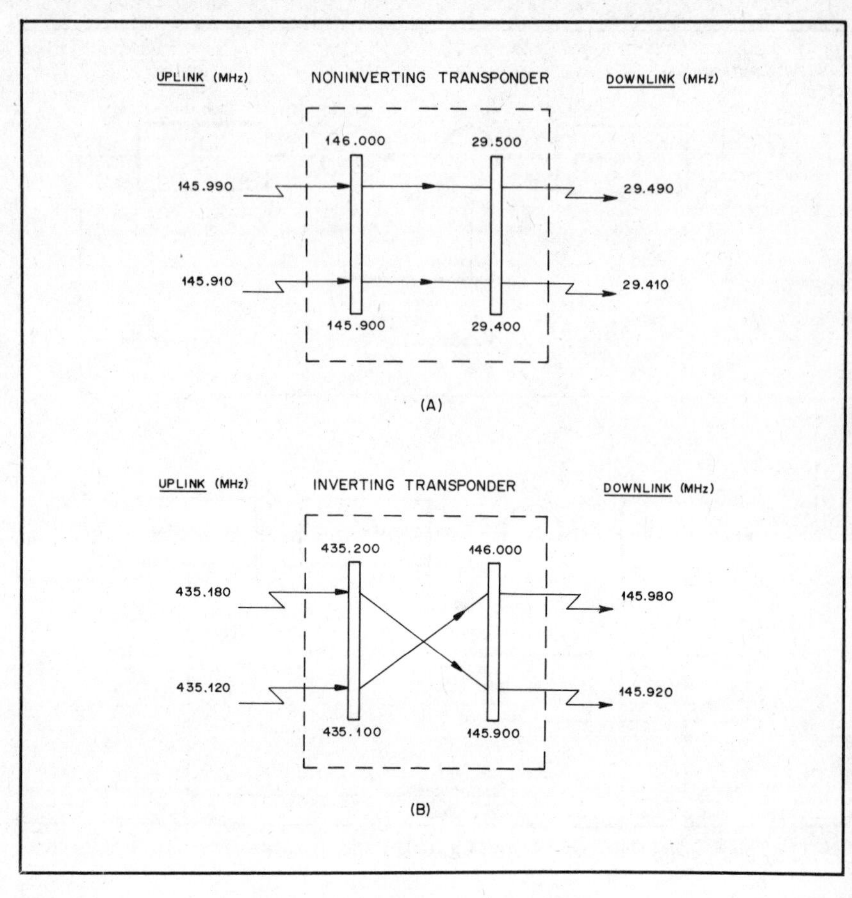

Fig. 2-8—Example of a noninverting transponder is shown at A, and an inverting transponder is shown at B.

Low-altitude (500-950 mi) satellites perform well with from 1 to 4 watts of PEP at frequencies between 29 and 435 MHz, using a 100-kHz-wide transponder. More sophisticated, high-altitude satellites provide acceptable performance with 35 watts of PEP using a 500-kHz-wide transponder downlink at 146 or 435 MHz.

Dynamic Range

Dynamic-range requirements for a transponder may appear to be less severe than for an HF receiver. That is not true! An HF receiver will handle signals differing in level by as much as 100 dB. That, however, is only part of the story. The HF receiver filters out all but the desired signal before introducing most of the gain. All transponder users must be accommodated simultaneously—quite a different situation. In one sense, all stations share a single channel. Transponder gain can, therefore, be limited by the strongest signal in the passband.

When a strong station is driving the satellite transponder to full power output, stations more than 22 dB lower than that level will not be heard on the downlink. That is true even if the weaker stations are perfectly readable when the strong signal is not present! User stations must cooperate by using only the minimum power necessary for communications through the satellites.

Using Amateur Radio Satellites

The first step in working through a satellite is to find out where it is. Since the spacecraft is constantly moving, some type of tracking calculator is needed. The simplest type of tracking device is an OSCARLOCATOR, available from ARRL. Briefly, you use one reference point each day. For low, nearly circular orbit satellites, that reference is usually the first time in a given (UTC) day that the satellite crosses the equator in a northerly direction. The equator crossing (EQX) time and longitude (when and where the satellite passes over the equator) is available from numerous sources such as W1AW bulletins and the various amateur publications. For satellites with high, elliptical orbits, the normal reference is time and location of apogee. Apogee and EQX data are available from the same sources. With this information, you can easily determine the satellite's approximate location and where to point your antenna at any given time from any point in the Northern Hemisphere.

More sophisticated calculating techniques are also possible. Actual antenna bearings can be mathematically calculated and recorded for a large number of points during a satellite pass. Programmable calculators and even computers facilitate this approach.

Regardless of the method used, it is a good idea to complete your calculations well in advance of the satellite pass. Jot down all the information on a work sheet, in a logical fashion that is easy to read. You will have enough to do during a pass, with tuning your receiver and transmitter, aiming antennas, logging and actually making contacts, without adding the burden of calculating the orbital data at the same time.

Operating Through Phase II Satellites

To use your system, whatever it may be, figure out when *AOS (acquisition of signal)* is for your location and the azimuth (compass heading) to where the *Phase II satellite* will rise. (Phase II satellites are low, circular-orbit satellites, such as OSCAR 6, 7 and 8, which are no longer operative, and most of the Russian satellites.) Since the satellite will come into "view" on the horizon, you know that the required antenna elevation at AOS is zero degrees; in other words the antenna points parallel to the ground. Record on the chart your calculated AOS time, and the associated antenna azimuth and elevation. Now, recalculate the antenna readings for the time one minute after AOS. Record these values on the chart. Continue for every one-minute interval until *LOS (loss of signal)*. If you later find that your antennas are very sharp in beamwidth and require greater pointing accuracy, reduce the interval to 30 seconds.

Using all this information is much simpler than it sounds. At the times indicated on your chart, simply point your antennas in the direction of the azimuth and elevation shown. It's that simple.

Now that the preliminaries are completed, the actual fun of working through the satellite can begin as soon as it's in range. The best procedure is to have your antennas pointed toward the AOS point and begin listening at the transponder beacon frequency several minutes in advance. You'll probably have to tune a little off the published frequency because Doppler shift makes the signal appear at a different frequency than expected. If you don't hear the beacon within a minute or so of when your chart predicts, something is wrong. Likely possibilities are miscalculations on your part, inaccurate orbital data, listening to the wrong transponder for that day, equipment failure or even an inaccurate clock. The keynote of satellites is reliability; satellite failure is rare compared with various equipment failures and mistakes encountered on the ground.

Once you hear signals from the satellite, you can set about the task of making

Table 2-4

Glossary of Satellite Terminology

AMSAT—The Amateur Radio Satellite Corporation, a nonprofit organization located in Washington, DC, has overseen the OSCAR program since the launch of OSCAR 5.

AOS—Acquisition of signal—The time you can first hear a satellite, usually just after it rises above the horizon.

Apogee—That point in a satellite's orbit where it is farthest above the earth.

Argument of Perigee—Polar angle that locates the perigee point in the orbital plane, measured counterclockwise from the ascending node.

Ascending Node—The point where the satellite crosses the equator traveling from the south to the north.

Ascending Pass—With respect to a particular ground station, a satellite pass during which the spacecraft is headed in a northerly direction while it is in range.

Az-el mount—Antenna mount that allows antenna positioning in both the azimuth and elevation planes.

Azimuth—Direction (side-to-side in the horizontal plane) from a given point on earth; usually specified in degrees (N = 0°, E = 90°, S = 180°, W = 270°).

Circular Polarization—A special case in which the electric field component of a radio wave is equal (but displaced 90°) in the vertical and horizontal planes and effectively rotates. The sense of polarization, whether right-hand circular or left-hand circular, is determined from behind the antenna, looking out along its axis of propagation.

Descending Node—The point where the satellite crosses the equator traveling from north to the south.

Descending Pass—With respect to a particular ground station, a satellite pass during which the spacecraft is headed in a southerly direction while it is in range.

Doppler Effect—An apparent shift in frequency caused by satellite movement toward or away from your location.

Downlink—The frequency of signals transmitted from the satellite to earth.

Eccentricity—The orbital parameter used to describe how much an elliptical orbit deviates from a circle; eccentricity values vary between 0 and 1: e = 0 for a circle.

EIRP—Effective isotropic radiated power—same as ERP except antenna reference is to an isotropic radiator.

Elliptical Orbit—Those orbits in which the satellite path traces an ellipse with the earth at one focus.

Elevation—Direction (up-and-down in the vertical plane) from a given point on earth usually specified in degrees (0° = plane of the earth's surface at your location; and 90° = straight up, perpendicular to the plane of the earth, overhead).

EQX—Equator crossing, usually specified in time (UTC) of crossing, and in degrees west longitude (0-360°).

ERP—Effective radiated power—System power output after transmission-line losses and antenna gain (referenced to a dipole) are considered.

Faraday Rotation—The rotation of the plane of polarization of an electromagnetic wave when it passes through the ionosphere.

Geostationary Orbit—An orbit at such an altitude (22,300 miles) and in such direction (W to E) over the equator that the satellite appears to be fixed above a given point.

Groundtrack—The imaginary line traced on the surface of the earth by a point directly below the satellite.

Inclination—The angle at which a satellite crosses the equator at its ascending node; also the highest latitude reached in an orbit. An orbit crossing directly over the North Pole would have an inclination of 90°, east of the pole less than 90° and west of the pole greater than 90°.

Increment—The number of degrees longitude that the satellite appears to move westward at the equator with each orbit, caused by the earth's rotation under the satellite during each orbit. (The earth rotates 360° in a 24-hour period.)

Keplerian Elements—Classical set of six numbers used as a standard to define orbits. The set is comprised of inclination, eccentricity, argument of perigee, Right Ascension of Ascending Node (RAAN), mean anomaly and semimajor axis of ellipse. For tracking purposes, the epoch, the date and time for which the elements are given, also must be specified.

LHCP—Left-hand circular polarization—counterclockwise.

LOS—Loss of signal—The time when the satellite passes out of range.

Mean Anomaly—An angle that increases uniformly with time that specifies where the satellite is in its orbit. In the case of OSCAR 10, the orbit is divided into 256 parts with 0/256 corresponding to perigee and 128/256 corresponding to apogee.

Mode A—Transponders with 2-meter uplink and 10-meter downlink.

Mode B—Transponders with 70-cm uplink and 2-meter downlink.

Mode J—Transponders with 2-meter uplink and 70-cm downlink.

Mode L—Transponders with 23-cm uplink and 70-cm downlink.

OSCAR—Orbiting Satellite Carrying Amateur Radio; there have been 10 Amateur Radio satellites named OSCAR as of the beginning of 1984 and nine Soviet Amateur Radio satellites, designated RS-1 through RS-8, and Iskra 2.

OSCARLOCATOR—A satellite tracking device consisting of a ranging oval and ground tracks superimposed on a polar projection map.

Pass—An orbit of the satellite.

Passband—The range of frequencies handled by a satellite transponder.

Perigee—The point in a satellite orbit where it passes closest to earth.

Period—The time it takes for a complete orbit, usually measured from one EQX to the next. The higher the altitude, the longer the period.

Phase I—The term given to the earliest, short-lived OSCAR satellites that were not equipped with solar cells. When their batteries were depleted, they ceased operating.

Phase II—The term given to low altitude, long-lived satellites. Equipped with solar panels that powered the spacecraft systems and recharge their batteries, these satellites have been shown to be capable of lasting up to five years (OSCARs 6, 7 and 8, for example).

Phase III—Extended-range, high-orbit satellites, typically in either elliptical orbit as AMSAT Phase IIIB, or in geostationary orbit.

Power Budget—A determination of how much power is actually available to operate the on-board satellite systems, taking into account such things as solar cell surface area, solar cell efficiency and angle toward the sun. A positive power budget means that ample power will be available to power the desired systems; a negative power budget means that periods of shutdown and recharge must be scheduled periodically.

Precession—An effect that is characteristic of AMSAT Phase III orbits. The satellite apogee SSP will gradually change over time.

RAAN—Right Ascension of Ascending Node the angular distance measured eastward along the celestial equator, between the vernal equinox and the hour circle of the ascending node of the spacecraft. This can be simplified to mean roughly the longitude of the ascending node.

Radio Sputnik—Soviet Amateur Radio satellites (see RS#).

Reference Orbit—The orbit beginning with the first ascending node during a given day UTC.

RHCP—Right-hand circular polarization—clockwise.

RS#—The designator used for most Soviet Amateur Radio satellites (e.g., RS-1, RS-5 and RS-8).

Semimajor Axis of Ellipse—One half the length of the major (long) axis of an ellipse, a Keplerian element that helps define an elliptical orbit.

Spin Modulation—Periodic amplitude fade-and-peak resulting from Phase III's 60 r/min spin; the effect is a 3-Hz "modulation" of the passband.

SSC—Special service channels—Frequencies in the downlink passband of AMSAT Phase III that are set aside for authorized, scheduled use in such areas as education, data exchange, scientific experimentation, bulletins and official traffic.

SSP—Subsatellite point—That point directly beneath a satellite on the surface of the earth at a given instant; usually defined in terms of latitude and longitude.

Sun-synchronous—A type of orbit that approximates the sun's apparent movement. For example, because its orbit is roughly sun-synchronous, OSCAR 8 was heard at a given location at about the same times each day.

Telemetry Beacon—The transmitters aboard each satellite that enable ground stations to monitor the satellite's vital functions.

Transponder—The repeater(s) aboard a satellite that retransmits on another frequency the signals it receives. Unlike terrestrial repeaters that operate on a fixed pair of frequencies, amateur satellite transponders translate an entire portion of one band (commonly 100 kHz bandwidth) to another band. Many signals can share a transponder simultaneously.

UoSAT-OSCAR—Amateur Radio satellite built under the coordination of radio amateurs and educators at the University of Surrey, England.

Uplink—The frequency on which radio signals are transmitted up to the satellites.

a contact. The toughest part is setting your transmitter frequency so that your signal is retransmitted on the proper downlink frequency. Calculating the expected downlink frequency for a given uplink frequency will put you fairly close. An alternative is to use a reference chart, which plots output frequency verses input frequency. With either of these methods, Doppler shifts and equipment frequency inaccuracies will only allow you to get close to the correct frequencies, but close enough so that sending a few words of sideband or a few dits will enable you to pinpoint your returning signal. The key word here is few. Unfortunately, you'll often hear a series of dits swishing up and down the passband. Not only is this inconsiderate, it also wastes time, which is in very short supply during a pass.

Initially, getting everything working and tuned properly may seem like an impossible task or one best suited to an octopus. With a little practice, however, the longing for an extra set of arms and hands gradually subsides. Since you're listening on an entirely different band from that on which you're transmitting, full duplex operation is possible: You hear what you're sending through the satellite as you send it! The ability of the station you're working to interrupt you while you're transmitting greatly improves the natural flow of the contact. It is even common for a CQ to be answered while it's being sent. Once again, anything to save time is to your advantage.

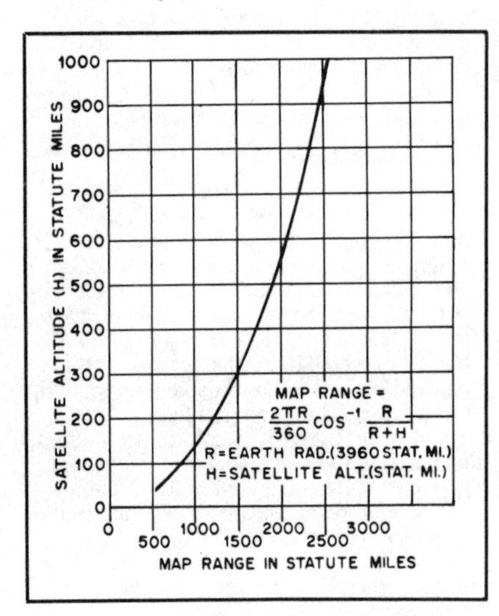

Fig. 2-9—Satellite altitude above earth versus ground station map range (statute miles).

The determining factor for maximum range of satellite communications is the height of the satellite. Fig. 2-9 can be used to determine range for the low, nearly circular orbit Phase II amateur satellites. Fig. 2-10 shows how to use range circles for two stations to determine when satellite communication is possible. The greater the range-circle overlap for the two stations, the longer the time that these stations can remain in contact.

Typical contacts through the satellite are short contest-type exchanges. Call signs, signal reports, locations and names are the extent of most contacts. Time is of the essence: Long-windedness doesn't work. Like an FM repeater, the satellite eventually times out, in this case, by going out of range.

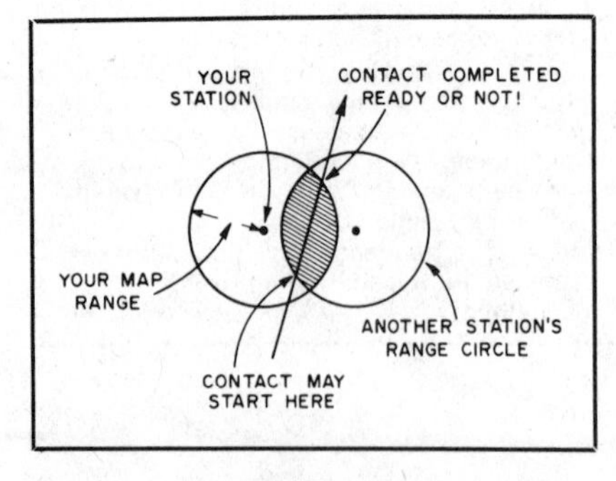

Fig. 2-10—Satellite passes through the range of two stations, enabling contact.

A typical contact might go like this:

CQ CQ CQ DE W1AW W1AW W1...
W1AW DE WA2LQQ 579 NY RIP
WA2LQQ DE W1AW 569 CONN CHUCK 73
W1AW DE KO5I...

There is one other very important aspect for the satellite user to consider. The transponder's internal automatic-gain-control system reduces the receiver sensitivity as the signal input level increases. This is done for two reasons. The first is to prevent receiver overload, which might produce spurious signals within the transmitter passband. Second, and more important, is that theoretically a very loud signal would be retransmitted with as much power as the transponder transmitter can muster. If this were to occur, serious damage could be sustained by the satellite. Even with the AGC system, transmitter current drain rises with very strong input signals, placing unnecessary and undesirable burden on the spacecraft batteries. It is every satellite user's responsibility to ensure that he or she does not transmit too great a signal to the satellite.

The proper transmitter power level is easy to determine. Since the downlink signal you receive indicates the effectiveness of your uplink, adjustments of your uplink power are reflected in the strength of the downlink signal. Your downlink signal should be no louder than the transponder beacon. Following this simple rule prevents receiver desense and guarantees easy transponder access for others and prevents satellite damage.

Operating Through Phase III Satellites

Tracking is not the problem with the higher flying *Phase III satellites* that it is with the Phase II variety. (Phase III satellites have extended communications range because they are in high, usually very elliptical orbits.) Instead of minutes, you may find hours of time available for QSOs. At times, the satellite will seem to "hang" in its elliptical orbit. Azimuth and elevation calculations for pointing your antenna can be made for 15- to 30-minute intervals. At times, you may not need to move your antennas for an hour or so. This makes operating considerably less complicated and contacts more leisurely.

With a Phase III satellite, there will be a significant path delay time—the time between transmitting an uplink signal to a satellite and receiving the downlink signal. This means that full duplex operation does not work very well.

Proper transmitter power level is determined in the same way as it is for Phase II. The downlink signal should be no stronger than the transponder beacon.

The key to more effective satellite communications is to improve your receiving capabilities. Start with a circularly polarized antenna with switching to change the polarity sense. A similar antenna should be used for transmitting. The proper polarity sense should be determined experimentally on each pass and checked occasionally during the pass. The next step is to install a low-noise preamp for receiving. For best results, mount the preamp at the antenna.

Table 2-4 is a glossary of common terms used in relation to satellites.

[Now turn to Chapter 10 and study FCC examination questions with numbers that begin 4BB-1. Review this section as needed.]

FAST-SCAN AMATEUR TELEVISION

With the most resemblance to broadcast-quality television, because it normally uses the same technical standards, *fast-scan TV* can be used by any amateur holding a Technician or higher-class license. Amateurs may use commercial transmission standards for TV. They are not limited to commercial standards, however.

Popularly known as *ATV (amateur television)*, this mode is permitted in the 420- to 450-MHz band and higher frequencies. Because the power density is comparatively low, typically 10 to 100 watts spread across 4 MHz, reliable amateur coverage is only on the order of 20 miles. Nevertheless, you might find yourself exchanging pictures with stations up to 200 miles away when tropospheric conditions are good.

Most wide-band A5 (A3F or C3F under the WARC-79 emission designators) activity occurs in the 420- to 450-MHz band. The exact frequency used depends on local custom. Some population centers have ATV repeaters. ATVers try to avoid interfering with the weak-signal work (moonbounce, for example) being done around 432 MHz and repeater operation above 442 MHz.

The Audio Channel

There are at least three ways to transmit voice information with a TV signal. The most popular method is by talking on another band, often 2-meter FM. This has the advantage of letting other local hams listen in on what you are doing—a good way to pick up some ATV converts! Rather than tie up a busy repeater for this, it is best to use a simplex frequency.

Commercial TV has an FM voice subcarrier 4.5 MHz above the TV picture carrier. If you provide FM audio at a frequency 4.5 MHz above the video frequency, the audio can easily be received in the usual way on a regular TV set. Many of the surplus FM rigs (tube type) that are available do not have enough bandwidth to pass both the picture carrier and the voice subcarrier at the same time, however, so other methods must be used.

Another way to go is to frequency modulate the video carrier. Since the video is amplitude modulated, it should not interfere with the FM audio, or vice versa. The usual way to receive this is with the FM receiver section of a UHF rig. It is also possible to use a lowband police/fire monitor coupled into the 44-MHz IF of the TV set.

Operating

Signal reporting on ATV differs from the RST system used on CW and phone. For example, "closed circuit" is the ATVer's term for "armchair copy"! The received signal strength is indicated by the amount of snow on the screen—the more snow, the weaker the signal. Some operators use a reporting system running from P0 to P5, roughly corresponding to readibility reports on voice or CW.

Signal reports are very useful for "talking the picture in" on the air while the sending station makes transmitter adjustments. In fact, this is the only reliable way to adjust the video modulator, since receiver overloading will give misleading results when you monitor your own signal. A picture that looks fine on your own TV receiver will likely have low contrast on a distant station's set.

Camera adjustments are best made with the output connected directly to a monitor or receiver video input. If you don't want to modify a TV set to serve as a monitor, most closed-circuit television (CCTV) cameras have provisions for generating a VHF signal on one of the lower TV channels.

The Scanning Process

A picture is divided sequentially into pieces for transmission or presentation (viewing); this process is called *scanning*. A total of 525 scan lines comprise a frame (complete picture) in the U.S. television system. Thirty frames are generated each second. Each frame consists of two fields, each field containing 262½ lines. Sixty fields are generated each second. Scan lines from one field fall between (interlace) lines from the other field. The scanned area is called the *television raster*. If all 525 scan lines are numbered from top to bottom, then one field scans all the even-numbered lines and the second field

Table 2-5

ATV Standards

Line rate	15,750 Hz
Field rate	60 Hz
Frame rate	30 Hz
Horizontal lines	262½/field 525/frame
Sound subcarrier	4.5 MHz
Channel bandwidth (VSB—C3F)	6 MHz

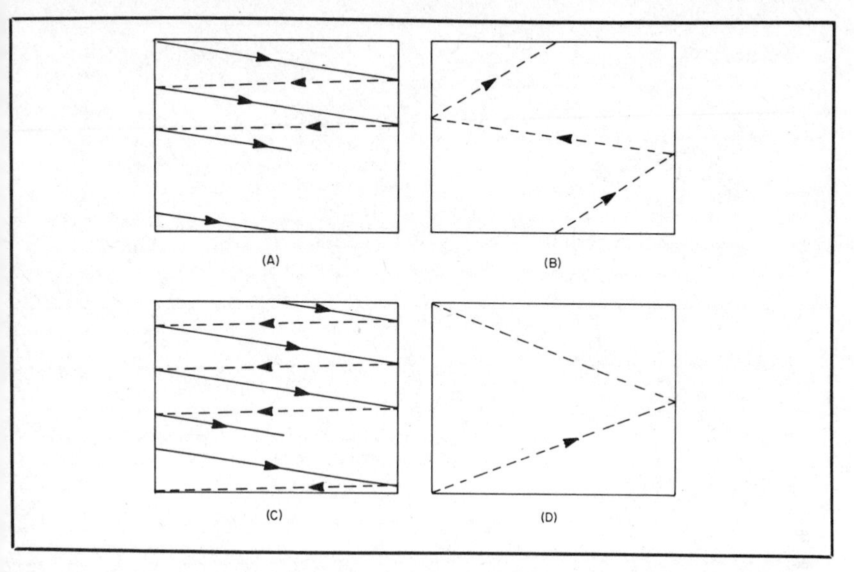

Fig. 2-11—Interlaced scanning used in TV. In field one, 262½ lines are scanned (A). At the end of field one, the electron scanning beam is returned to the top of the picture area (B). Scanning lines in field two (C) fall between the lines of field one. At the end of field two, the scanning beam is again returned to the top where scanning continues with field one (D).

scans all the odd-numbered lines. Table 2-5 lists some common ATV standards.

Fig. 2-11 illustrates the principle involved in scanning an electron beam (in the pickup or receiving device) across and down to produce the television raster. Field-one scanning begins in the upper left corner of the picture area. The electron beam is swept across the picture to the right side. At the end of the line, the beam is "turned off" or blanked and returned to the left side where the process repeats. In the meantime, the beam has also been moved slightly downward. At the end of 262½ horizontal scans (lines), the beam is blanked and rapidly returned to the top of the picture area. At that point, scanning of field two begins. Notice that this time the beam starts scanning from the middle of the picture. For that reason, the scanning lines of field two will fall halfway between the lines of field one. At the end of 262½ lines the beam is rapidly returned to the top of the picture again. Scanning continues—this time field one of the next frame.

The picture area is scanned 60 times each second. That is frequent enough to avoid visible flicker. Because the entire picture area is scanned only 30 times each second, bandwidth is reduced.

It should be obvious that the horizontal and vertical oscillators that control the electron-beam movement must be "locked together" for the two fields to interlace properly. If the frequencies of these oscillators are not locked, proper interlace will be lost and vertical resolution or detail will be degraded.

Deflection

In most TV applications, the electron beam is scanned (deflected) by means of two coil pairs. Because the deflection of the electron beam is accomplished magnetically, coils for horizontal deflection are located above and below the beam. Vertical deflection coils are located on either side of the beam. See Fig. 2-12.

The electron beam is deflected as a result of a "sawtooth" current passing through

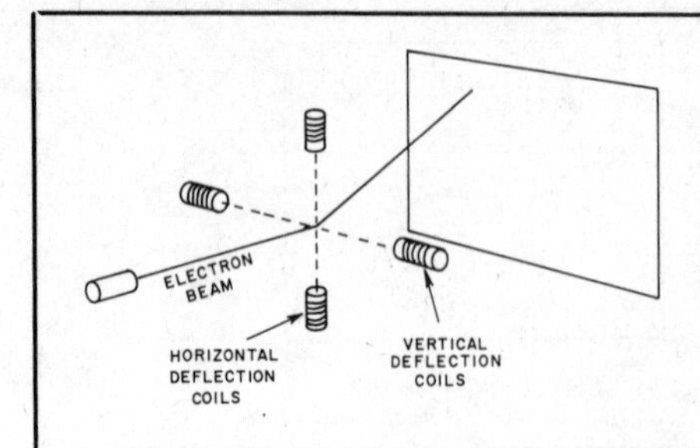

Fig. 2-12—Electron beam deflection in TV cameras and receivers is usually accomplished by using two sets of coils. This is called electromagnetic deflection.

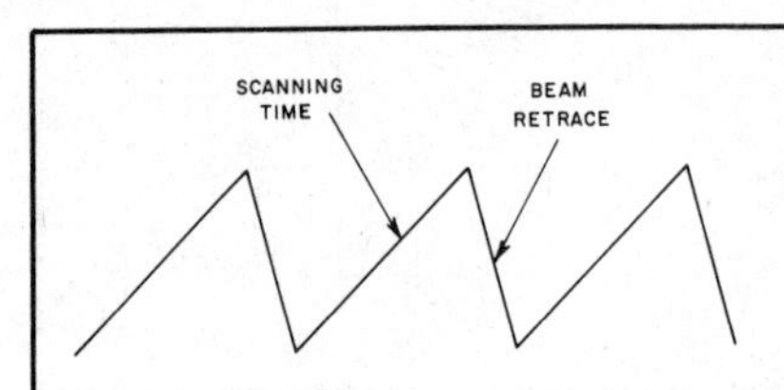

Fig. 2-13—Current waveform used in electromagnetic deflection coils for deflection of the electron scanning beam.

the deflection coils (Fig. 2-13). The frequency of the horizontal sawtooth current is 15,750 Hz. A similar waveform with a frequency of 60 Hz causes vertical deflection. The electron beam is turned off during beam retrace by a process called *blanking*. Blanking is only associated with the electron beam and does not affect the deflection coil current or the resulting magnetic fields.

Synchronizing Pulses

To ensure a stable picture at the receiver, the scanning process at the transmitting and receiving ends must be synchronized. The pulses used to control the horizontal scan are called *horizontal sync pulses*, and occur once per line—a frequency of 15,750 Hz. Horizontal sync pulses occur during the horizontal blanking interval.

Vertical scanning is controlled by *vertical sync pulses*, which can be seen in Fig. 2-14. The vertical sync pulse is preceded and followed by six equalizing pulses. The vertical sync pulse has a duration of three line times and is slotted or serrated at the end of each half-line time. Blanking turns off the electron beam during vertical retrace and sync.

The equalizing pulses are twice the frequency and approximately half the duration of horizontal sync pulses (2 equalizing pulses per line). The equalizing pulses ensure that there is always a pulse to lock the horizontal sweep to at the beginning of each line. Line 1 of field 1 begins at the first equalizing pulse and vertical sync starts at line 3½. This creates the ½-line offset that generates the interlaced scanning pattern shown in Fig. 2-11.

Composite Video

Video, as it comes from the *vidicon tube*, is not suitable for transmission. (See Chapter 6 for more information about vidicon tubes.) Horizontal and vertical sync must be added first. The *composite video signal* contains video, blanking and sync. Fig. 2-15 shows these three elements combined for two horizontal lines.

Today's compact, hand-held cameras require only power and light to generate a composite video signal. Some older cameras and most TV studio cameras require

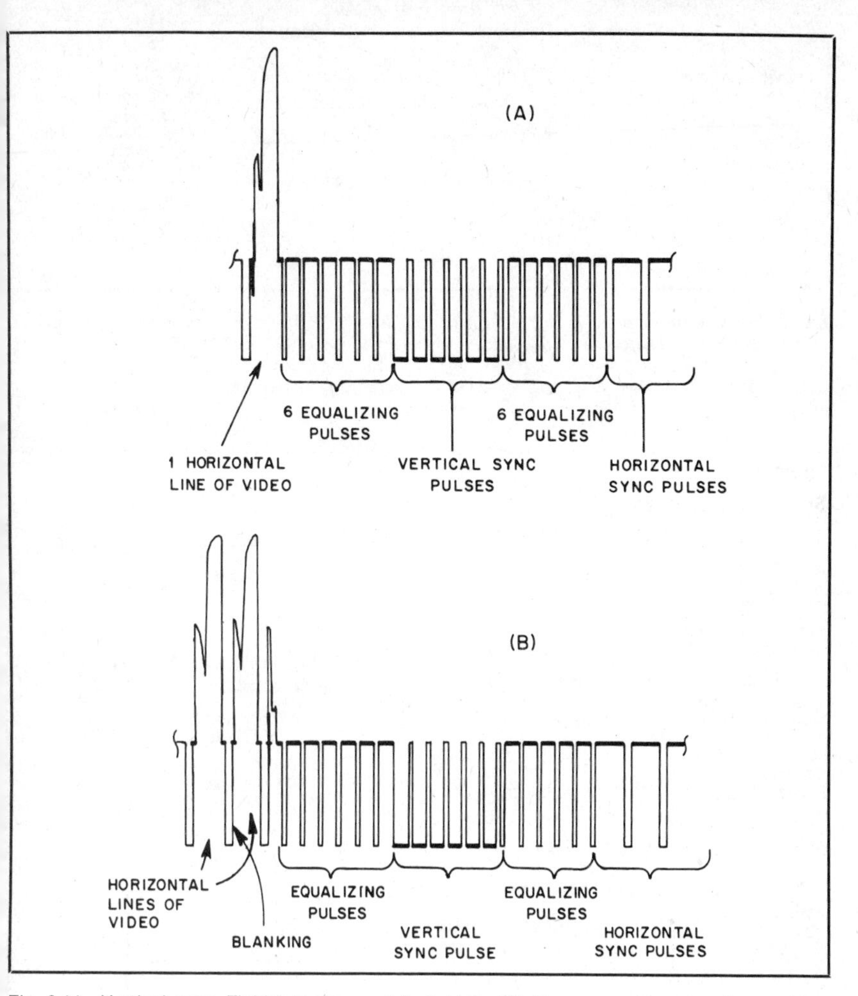

Fig. 2-14—Vertical sync. Field 1 is shown at A, field 2 at B. These waveforms are inverted for transmission.

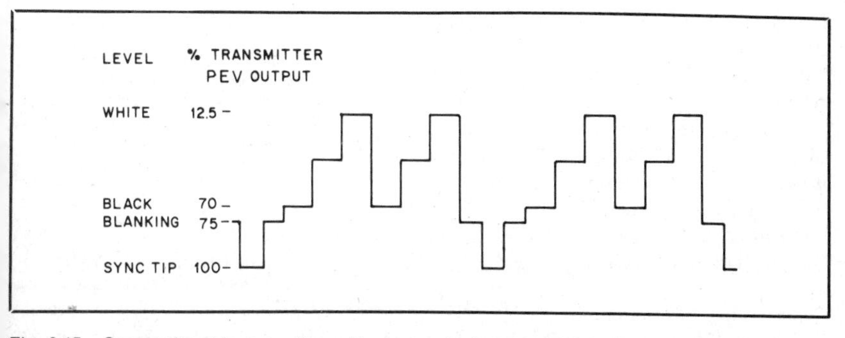

Fig. 2-15—Composite video waveform. For transmission, sync is positive and white is negative.

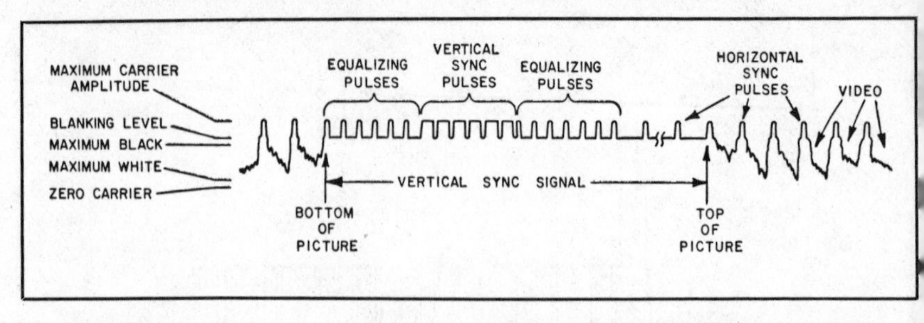

Fig. 2-16—A composite video signal as it is sent from a TV transmitter. The horizontal sync pulses, superimposed on the video (picture information) signal "tell" the TV receiver when to start a new horizontal line. The vertical sync signal is sent at the bottom of each vertical sweep to tell the TV set to go to the top of the screen to start a new field.

Table 2-6

Standard Video Levels

	IEEE Units	*%PEV*
Zero carrier	120	0
White	100	12.5
Black	7.5	70
Blanking	0	75
Sync tip	−40	100

PEV is peak envelope voltage and corresponds to levels as seen on an oscilloscope.

horizontal and vertical sync signals from an external source. Those signals can be provided by a sync generator.

Video Transmission

Video from the TV camera normally has white positive and sync negative. The standard video voltage level between white and the sync tip is 1-V peak to peak. Monitors are made for sync-negative video. For transmission, however, the polarity or sense of the video is inverted—the sync is positive. That puts the sync tip at peak power output from the transmitter. The video sense is inverted again in the receiver. See Table 2-6 and Fig. 2-16.

Because the sync tip corresponds to maximum transmitter output, a receiver is better able to hold a stable picture. This is especially true under adverse reception conditions such as noise or weak signals. It is far better to have a bit more noise in the picture than to have it be unstable.

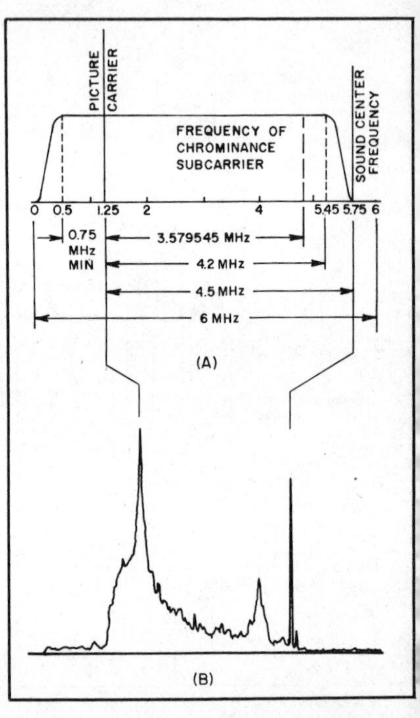

Fig. 2-17—Frequency spectrum of a color TV signal shown in diagram at A. B represents a spectrum analyzer display. Spectrum power density will vary with picture content, but typically 90% of sideband power is within the first megahertz.

Channel Bandwidth

Color TV pictures require about 4-MHz bandwidth. Satisfactory black and white pictures can be realized with less bandwidth. If a color TV picture is to be trans-

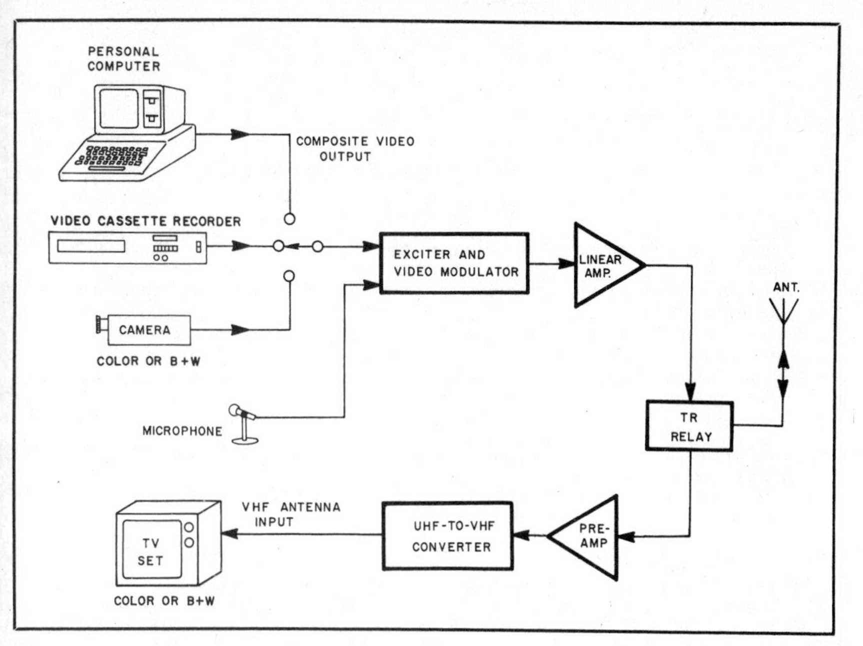

Fig. 2-18—Block diagram of a fast-scan ATV station.

mitted with double sideband AM (DSB), more than 8 MHz of spectrum space is required. It is not necessary to use that much spectrum—although amateurs can, and some do, use DSB. The WARC-79 emission designator for double-sideband emissions on a single channel containing analog video information is A3F.

Commercial TV stations and many amateurs use *vestigial sideband (VSB)* for transmission. VSB is like SSB with full carrier except a portion (vestige) of the unwanted sideband is retained. In the case of VSB TV, approximately 1 MHz of the lower sideband and all of the upper sideband plus full carrier comprise the transmitted picture signal. The WARC-79 emission designator for vestigial-sideband emissions on a single channel containing analog video information is C3F. Both A3F and C3F emissions are designated A5 under the old system of emissions designators.

The sound carrier is 4.5 MHz above the picture carrier, as shown in Fig. 2-17. You can also see in Fig. 2-17 that the channel bandwidth for VSB TV with sound carrier is 6 MHz.

Bandwidth of an ATV signal is usually said to be 6 MHz. It may be less for black-and-white pictures with no sound. Bandwidth will be greater if DSB transmission is used. As with voice transmission, it is not necessary to transmit both video sidebands.

A block diagram of a complete ATV station is shown in Fig. 2-18. The standard 4.5-MHz FM audio subcarrier is being used. Both the linear amplifier and the antenna should have sufficient bandwidth for the 6-MHz-wide TV signal.

[Before proceeding to Chapter 3, turn to Chapter 10 and study FCC examination questions with numbers that begin 4BB-2. Review this section as needed.]

Key Words

Apogee—That point in a satellite's orbit (such as the moon) when it is farthest from the earth.

Crooked-path propagation—Radio wave propagation along any direction other than a great-circle path.

Declination angle—The angle measured north or south from the celestial equator to an object in the sky.

Earth-moon-earth (EME)—A method of communicating with other stations by reflecting radio signals off the moon's surface.

Galactic plane—An imaginary plane surface extending through the center of the Milky Way galaxy. This plane is used as a reference for some astronomical observations.

GaAsFET—Galium-arsenide field-effect transistor. A low-noise device used in UHF and higher frequency amplifiers.

Gray line—A band around the earth that is the transition region between daylight and darkness.

Gray-line propagation—A generally north-south enhancement of propagation that occurs along the gray line, when D layer absorption is rapidly decreasing at sunset, or has not yet built up around sunrise.

Great circle—An imaginary circle around the surface of the earth formed by the intersection of the surface with a plane passing through the center of the earth.

Great-circle path—The shortest distance between two points on the surface of the earth, which follows the arc of a great circle passing through both points.

Greenwich Hour Angle (GHA)—The angle measured east or west from the prime meridian to an object in the sky.

Libration fading—A fluttery, rapid fading of EME signals, caused by short-term variations on the aspect of the moon relative to earth.

Long-path propagation—Propagation between two points on the earth's surface that follows a path along the great circle between them, but is in a direction opposite from the shortest distance between them.

Meteor—A particle of mineral or metallic material that is in a highly elliptical orbit around the sun. As the earth's orbit crosses the orbital path of a meteor, it is attracted by the earth's gravitational field, and enters the atmosphere. A typical meteor is about the size of a grain of sand.

Meteor-burst communication—A method of radio communication that uses the ionized trail of a meteor that has burned up in the earth's atmosphere to refract (bend) radio signals back to earth.

Moonbounce—A common name for EME communication.

Path loss—The total signal loss between transmitting and receiving stations relative to the total radiated signal energy.

Perigee—That point in a satellite's orbit (such as the moon) when it is closest to the earth.

Semidiameter—The apparent radius of a generally spherical object in the sky. Semidiameter is usually expressed in degrees of arc as measured between imaginary lines drawn from the observer to the center of the object (such as the moon) and from the observer to one edge.

Sky temperature—A measure of relative background noise coming from space, often expressed in kelvins. The sun is a source of much background noise, and so has a high noise temperature.

Transequatorial propagation—A form of F-layer ionospheric propagation, in which signals of higher frequency than the expected MUF are propagated across the earth's magnetic equator.

Chapter 3

Radio-Wave Propagation

By the time you are ready to study for an Amateur Extra Class license, you should have a pretty good understanding of the basic modes of propagation for radio waves. Now it is time to learn about ways to take advantage of some exotic propagation methods, such as bouncing signals off the moon or using the ionized trail from a meteor to refract (bend) signals back to earth. You will also learn about some of the directional characteristics of ionospheric propagation, so you can determine the best direction to point your beam antenna to contact a desired part of the world. When you have studied the information in each section of this chapter, use the examination questions to check your understanding of the material. If you are unable to answer a question correctly, go back and review the appropriate part of this chapter.

EARTH-MOON-EARTH

The concept of *EME*, popularly known as *moonbounce*, is straightforward: Stations that can simultaneously see the moon communicate by reflecting VHF or UHF signals off the lunar surface. Those stations may be separated by nearly 180° of arc on the earth's surface—a distance of more than 11,000 miles.

There is a drawback, though; since the moon's mean distance from earth is 239,000 miles, path losses are huge when compared to "local" VHF paths. *Path loss* refers to the total signal loss between the transmitting and receiving stations relative to the total radiated signal energy. Thus, each station on an EME circuit demands the most out of the transmitter, antenna, receiver and operator skills. Even when all those factors are optimized, the signal in the headphones may be barely perceptible above the noise. Nevertheless, for any type of amateur communication over a distance of 500 miles or more at 432 MHz, for example, moonbounce comes out the winner over terrestrial propagation paths when various factors are weighed on a balance sheet.

EME thus presents amateurs with the ultimate challenge in radio system performance. Today, most of the components for an EME station on 144, 220, 432 or 1296 MHz are commercially available. Whether one chooses to buy or build station equipment, some system design requirements must be met, because this is extremely weak-signal work.

1) Transmissions must be made on CW or SSB with as close to the maximum power level as possible.

2) The antenna should have at least 20 dB gain over a dipole. An array of four or more antennas is usually required to realize this amount of gain. Chapter 9 has more information about antenna-system gain.

3) Antenna rotators are needed for both azimuth and elevation. Since the half-power beamwidth of a high-gain antenna is quite sharp, the rotators must have an appropriate accuracy.

4) Transmission-line losses should be held to a minimum.

5) The receiving system should have a very low noise figure.

EME Scheduling

The best days to schedule an EME contact are usually when the moon is at *perigee* (closest to the earth) since the path loss is typically 2 dB less than when the moon is at *apogee* (farthest from the earth). The moon's perigee and apogee dates may be determined from publications such as *The Nautical Almanac* by inspecting the section of the tables headed "S.D." (semidiameter of the moon in minutes of arc). An S.D. of 16.53 equates to an approximate earth-to-moon distance of 225,000 miles, typical perigee, and an S.D. of 14.7 to an approximate distance of 252,500 miles, typical apogee. Once you know the semidiameter of the moon for a given date, the value is located on the graph of Fig. 3-1, and the EME path loss in decibels may be determined for the most popular amateur frequencies.

The moon's orbit is slightly elliptical. Hence, the day-to-day path-loss changes at apogee and perigee are minor. The greatest changes take place at the time when the moon is traversing between apogee and perigee. However, several other factors must be considered for optimum scheduling, aside from the path losses.

If perigee occurs near the time of a new moon, one to two days will be unusable since the sun is behind the moon and will cause increased sun-noise pickup. This noise will mask weak signals. Therefore, avoid schedules when the moon is within 10 degrees of the sun (and farther if your antenna has a wide beamwidth or strong side lobes). The moon's orbit follows a cycle of 18 to 19 years, so the relationships between perigee and new moon will not be the same from one year to the next.

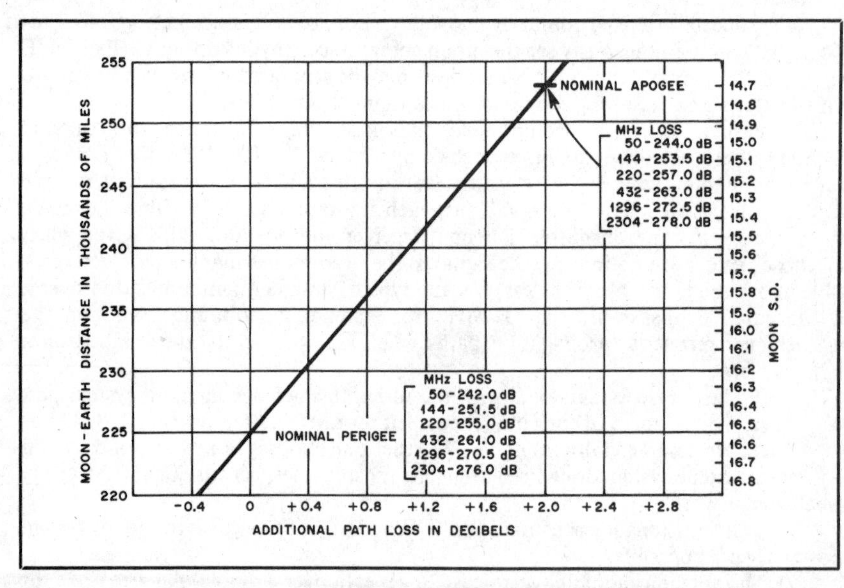

Fig. 3-1—Variations in EME path loss can be determined from this graph. S.D. refers to the semidiameter of the moon, which is indicated for each day of the year in *The Nautical Almanac*.

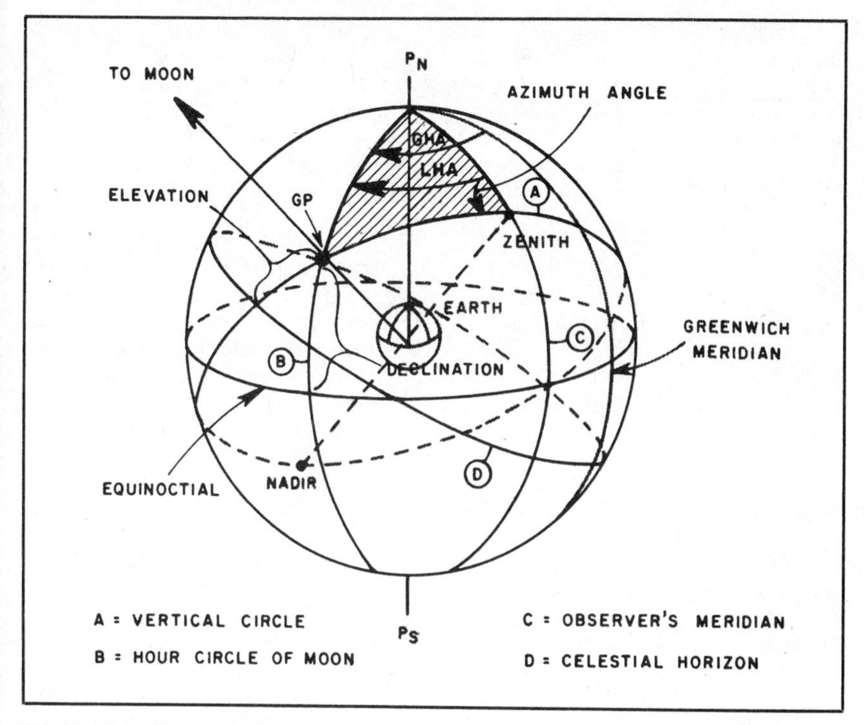

Fig. 3-2—Spherical coordinate system used to define azimuth and declination or elevation angles for celestial objects.

In order to calculate the position of the moon to determine the times you can work moonbounce, and for pointing your antenna system accurately, we must define a reference coordinate system. Astronomers use a set of coordinates called spherical coordinates. There are a number of spherical coordinate systems in common use, but the one that is easiest to understand, and that is entirely adequate for our purpose, assumes the earth to be at the center of the universe. Fig. 3-2 illustrates the terms associated with this coordinate system. The sky is like a large balloon, completely surrounding the earth. If we extend a line through the north and south poles, so it touches the "celestial sphere," then we have defined the celestial poles. The earth's equator is a plane that passes through the center of the earth, and is perpendicular to the line passing through the poles. If we extend that plane to meet the celestial sphere, then we have defined the celestial equator, or the equinoctial.

Any plane that passes through the center of a sphere marks off a *great circle* on that sphere. So the equator is a great circle on the earth and the equinoctial is a great circle on the celestial sphere. The diameter of a great circle is equal to the diameter of the sphere. Note that you can draw other planes through the circle, and if they do not pass through the center of the sphere, the circles they define will have diameters smaller than the diameter of the sphere. Lines of latitude on the earth are examples of non-great-circle planes. You can also define any number of great circles around the sphere. Besides the equator, sometimes called the primary great circle, we are interested in a set of great circles that go through the poles. These correspond to lines of longitude on the earth, and are referred to as hour circles or secondary great circles on the celestial sphere.

Imagine a secondary great circle defined by the plane that passes through the prime meridian in Greenwich, England. Since we use this circle as a reference for measurements on earth, we can also use it as a reference for measurements on our celestial sphere. You should also imagine a plane extending through your location to define a great circle on the celestial sphere, and one passing through the moon and the center of the earth. The equinoctial, the Greenwich meridian and the moon's hour circle will allow you to define the position of the moon in such a way that it can be located from any place on earth.

The angle measured from the Greenwich meridian to the moon's hour circle is called the *Greenwich hour angle (GHA)*. The local hour angle is measured from your hour circle to the moon's hour circle. The *declination angle* is measured from the equinoctial, north or south along the moon's hour circle, to the moon.

Also shown on Fig. 3-2 are two points labeled zenith and nadir, along with a great circle labeled celestial horizon. Zenith refers to a point directly overhead at your location, nadir is the point on the celestial sphere directly under you, and the celestial horizon is the extension of a plane passing through your horizon and the center of the earth. These points define another spherical coordinate system, but one which is specific only to your location, and would not be useful for stations at other locations. It is, however, more convenient for you to calibrate your antenna system based on elevation angle and azimuth.

High moon declinations and high antenna elevation angles should yield best results. By contrast, low moon declinations and low aiming elevations generally produce poor results and should be avoided if possible. Generally, low elevation angles increase antenna-noise pickup and increase tropospheric absorption, especially above 420 MHz, where the galactic noise is very low. This situation cannot be avoided when one station is unable to elevate the antenna above the horizon or when there is a great terrestrial distance between stations. Ground gain (gain obtained when the antenna is aimed at the horizon) has been used very effectively at 144 MHz. While potentially useful, ground gain is a complex phenomenon and difficult to predict.

Usually, signals are stronger in the fall and winter months and weaker in the summer. Also, signals are generally better at night than during the day. This may be attributable to decreased ionization or less Faraday rotation.

Whenever the moon crosses the *galactic plane* (twice a month for three to five days each occurrence), the *sky temperature* will be higher (more noise). The galactic plane is an imaginary plane surface extending through the center of the Milky Way galaxy. Other areas of the sky to avoid are the constellations Orion and Gemini at northern declinations and Scorpius and Sagittarius at southern declinations. Positions of the moon with respect to these constellations can be checked with *Sky and Telescope* magazine or *The Nautical Almanac*. The galactic plane is biased toward southern declinations, which will cause southerly declinations to be less desirable (with respect to noise) than are northerly declinations.

Finally, the time of the day and the day of the week must be considered since most of us have to work for a living and cannot always be available for schedules. Naturally, weekends and evenings are preferred, especially when perigee occurs on a weekend.

General Considerations

It helps to know your own EME window as accurately as possible. The term "window" means the period of time that a station can "see" the moon. This can be determined with the help of information contained in a later section of this chapter. Most EME operators determine their local window and translate it into GHA (Greenwich hour angle) and declination. This information is a constant, so once determined, it is usable by other stations just as one would use UTC. Likewise, it helps to know the window of the station to be scheduled. Most EME stations are limited in some way by local obstructions, antenna-mounting constraints, geographical considerations, and the like. Therefore, the accuracy of each station's EME window is very important for locating common windows and setting schedule times.

Most multiple-antenna arrays exhibit some pattern skewing. Skewing means that the main antenna lobe does not coincide with the "front" of the mechanical antenna. In those cases, a boresight of some type is practically mandatory in order to align your antenna accurately with the moon. A simple calibration method is to peak your antenna on received sun noise and then align the boresight tube on the sun. The boresight of the antenna is now calibrated and can be used to aim the antenna at the moon. Do not use a telescope or other device employing lenses as a boresight device! Even the best optical filters will not eliminate the hazard from solar radiation when viewed directly. A simple piece of small-diameter tubing, 2 or 3 feet long, can serve this purpose. A symmetrical spot of light cast upon a piece of paper near the back end of the tube will indicate alignment. With this type of boresight system, you should never look through the tube to align the antenna with the sun.

An accurate rotator readout (±2°) is highly recommended. A remote readout is particularly important for scheduling when the moon is within 45 degrees of the sun or when the sky is overcast. There are very few areas where the moon is not occasionally obscured by cloud cover. Aiming an antenna blindly seldom pays off.

Libration Fading of EME Signals

One of the most troublesome aspects of receiving a moonbounce signal, besides the enormous path loss and Faraday-rotation fading, is *libration* (pronounced lie-bray-shun) *fading*. Libration fading of an EME signal is characterized in general as fluttery, rapid, irregular fading not unlike that observed in tropospheric-scatter propagation. Fading can be very deep, 20 dB or more, and the maximum fading will depend on the operating frequency. At 1296 MHz the maximum fading rate is about 10 Hz, and is directly proportional to frequency.

On a weak CW EME signal, libration fading gives the impression of a randomly keyed signal. In fact on very slow CW telegraphy the effect is as though the keying is being done at a much faster speed. On very weak signals only the peaks of libration fading are heard in the form of occasional short bursts or "pings."

Fig. 3-3 shows samples of a typical EME echo signal at 1296 MHz. These recordings show the wild fading characteristics with sufficient S/N ratio to record the deep fades. Circular polarization was used to eliminate Faraday fading; thus, these record-

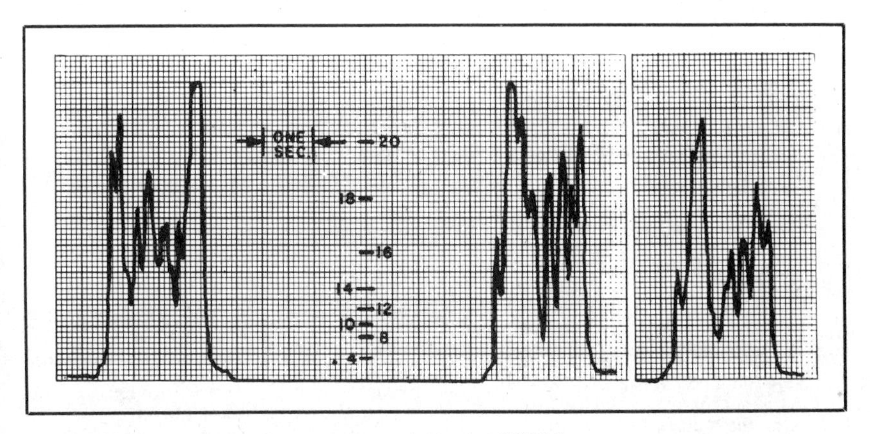

Fig. 3-3—Chart recording of moon echoes received at W2NFA.

ings are of libration fading only. The recording bandwidth was limited to about 40 Hz to minimize the higher sideband-frequency components of libration fading that exist but are much smaller in amplitude. In the recordings shown by Fig. 3-3, the average signal-return level computed from path loss and mean reflection coefficient of the moon is at about the +15 dB S/N level.

It is clear that enhancement of echoes far in excess of this average level is observed. This point should be kept clearly in mind when attempting to obtain echoes or receive EME signals with marginal equipment. The probability of hearing an occasional peak is quite good since random enhancement as much as 10 dB is possible. Under these conditions, however, the amount of useful information that can be copied will be near zero. Enthusiastic newcomers to EME communications will be frustrated by this effect. They hear the signal strong enough to copy on peaks but they can't copy enough to make sense.

What causes libration fading? Very simply, multipath scattering of the radio waves from the very large (2000-mile diameter) and rough moon surface combined with the relative short-term motion between earth and moon. These short-term oscillations in the apparent aspect of the moon relative to earth are called librations.

To understand these effects, assume first that the earth and moon are stationary (no libration) and that a plane wave front arrives at the moon from your earth-bound station as shown in Fig. 3-4A.

The reflected wave shown in Fig. 3-4B consists of many scattered contributions from the rough moon surface. It is perhaps easier to visualize the process as if the scattering were from many small individual flat mirrors on the moon that reflect small portions (amplitudes) of the incident wave energy in different directions (paths) and with different path lengths (phase). Those signals reflected from the moon arrive at your antenna as a collection of small wave fronts (field vectors) of various amplitudes and phases. All these returned waves (we can consider their number to be infinite) are combined at the feed point of your antenna. The level of the final addition, as measured by a receiver, can have any value from zero to some maximum. Remember that we assumed the earth and moon were not moving with respect to one another.

Consider now that the earth and moon are moving relative to each other (as they are in nature), so the incident radio wave "sees" a slightly different surface of the moon from moment to moment. Since the lunar surface is very irregular, the reflected wave will be equally irregular, changing in amplitude and phase from moment to moment. The result is a number of continuously varying multipath signals at your antenna feed point, which produces the effect called libration fading of the moon-reflected signal.

The term libration is used to describe small irregularities in the movement of celestial bodies. Lunar libration consists mainly of its 28-day rotation, which appears as a very slight rocking motion with respect to an observer on earth. This rocking motion can be visualized as follows: Place a marker on the surface of

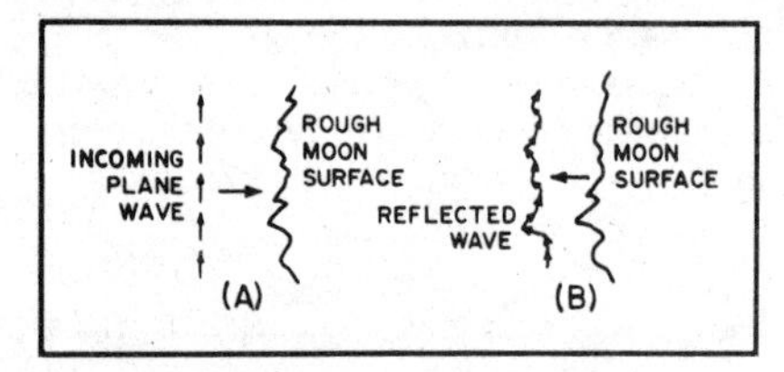

Fig. 3-4—How the rough surface of the moon reflects a radio wave.

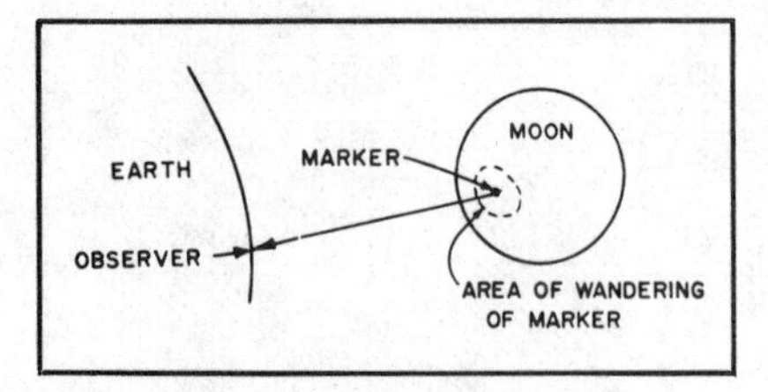

Fig. 3-5—The moon appears to "wander" in its orbit about the earth. Thus, a fixed marker on the moon's surface will appear to move about in a circular area.

the moon at the center of the moon disc, which is the point closest to the observer, as shown in Fig. 3-5. Over time, we will observe that this marker wanders around within a small area. This means the surface of the moon as seen from the earth is not quite fixed but changes slightly as different areas of the periphery are exposed because of this rocking motion. Moon libration is very slow. Although the libration motions are very small and slow, the larger surface area of the moon has an infinite number of scattering points, each with a very small area. This means that even slight movements can alter the returned multipath echo by a significant amount.

Frequencies

EME contacts are generally made randomly or by prearranged schedule. Many stations, especially those with marginal capability, prefer to set up a specific time and frequency in advance so that they will have a better chance of finding each other. The larger stations, especially on 144 and 432 MHz where there is a good amount of activity, often call CQ during evenings and weekends when the moon is at perigee and listen for random replies. Most of the work on 220, 1296 and 2304 MHz, where activity is light, is done by schedule.

An EME net meets on weekends on 14.345 MHz for the purpose of arranging schedules and exchanging pertinent information. Those operating EME at 432 MHz and above meet at 1600 UTC, followed by the 144-MHz operators at 1700 UTC. Both nets carry information on 220-MHz EME operation.

Most amateur EME work on 144 and 220 MHz takes place near the low edge of the band. Activity is found 50 kHz or higher in the band during peak hours. Generally, random activity and CQ calling take place in the lower 10 kHz or so, and schedules are run higher in the band.

Formal schedules (that is, schedules arranged well in advance and published in the *432 and Above EME Newsletter* published by Al Katz, K2UYH) are run on 432.000, 432.025 and 432.030 MHz. Other schedules are normally run on 432.035 and up to 432.070. For this band, the EME random-calling frequency is 432.010, with random activity spread out between 432.005 and 432.020. Random SSB CQ calling is at 432.015. Terrestrial activity is centered on 432.100 and is, by agreement, limited to 432.075 and above in North America.

Moving up in frequency, formal schedules are run on 1296.000 and 1296.025 MHz. The EME random calling frequency is 1296.010 with random activity spread out between 1296.005 and 1296.020. Terrestrial activity is centered at 1296.100. There is some EME activity on 2304 MHz. Specific frequencies are dictated by equipment availability and are arranged by the stations involved.

For EME SSB contacts on 144 and 432 MHz, contact is usually established on CW, and then the stations move up 100 kHz from the CW frequency. (This method was adopted because of the U.S. requirement for CW only below 144.1 MHz.)

Of course, it is obvious that as the number of stations on EME increases, the frequency spread must become greater. Since the moon is in convenient locations only a few days out of the month, and only a certain number of stations can be scheduled for EME during a given evening, the answer will be in the use of simultaneous schedules, spaced a few kilohertz apart. The time may not be too far away—QRM has already been experienced on each of our three most active EME frequencies.

Receiver Requirements

A low-noise receiving setup is essential for successful EME work. Since many of the signals to be copied on EME are barely, but not always, out of the noise, a low-noise-figure receiver is a must. The mark to shoot for at 144 MHz is something under 0.5 dB, as the cosmic noise will then be the limiting factor in the system. Noise figures of this level are relatively easy to achieve with inexpensive modern devices. As low a noise figure as can be attained will be usable at 432 and 1296 MHz. Noise

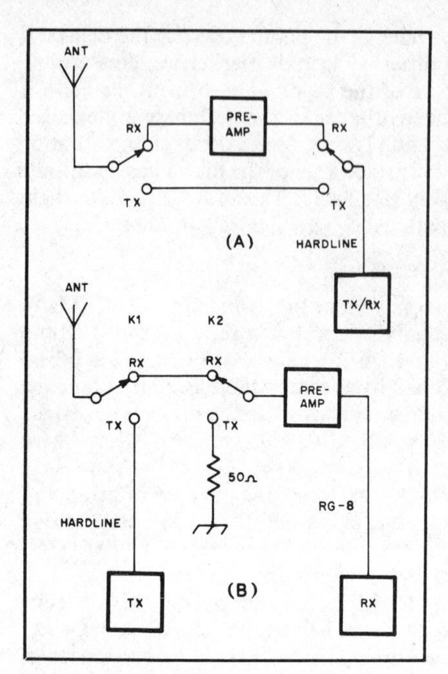

Fig. 3-6—Two systems for switching a preamplifier in and out of the receive line. At A, a single length of cable is used for both the transmit and receive line. At B is a slightly more sophisticated system that uses two separate transmission lines. See text for details.

figures on the order of 0.5 dB are possible with amplifiers that use *GaAsFET* devices.

The loss in the transmission line that connects the antenna to the preamplifier adds directly to the system noise figure. Not only does the line have loss, it also is a source of noise, which further deteriorates the system noise figure. For those reasons, most serious EME operators mount a preamplifier at the top of the tower or directly at the antenna. If an exceptionally good grade of transmission line is available, it is possible to obtain almost as good results with the preamplifier located in the shack. Two relay/preamplifier switching systems are sketched in Fig. 3-6. The system at A makes use of two relays and a single transmission line for both transmit and receive. The preamplifier is simply switched "in" for receive and "out" for transmit.

The system outlined at Fig. 3-6B also uses two relays, but the circuit is somewhat more sophisticated. Two transmission lines are used, one for the receive line and one for the transmit line. In addition, a 50-ohm termination is provided for the receiver during transmit times. Since relays with high isolation in the VHF/UHF range are expensive, and difficult to obtain, two relays with a lower isolation factor may be used. When the relays are switched for the transmit mode, K1 connects the antenna to the transmit line, K2 switches the preamplifier into the 50-ohm termination. Hence, two relays provide the isolation between the transmitter connection and the preamplifier.

If independent control of K2 is provided, the preamplifier can be switched between the 50-ohm termination and the antenna during receive. This feature is especially useful when making sun-noise measurements to check system performance. For this measurement, the antenna is directed toward the sun and the preamplifier is alternately switched between the 50-ohm load and the antenna. The decibel difference can be recorded and used as a reference when checking system improvements. A less convenient, but better, check is to compare sun noise with cold sky noise.

Since the preamplifier is mounted ahead of the transmission line to the receiver, a cable of good (as opposed to excellent) performance can be used. The loss in the cable, as long as it is within reason, will not add appreciably to the system noise figure. Foam-type RG-8 cable is acceptable for runs up to 100 feet at 144 MHz.

It is important to get as much transmitter power as possible to the antenna. For this reason rigid or semirigid low-loss cable (hardline) is specified for the transmit line.

Transmitter Requirements

In many EME installations the antenna gain is not much above the minimum required for communications. It is highly likely that the maximum legal power limit will be required for successful EME work on 432 MHz and lower frequencies.

Since many contacts may require long, slow sending, the transmitter and amplifier should have adequate cooling. An amplifier with some power to spare rather than an amplifier running "flat out" is desirable. This is especially important should SSB communication be attempted. An amplifier run all out on SSB will likely produce large amounts of odd-order IMD products that fall within the band. While the splatter produced will not affect your communications, it will certainly affect that of others close in frequency!

[Study FCC examination questions with numbers that begin 4BC-1. Review this section as needed.]

METEOR-BURST PROPAGATION

Meteors are particles of mineral or metallic matter that travel in highly elliptical orbits about the sun. Most of these are microscopic in size. Every day hundreds of millions of these meteors enter the earth's atmosphere. Drawn by the earth's gravitational field, they attain speeds from 6 to 60 mi/s (22,000 to 220,000 mi/h).

As a meteor speeds through the upper atmosphere, it begins to burn or vaporize as it collides with air molecules. This action creates heat and light and leaves a trail of free electrons and positively charged ions behind as the meteor races along its parabolic path. Trail size is directly dependent on meteor size and speed. A typical meteor is the size of a grain of sand. A particle of this size creates a trail about 3 feet in diameter and 12 to 40 miles long, depending on speed.

Duration of meteor-produced ionization is directly related to electron density. Ionized air molecules contact and recombine with free electrons over time, gradually lowering the electron density until it returns to its previous state.

Radio waves can be refracted as they encounter the ionized trail of a meteor. The ability of a meteor trail to refract radio signals depends on electron density—greater density causing greater refracting ability and refraction at higher frequencies. The electron density in a typical meteor trail will strongly affect radio waves between 24 and 60 MHz.

The signals refracted by a meteor trail propagate just as they do for regular ionospheric propagation. Meteor trails are formed at approximately the altitude of the ionospheric E layer, 50 to 75 miles above the earth. That means that the range for meteor-burst propagation is about the same as for single-hop E (or sporadic-E) skip, a maximum of approximately 1200 miles (Fig. 3-7).

A few meteors create enough ionization to refract 144-MHz signals. Fewer yet can refract 220-MHz signals. Rarely do meteor trails have enough ionization to refract

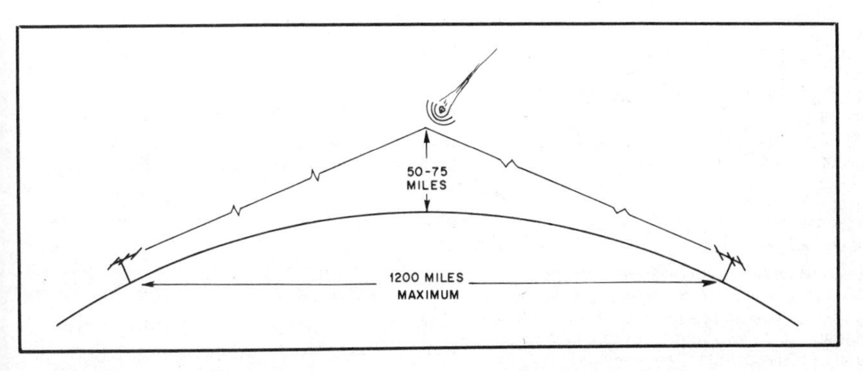

Fig. 3-7—Meteor-burst communication makes extended-range VHF communications possible.

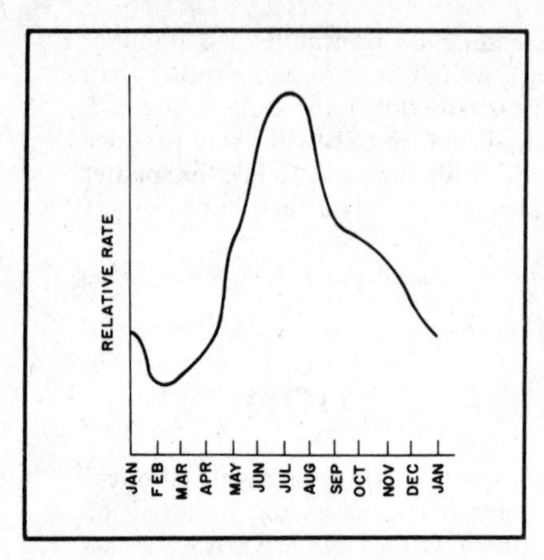

Fig. 3-8—Seasonal variation of sporadic meteor rates. This graph does not include shower meteors. See text and Table 3-1.

432-MHz signals, although meteor-burst contacts have been made on that frequency.

Propagation potential of a "meteor trail" is highly frequency dependent. For example, consider the result of a relatively large (about the size of a peanut) meteor entering the earth's atmosphere. Using the ionized trail of that meteor, no 432-MHz propagation is possible. At 220 MHz, moderately strong signals are heard over a 1200-mile path for perhaps 12 seconds. The signals gradually fade into the noise over the next four or five seconds. At 144 MHz, the same conditions might result in very strong signals for 20 seconds that slowly fade into the noise during the next 20 seconds. At 28 MHz, propagation might well last for a couple of minutes.

Meteors the size of a peanut are relatively rare; most are much smaller. As a result, the typical meteor burst is much shorter than that cited in the example. Only a few meteors will generate a trail strong enough to propagate 220-MHz signals.

The number of meteors the earth encounters as it sweeps through its orbit around the sun varies from month to month as shown in Fig. 3-8. The curve in Fig. 3-8 shows only sporadic or random meteors as contrasted with shower meteors.

Meteor Showers

At certain times of the year, the earth encounters greatly increased numbers of meteors. Great swarms of meteors, probably the remnants of a comet, orbit the sun. Each year the earth passes through these swarms causing the so-called meteor showers. These showers greatly enhance meteor-burst communication at VHF. The degree of enhancement depends on the time of day, shower intensity and the frequency in use. The largest meteor showers of the year are the Perseids and the Geminids. The Perseids appear to come from the constellation Perseus; this shower occurs in August. December 11 and 12 is when the Geminids peak; these seem to come from Gemini. Table 3-1 is a partial list of meteor showers throughout the year.

Table 3-1
Major Meteor Showers

Date	*Name*
January 3-5	Quadrantids
April 19-23	Lyrids
*May 19-21	Cetids
*June 4-6	Perseids
*June 8	Arietids
*June 30-July 2	Taurids
July 26-31	Aquarids
July 27-August 14	Perseids
October 18-23	Orionids
October 26-November 16	Taurids
November 14-18	Leonids
December 10-14	Geminids
December 22	Ursids

All showers occur in evening except those marked (*), which are daytime showers. Evening showers begin at approximately 2300 local standard time, daylight showers at approximately 0500 local standard time.

Meteor-burst communication is best between midnight and dawn. The part of the earth that is between those hours, local time, is always on the leading edge as the earth travels along its orbit. It is at that leading edge where most meteors enter the earth's atmosphere.

Conditions peak around dawn; that is when the relative velocity between meteors and earth is the greatest. See Figs. 3-9 and 3-10. At the leading edge of the earth, the earth's orbital velocity is

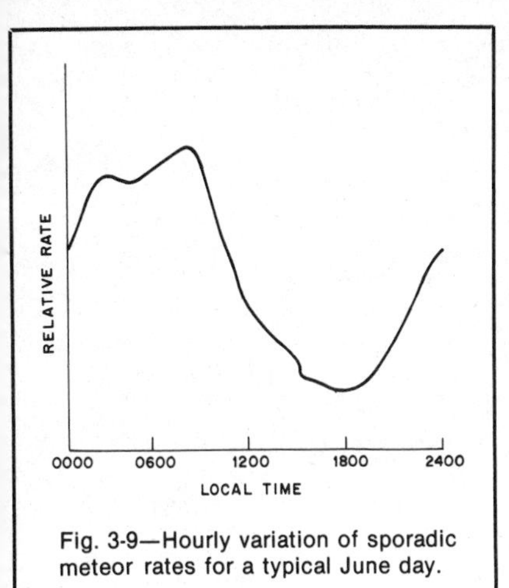

Fig. 3-9—Hourly variation of sporadic meteor rates for a typical June day.

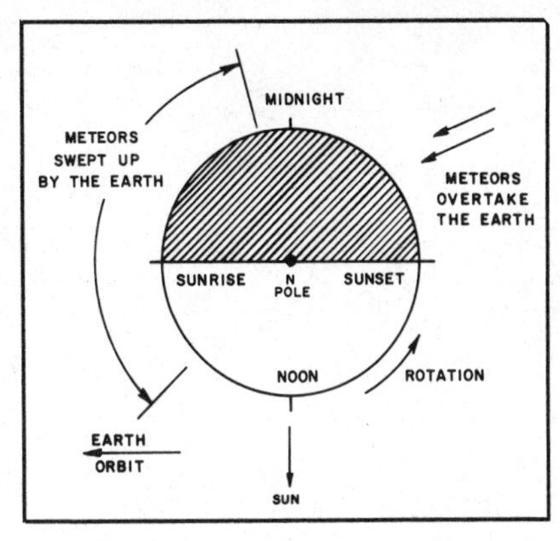

Fig. 3-10—Best meteor rates occur around dawn. During afternoon and evening hours only meteors overtaking the earth enter the atmosphere.

added to the meteor velocity—like two vehicles involved in a head-on crash. Because burst duration is related to meteor velocity, the relative velocity improvement also results in an extended burst duration.

Operating Hints

The secret to successful meteor-burst communication is short transmissions. To call CQ, for example, say (or send) CQ followed by your call sign, repeated two or three times. To answer a CQ, give the other station's call once followed by your call phonetically. Example:
"CQ from Whiskey Bravo Five Lima Uniform Alfa." "WB5LUA this is Whiskey One Juliett Romeo." WB5LUA responds: "W1JR 59 Texas." W1JR immediately responds: "Roger 59 Massachusetts." WB5LUA says: "Roger QRZ from WB5LUA." The entire QSO with information exchanged and confirmed in both directions lasted only 12 seconds!

A single meteor may produce a strong enough path to sustain communication long enough to complete a short QSO. At other times multiple bursts are needed to complete the QSO, especially at higher frequencies. Remember, the key is short, concise transmissions. Do not repeat information unnecessarily; that is a waste of time and propagation.

In the United States there is an accepted convention for transmission timing. Each minute is broken up into four 15-second periods. The station at the western end of the path transmits during the first and third period of each minute. During the second and fourth 15-second periods, the eastern station transmits.

Meteor-burst propagation and packet radio seem to be made for each other. High-speed data transfers can take place even during short bursts. Commercial and military meteor-scatter communication using packet radio takes place on the 40-50 MHz band. Some experimentation has been done by amateurs using this combination, but more can be learned by further work in this promising field.

[Turn to Chapter 10 and study FCC examination questions with numbers that begin 4BC-2. Review this section as needed.]

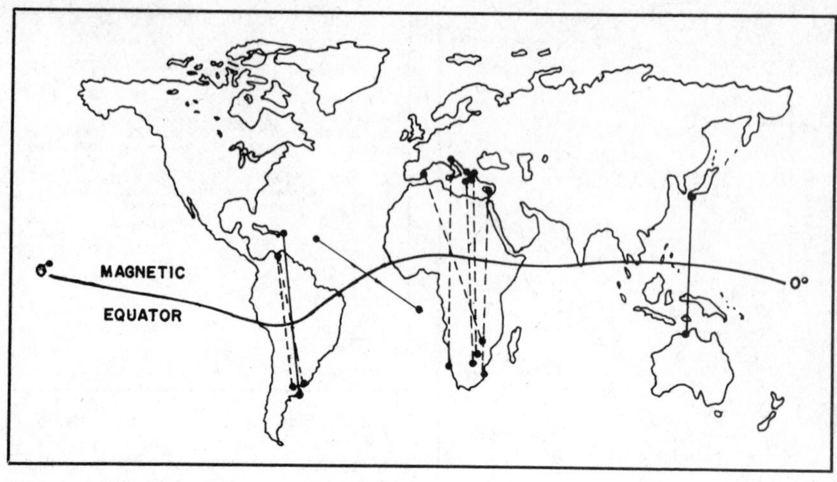

Fig. 3-11—TE paths worked by amateurs on 144 MHz, showing the symmetrical distribution of stations with respect to the magnetic equator.

TRANSEQUATORIAL PROPAGATION

Transequatorial propagation (TE) is a form of F-layer ionospheric propagation that was discovered by amateurs in the late 1940s. Amateurs on all continents reported the phenomenon almost simultaneously on various north-south paths. Those amateurs were communicating successfully on 50 MHz during evening hours. At that time, the predicted MUF was around 40 MHz for daylight hours, a time when the MUF should have been maximum. Since that time, research carried out by amateurs has shown that the TE mode works on 144 MHz and even to some degree at 432 MHz.

The ionosphere is directly influenced by solar radiation. One might expect that the density of ionization would show a maximum over the equator at the equinoxes. In fact, at the equinoxes there is not one area of maximum ionization but two. These maxima form in the morning, are well established by noon and last until after midnight. The high-density-ionization regions form approximately between 10 and 15 degrees on either side of the earth's magnetic equator—not the geographic equator. The system does not move north and south with the seasons. As the relative position of the sun moves away from the equator, the ionization levels in the northern and southern hemispheres become unbalanced.

At HF, TE may provide very strong signals during the afternoon and early evening. Later at night, and sometimes in the early morning as well, only weak and watery signals can be heard arriving by TE.

As the signal frequency is increased, the communication zones become more restricted to those equidistant from and perpendicular to the magnetic equator (Fig. 3-11). Further, the duration of the opening tends to be shorter and closer to 8 P.M. local time. The rate of flutter fading and the degree of frequency spreading increase with signal frequency. TE range extends to approximately 5000 miles—2500 miles on each side of the magnetic equator.

Greatest MUFs will be experienced on TE during solar activity peaks. Best conditions exist when the earth's magnetic field is quiet. Seasonal variations favor the periods around the equinoxes.

[Now study FCC examination questions with numbers that begin 4BC-3. Review this section as needed.]

LONG-PATH PROPAGATION

Propagation between any two points on the earth's surface is usually by the shortest direct route, which is a *great-circle path* between the two points. Remember that a great circle is an imaginary line drawn around the earth, formed by a plane passing through the center of the earth. The diameter of a great circle is equal to the diameter of the earth. You can find a great-circle path between two points by stretching a string tightly between those two points on a globe. If a rubber band is used to mark the entire great circle, by stretching it around the globe, then you can see that there are really two great-circle paths. See Fig. 3-12.

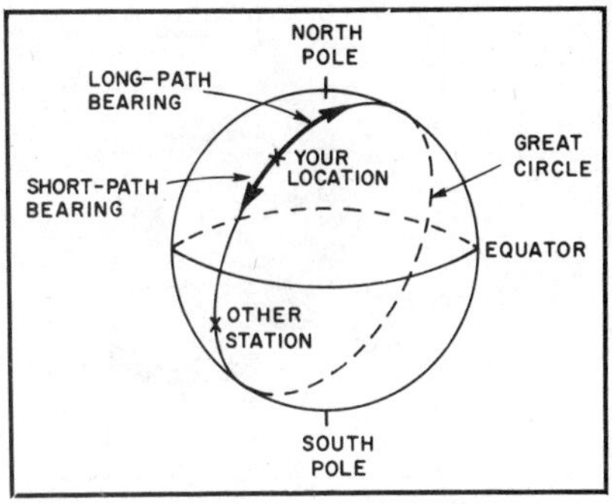

Fig. 3-12—Sketch of the earth, showing a great circle drawn between two stations. Short-path and long-path bearings are shown from the northern-hemisphere station.

One of those paths will usually be longer than the other, and it may be useful for communications when conditions are favorable. Of course you must have a beam antenna that you can point in the desired direction to make effective use of *long-path propagation*. The station at the other end of the path must also point his or her antenna in the long-path direction to your station to make the best use of this propagation.

The long- and short-path directions always differ by 180°. Since the circumference of the earth is 24,000 miles, short-path propagation is always over a path length of less than 12,000 miles. The long-path distance is 24,000 miles minus the short-path distance for a specific communications circuit. For example if the distance from Gordon, PA to the Canary Islands is 3510 miles at a bearing of 85°, then the long-path circuit would be a distance of 20,490 miles at a bearing of 265°. (If you want to know how to perform distance and bearing calculations, see Chapter 16 of *The ARRL Antenna Book*).

Normally, radio signals propagate most effectively along the great-circle path that provides the least absorption. Since the D layer of the ionosphere is most responsible for signal absorption, the amount of sunlight that the signal must travel through helps determine how strong the signals will be. So at times, your signals will be stronger at the receiving station when they travel over the nighttime side of the earth.

For paths less than about 6000 miles, the short-path signal will almost always be stronger, because of the increased losses caused by multiple-hop ground-reflection losses and ionospheric absorption over the long path. When the short path is more than 6000 miles, however, long-path propagation may be more efficient than that over the short path. Long-path propagation will usually be observed either along the "gray line" between darkness and light, or over the nighttime side of the earth.

While signals will generally travel best along the great-circle path, there is often some deviation from the exact predicted beam heading. Ionospheric conditions can cause radio signals to reflect or refract in unexpected ways. So it is always a good idea to rock your rotator control back and forth around the expected beam heading, and listen for the peak signals. If the strongest signals come from a direction that is significantly different from the great-circle headings, then we say the signals are following a crooked path. Anytime signals are not following a great-circle path, you have *crooked-path propagation*.

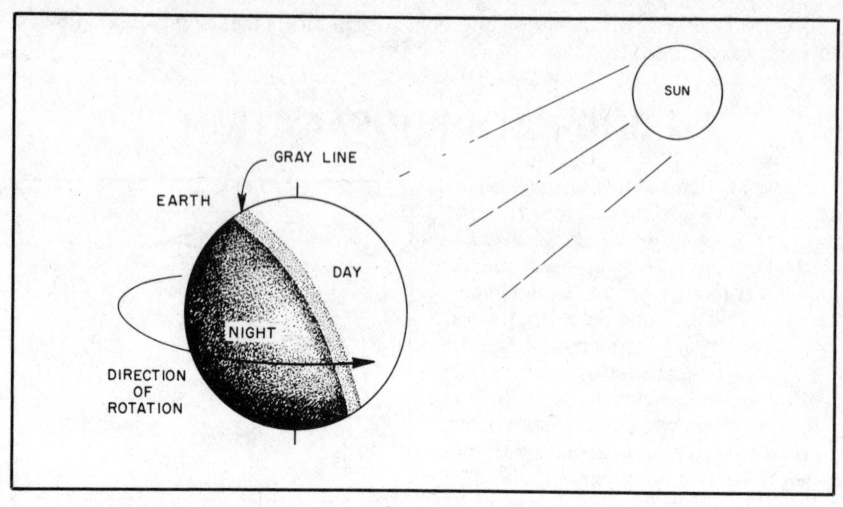

Fig. 3-13—The gray line is a transition region between daylight and darkness. One side of the earth is coming into sunrise, and the other side is just past sunset.

GRAY-LINE PROPAGATION

The *gray line* (sometimes called the "twilight zone") is a band around the earth that separates the daylight from the darkness. Astronomers call this the terminator. It is a somewhat fuzzy region because the earth's atmosphere tends to diffuse the light into the darkness. Fig. 3-13 illustrates the gray line around the earth. Notice

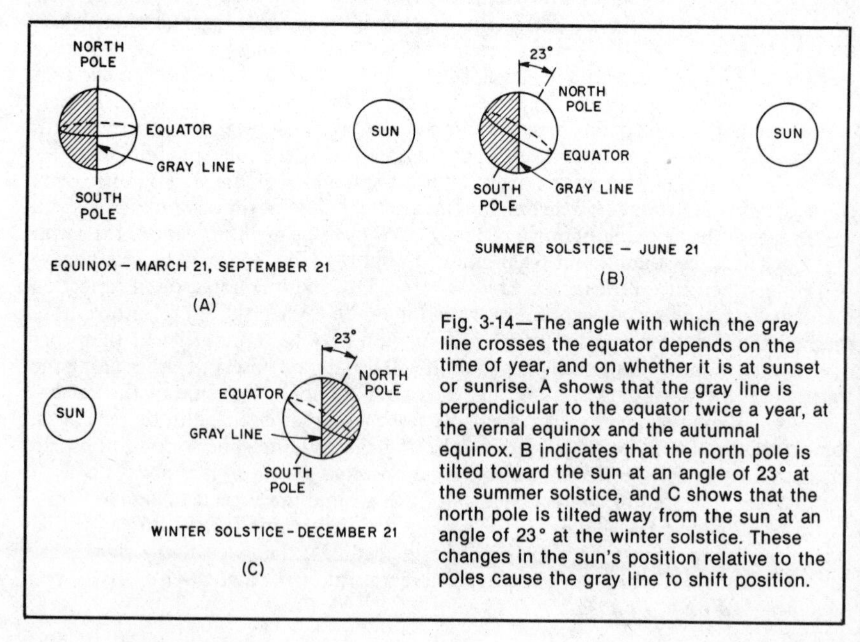

Fig. 3-14—The angle with which the gray line crosses the equator depends on the time of year, and on whether it is at sunset or sunrise. A shows that the gray line is perpendicular to the equator twice a year, at the vernal equinox and the autumnal equinox. B indicates that the north pole is tilted toward the sun at an angle of 23° at the summer solstice, and C shows that the north pole is tilted away from the sun at an angle of 23° at the winter solstice. These changes in the sun's position relative to the poles cause the gray line to shift position.

that on one side of the earth, the gray line is coming into daylight (sunrise), and on the other side it is coming into darkness (sunset).

Propagation along the gray line is very efficient. One major reason for this is that the D layer, which absorbs HF signals, disappears rapidly on the sunset side of the gray line, and it has not yet built up on the sunrise side.

The gray line runs generally north and south, but varies as much as 23° either side of the north-south line. This variation is caused by the tilt of the earth's axis relative to its orbital plane around the sun. The gray line will be exactly north and south at the equinoxes (March 21 and September 21). On the first day of summer in the northern hemisphere, June 21, it is tilted a maximum of 23° one way, and on the first day of winter, December 21, it is tilted a maximum of 23° the other way. Fig. 3-14 illustrates the changing gray-line tilt. The tilt angle will be between these extremes during the rest of the year.

One way to describe the gray-line tilt is as an angle measured upward from the equator, looking east. It is important to note that if you measure an angle greater than 90° at sunrise, then you will measure an angle less than 90° at sunset on the same day. For example, at sunrise on April 16 the gray line makes an angle of approximately 99° with the equator, and at sunset it makes an angle of 81°. This means you can work into a different area of the world using *gray-line propagation* at sunset than you could at sunrise of the same day.

Key Words

Adcock array—A radio direction finding antenna array consisting of two vertical elements fed 180° apart and capable of being rotated.

Alternator whine—A common form of conducted interference typified by an audio tone being induced onto the received or transmitted signal. The pitch of the noise varies with alternator speed.

Antenna effect—One of two operational modes of a simple loop antenna wherein the antenna exhibits the characteristics of a small, nondirectional vertical antenna.

Cardioid pattern—A heart-shaped antenna pattern characterized by a single, large lobe in one direction, and a deep, narrow null in the opposite direction.

Coaxial capacitor—A cylindrical capacitor used for noise-suppression purposes. The line to be filtered is through-connected to the two ends of the capacitor, and a third connection is made to electrical ground.

Conducted noise—Electrical noise that is imparted to a radio receiver or transmitter through the power connections to the radio.

Corona discharge—A condition when a static-electricity charge builds up on an antenna, usually a mobile antenna, and then discharges as the air insulation around the antenna becomes ionized and glows light blue.

Directive antenna—An antenna that concentrates the radiated energy to form one or more major lobes, in specific directions. The receiving pattern is the same as the transmitting pattern.

Ferrite loop (loopstick)—A loop antenna wound on a ferrite rod to increase the magnetic flux. Also called a loopstick antenna.

Frequency domain—A time-independent way to view a complex signal. The various component sine waves that make up a complex waveform are shown by frequency and amplitude on a graph or the CRT display of a spectrum analyzer.

Ground-wave signals—Radio signals that are propagated along the ground rather than through the ionosphere or by some other means.

Intermodulation distortion (IMD)—A type of interference that results from the mixing of integer multiples of signal frequencies in a nonlinear stage or device. Mixing products result, which can interfere with desired signals on the mixed frequencies.

Logic probe—A simple piece of test equipment used to indicate high or low logic states (voltage levels) in digital-electronic circuits.

Loop antenna—An antenna configured in the shape of a loop. If the current in the loop, or in multiple parallel turns, is essentially uniform, and if the loop circumference is small compared with a wavelength, the radiation pattern is symmetrical, with maximum response in either direction of the loop plane.

Night effect—A special type of error in a radio direction-finding system, occurring mainly at night, when sky-wave propagation is most likely.

Radiated noise—Usually referring to a mobile installation, noise that is being radiated from the ignition system or electrical system of a vehicle and causing interference to the reception of radio signals.

Sensing antenna—An omnidirectional antenna used in conjunction with an antenna that exhibits a bidirectional pattern to produce a radio direction-finding system with a cardioid pattern.

Sky-wave signals—Radio signals that travel through the ionosphere to reach the receiving station. Sky-wave signals will cause a variation in the measured received-signal direction, resulting in an error with a radio direction-finding system.

Spectrum analyzer—A test instrument generally used to display the power (or amplitude) distribution of a signal with respect to frequency.

Time domain—A method of viewing a complex signal. The amplitude of the complex wave is displayed over changing time. The display shows only the complex waveform, and does not necessarily indicate the sine-wave signals that make up the wave.

Triangulation—A radio direction-finding technique in which compass bearings from two or more locations are taken, and lines are drawn on a map to predict the location of a radio signal source.

Chapter 4

Amateur Radio Practice

This chapter contains material on Amateur Radio practices that you must be familiar with to pass your Amateur Extra Class Amateur Radio license examination. The information on test equipment covers the theory and use of the spectrum analyzer in testing of radio equipment, and the use of the logic probe for determining the state of digital logic circuits. The sections dealing with electromagnetic compatibility explain methods that have proven to be effective in suppressing vehicular noise. Direction-finding techniques are covered at the end of the chapter.

USE OF TEST EQUIPMENT

Spectrum Analyzer

The *spectrum analyzer* is similar to an oscilloscope in that it permits a graphic representation of a dynamic, or changing, electrical signal. The oscilloscope presents complex signals in the *time domain*. That is, it shows amplitude as a function of time. There are, however, signals that cannot be represented properly in the time domain. Amplifiers, oscillators, detectors, modulators, mixers, and filters are best characterized in terms of their frequency response. We must observe these signals in the *frequency domain* (amplitude as a function of frequency.) The spectrum analyzer is one instrument that can display the frequency domain.

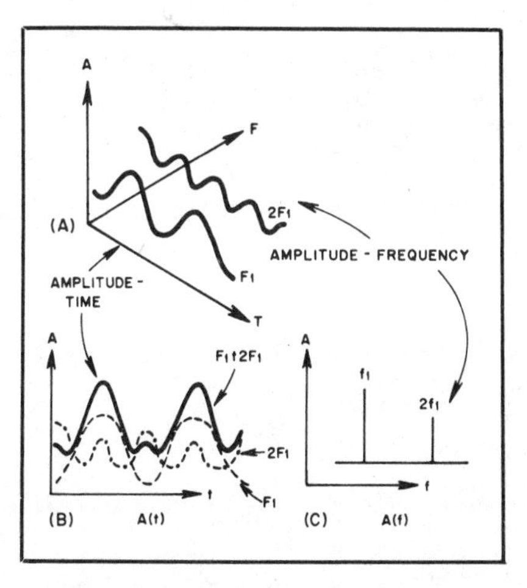

Fig. 4-1—Different ways in which a complex signal may be characterized. At A is a three-dimensional display of amplitude, time and frequency. At B this information is shown only in the time domain as would be seen on an oscilloscope. At C the same information is shown in the frequency domain as it would be viewed on a spectrum analyzer.

Understanding the Time and Frequency Domains

To better understand the concepts of time and frequency domains, refer to Fig. 4-1. In Fig. 4-1A, the three-dimensional coordinates show time (as the line sloping toward the bottom right), frequency (as the line sloping toward the top right), and amplitude (as the vertical axis). The two frequencies shown are harmonically related (f1 and 2f1). The time domain is represented in Fig. 4-1B, where all frequency components are added together.

If the two frequencies were applied to the input of an oscilloscope, we would se
the solid line (which represents f1 + 2f1) on the display.

The display shown in Fig. 4-1C is typical of a spectrum analyzer presentatio
of a complex signal (a signal composed of more than one frequency). Here the sign
is separated into the individual frequency components, and a measurement may b
made of the power level at each frequency.

The frequency domain contains information not found in the time domain, an
vice versa. Hence, the spectrum analyzer offers advantages over the oscilloscope fc
certain measurements, but for measurements in the time domain, the oscilloscop
is an invaluable instrument.

Spectrum Analyzer Basics

There are several types of spectrum analyzers. The most popular type is the swe
superheterodyne. A simplified block diagram of such an analyzer is shown in Fig. 4-
The analyzer is basically a narrow-band receiver that is electronically tuned in fre
quency. Tuning is accomplished by applying a linear ramp voltage to the frequency
controlling element of a voltage-controlled oscillator. The same ramp voltage i
simultaneously applied to the horizontal deflection plates of the cathode ray tub
(CRT). The receiver output is synchronously applied to the vertical deflection plate
of the CRT, resulting in an amplitude-versus-frequency display.

Spectrum analyzers are calibrated in both frequency and amplitude for relativ
and absolute measurements. The frequency range, controlled by the scan-width cor
trol, is calibrated in hertz, kilohertz or megahertz per division (graticule marking)
Each horizontal division might thus correspond to 10 MHz, 1 MHz or 10 kHz. Th
vertical axis of the display is calibrated for amplitude. Common calibrations fo
amplitude are 1 dB, 2 dB or 10 dB per division. For transmitter testing the 10-dB
per-division range is commonly used, because it allows you to view a wide range c
signal strengths, such as those of the fundamental signal, harmonics and spuriou
signals.

Transmitter Testing with the Spectrum Analyzer

The spectrum analyzer is ideally suited for checking the output from a trans
mitter or amplifier for spectral purity. There are many other practical uses, also

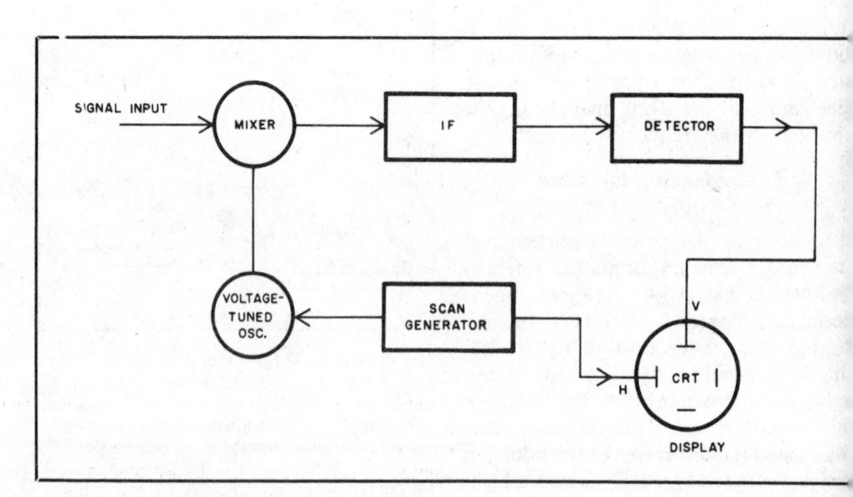

Fig. 4-2—A simplified block diagram of a spectrum analyzer.

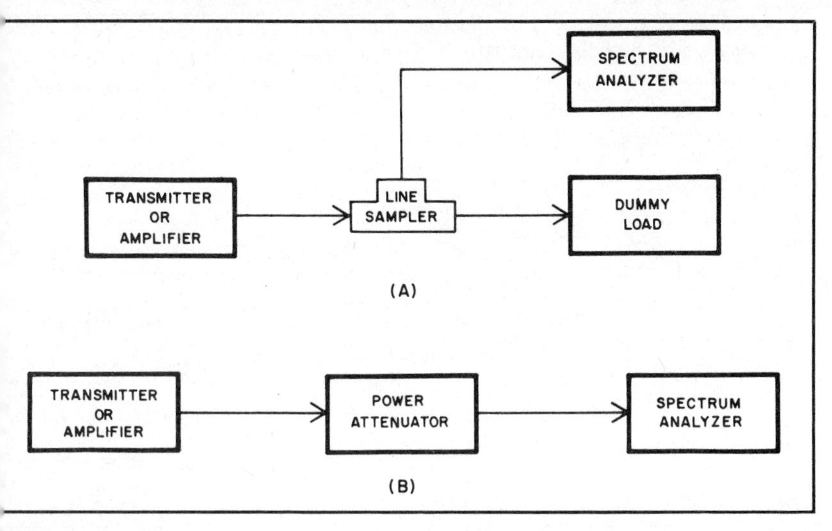

ig. 4-3—Two commonly used test setups to observe the output of a transmitter or amplifier n a spectrum analyzer. The system at A uses a line sampler to pick off a small amount of e transmitter or amplifier power. At B the majority of the transmitter power is dissipated in e power attenuator.

ig. 4-3 shows two test setups commonly sed for transmitter testing. The setup at is the better and more accurate approach for broadband measurements because most line-sampling devices do not xhibit a constant-amplitude output cross a broad frequency spectrum. The aboratory at ARRL Hq. uses the setup hown in B. The power attenuator may e a single, high-power type, or it may onsist of a string of lower-power attenuators connected in series to reduce the ated output of the transmitter to a level uitable for the analyzer (typically 0 mW, or less).

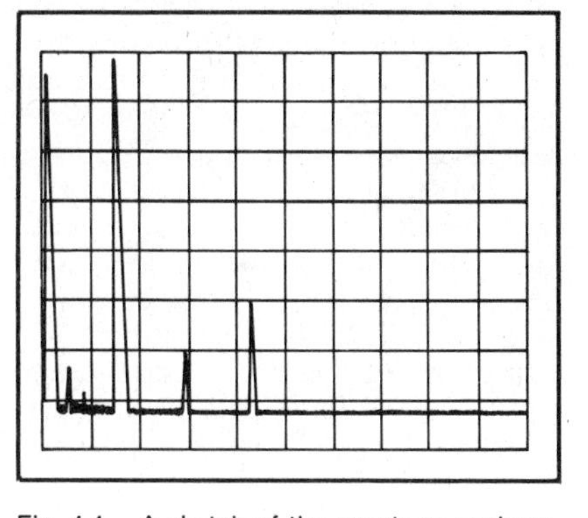

Fig. 4-4— A sketch of the spectrum analyzer display for a well-designed transmitter operating key down on the 40-meter band. Each horizontal division represents 5 MHz and each vertical division is 10 dB.

Fig. 4-4 is a sketch showing a typical pectrum analyzer display for a transmitter operating key down on the 40-meter and. The full-scale pip at the far left of he display is generated within the spectrum analyzer and represents "zero" frequency. The horizontal scale is 5 MHz per division, and the vertical scale is 10 dB er division. Moving to the right, the next tall pip is seen at roughly 7 MHz. This ignal is the fundamental frequency. When the spectrum analyzer is adjusted so that he top of this signal touches the top (reference) line of the display, all other signal evels can be referenced to the power of the fundamental. Moving farther right, the ext signal, at 14 MHz, is the second harmonic of the fundamental. Its level is 0 dB down from the fundamental. Even farther to the right is the third harmonic, t 21 MHz, which is 50 dB down. To the left of the fundamental are a couple of mall pips. These signals, probably spurious mixing products or oscillator leakage, are more than 60 dB below the fundamental, an acceptable level. This spectrograph is typical of a well-designed multiband rig.

The spectrum shown in Fig. 4-5 represents the spectrum output from a not-so-well-designed rig. The horizontal and vertical calibration is the same as shown in Fig. 4-4. In addition to the higher order harmonics, at about 28 MHz and 35 MHz, a number of mixing products are visible above and below the fundamental. The chances of causing interference are greater with this transmitter. Both transmitters, however, are legal according to the FCC rules and regulations.

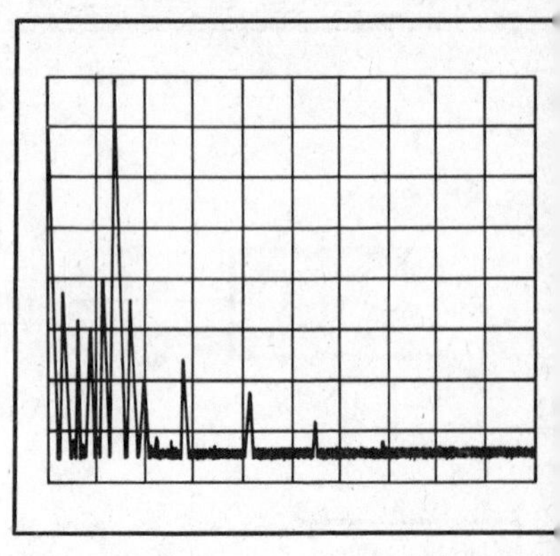

Fig 4-5—A sketch of the spectrum analyzer display for a not-so-well-designed transmitte operating key down on the 40-meter band. Compare with Fig. 4-4.

Another area of concern in the realm of transmitter spectral purity has to do with the *intermodulation distortion (IMD)* levels associated with SSB transmitters and amplifiers. The test setup shown in Fig. 4-6 is used for transmitter IMD testing. Two equal-amplitude, but not harmonically related, audio tones ar fed into the transmitter. (In the ARRL lab we use 700- and 1900-Hz tones.) Th transmitter is first adjusted for rated PEP output using just a single tone, then th single tone is replaced by the two equal-amplitude tones and the output from th two-tone generator and microphone-level control on the transmitter are adjusted fo best IMD performance while maintaining each tone at a level 6 dB below the to line (PEP output). Fig. 4-7 shows a typical display. Responses other than the tw individual tones, which appear near the center of the display, are distortion products Third-order products are down 30 dB, fifth-order products are down 37 dB, an seventh-order products are down 44 dB from the PEP output. The two individua tones are 6 dB below PEP output because they are displayed as two discrete frequen cies. At the instant when the voltages of the individual tones are in phase, they ad to produce a peak in the envelope-waveform pattern which is twice the voltag amplitude of a single tone alone. The power at the peaks of the envelope (PEP) i therefore four times that of a single tone—a 4:1 power ratio being equal t

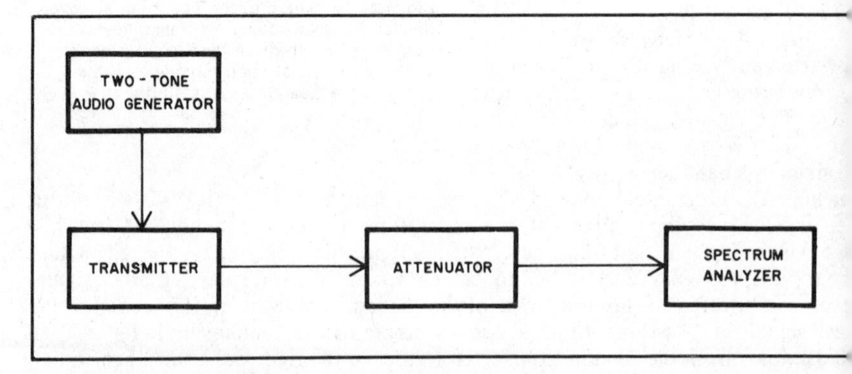

Fig. 4-6—Test setup used in the ARRL laboratory for measuring the IMD performance of transmitters and amplifiers.

dB. The power ratio is four because loubling the applied voltage will double he current, assuming a constant load esistance.

A more detailed discussion of spectrum inalyzer measurement techniques may be ound in *The ARRL Handbook*.

[Turn to Chapter 10 and study FCC xam questions 4BD-1. Review this secion if you have any difficulty with these questions.]

The Logic Probe

Most amateurs are familiar with the use of the standard multimeter for troubleshooting, but how do you go about testng digital-logic circuits? In digital circuitry there are only two states to worry about—the logical "one" and the logical 'zero." A voltmeter or oscilloscope could be used to monitor these logic states on a

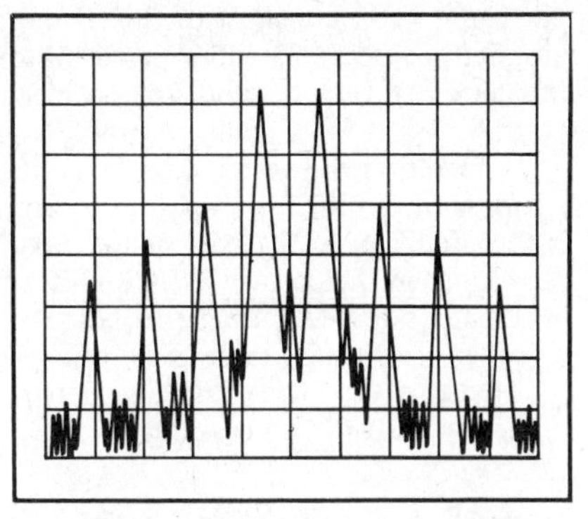

Fig. 4-7—Sketch of a spectrum analyzer display showing the result of a two-tone test on an SSB transmitter. Each horizontal division is equal to 1 kHz and each vertical division is 10 dB. Third-order products are 30 dB below PEP output (top line), fifth-order products are down 37 dB, and seventh-order products are down 44 dB. This represents acceptable, but not ideal, performance.

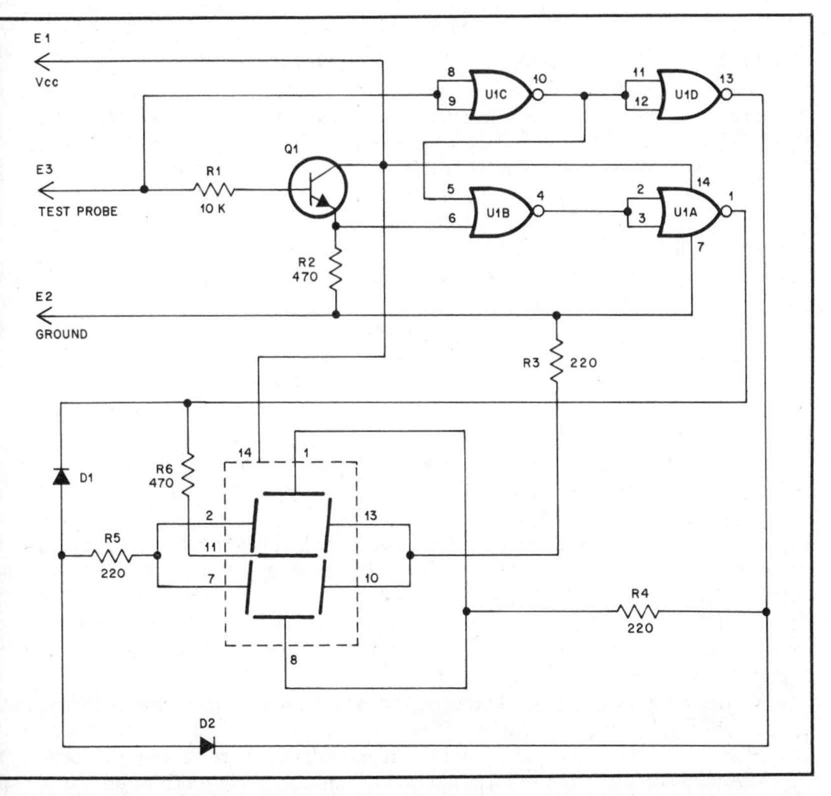

ig. 4-8—Schematic diagram of a simple TTL logic probe. The readout will indicate "0" for a ero state, "1" for a one state, and "H" for a high-impedance state.

particular gate, but most of the time it is necessary to know only if there is eith a one or a zero at the input or output of a gate. The *logic probe* provides a mear of doing just that, without the problem of carrying a scope or VOM around wit you. A circuit for a small, portable logic probe is shown in Fig. 4-8.

Operation of this logic probe is simple—connect the $+V_{CC}$ lead to the positiv terminal of the IC or power supply, and the negative lead to ground. The probe ti is then touched to the part of the circuit you wish to check for a logic state. Th readout then indicates 1 for a logical one, 0 for a logical zero, and H for a higl impedance point. Other logic probes may use LEDs to indicate the high and low state or other conditions, such as a pulsed signal.

[Now study FCC exam questions 4BD-2. Review this section if you have an difficulty with these questions.]

ELECTROMAGNETIC COMPATIBILITY

Vehicle Noise Suppression

One of the most significant deterrents to effective signal reception during mobil or portable operation is electrical impulse noise from the automotive ignition systen The problem also arises during use of gasoline-powered portable generators. Thi form of interference can completely mask a weak signal. Other sources of noise ir clude conducted interference from the vehicle battery-charging system, instrument caused interference, static and corona discharge from the mobile antenna.

Ignition Noise

Most electrical noise can be eliminated by taking logical steps to suppress it. Th first step is to clean up the noise source itself, then to utilize the receiver's built-i noise-reducing circuit as a last measure to minimize any noise impulses from passin cars or other man-made sources. Most vehicles manufactured prior to 1975 were equir ped with inductive-discharge ignition systems. A variety of noise-suppression method were devised for these systems, such as: resistor spark plugs, clip-on suppressors resistive high-voltage spark-plug cable, and even complete shielding. Resistor spar plugs and resistive high-voltage cable provide the most effective noise reduction fc inductive-discharge systems at the least cost and effort. Almost all vehicles produce after 1960 had resistance cable as standard equipment. Such cable develops crack in the insulation after a few years of service, and should be replaced on a regula schedule. Two years is a reasonable replacement schedule.

Some later model automobiles employ sophisicated, high-energy, electronic ignition systems to reduce exhaust pollution and increase fuel mileage. Solutions t noise problems that are effective for inductive-discharge systems cannot uniforml be applied to the modern electronic systems. Such fixes may be ineffective at best and at worst may impair engine performance. One significant feature of capacitive discharge systems is extremely rapid voltage rise, which combats misfiring cause by fouled plugs. Rapid voltage rise depends on a low RC time constant being presente to the output transformer. High-voltage cable designed for capacitive-discharge system exhibits a distributed resistance of about 600 ohms per foot, compared with 10 kilohm per foot for cable used with inductive-discharge systems. Increasing the RC produc by shielding or installing improper spark-plug cable could seriously affect th capacitive-discharge-circuit operation.

Ferrite beads are a possible means for RFI reduction in newer vehicles. Bot primary and secondary ignition leads are candidates for beads. Install them liber ally, then test the engine under load to ensure adequate spark-plug performance

Electrical bonding can reduce the level of ignition noise, both inside the vehicl and out. The plane sheet metal surfaces and cylindrical members often exhibi resonance in one of the amateur bands, and such resonances encourage the reradiatio f spark impulse energy. Other types of noise, described later, can also be helped y bonding. Bond the following structural members with heavy metal braid: (1) engine frame; (2) air cleaner to engine block; (3) exhaust lines to frame; (4) battery ground rminal to frame; (5) hood to fire wall; (6) steering column to frame; (7) bumpers frame; and (8) trunk lid to frame.

Charging-System Noise

Noise from the vehicle battery-charging system can interfere with both reception nd transmission of radio signals. The charging system of a modern automobile consts of a belt-driven, three-phase alternator and a solid-state voltage regulator. Interrence from the charging system can affect receiver performance in two ways: RF diation can be picked up by the antenna, and noise can be conducted directly into e circuitry through the power cable. *Alternator whine* is a common form of conucted interference, and can affect both transmitting and receiving. VHF FM comunications are the most affected, since synthesized carrier generators and local scillators are easily frequency modulated by power-supply voltage fluctuations. The ternator ripple is most noticeable when transmitting, because the alternator is most eavily loaded in that condition.

Conducted noise can be minimized by connecting the radio power leads directly the battery, as this point is the lowest impedance point in the system. If the voltage gulator is adjustable, set the voltage no higher than necessary to ensure complete attery charging. *Radiated noise* and conducted noise can be suppressed by filtering e alternator leads. *Coaxial capacitors* (about 0.5 μF) are suitable, but never conect a capacitor to the field lead. The field lead can be shielded, or loaded with ferte beads. A parallel-tuned LC trap in this lead may also be effective against radiated oise. A parallel-tuned trap in the output lead should be made with no. 10 wire, or rger, as some alternators conduct up to 100 amperes for short times. Keep the alterator slip rings clean to prevent excess arcing. An increase in "hash" may indicate at the brushes need replacement.

Instrument Noise

Some automotive instruments can create noise. Among these gauges and senders e the engine-heat and fuel-level indicators. Ordinarily the installation of a 0.5-μF oaxial capacitor at the sender element will cure this problem.

Other noise-generating accessories include turn-signals, window-opener motors, eating-fan motors, and electric windshield-wiper motors. The installation of a 0.25-μF apacitor across the motor winding will usually eliminate this type of interference.

Corona-Discharge Noise

Some mobile antennas are prone to static build up and *corona discharge*. This atic electricity build up is because of the motion of the vehicle and antenna through e air. The corona discharge occurs when the air insulation around the antenna (usully at the tip) breaks down and becomes ionized. Then it begins to glow as the static arge leaks off. There is no cure for this condition, but it can be reduced by atching static-discharge strips to the frame and allowing them to drag or almost touch e ground. Corona-discharge noise does not originate in the automobile electrical stem.

Whip antennas that come to a sharp point will sometimes be more prone to this pe of noise. Most mobile whips have steel or plastic balls at their tips to help prent the discharge, but regardless of the antenna structure, corona buildup can occur. will be more prevalent during, or just before, a severe electrical storm. The symptoms e a high-pitched "screaming" noise in the receiver, which comes in cycles of one two minutes' duration, then changes pitch and dies down as it discharges through e receiver front end. The condition will repeat itself as soon as the antenna system arges up again.

[Now study FCC exam questions 4BD-3. Review this section if you have an difficulty with these questions.]

DIRECTION-FINDING TECHNIQUES

Radio direction finding (RDF) is as old as radio itself. RDF is just what the nam implies—finding the direction or location of a transmitted signal. The practical aspec of RDF include radio navigation, location of downed aircraft, and identification o sources of malicious jamming or radio-frequency interference. In many countries, th hunting of hidden transmitters has become a sport, with participants in automobil or on foot dashing about the countryside in search of a transmitter. The equipme required for any RDF system is a *directive antenna* and a device for detecting the radi signal. In Amateur Radio applications, the signal detector is usually a receiver wi a meter to indicate signal strength. Some form of RF attenuation is desirable to allo proper operation of the receiver under high signal conditions, such as when "zeroir in" on the transmitter at close range. The directive antenna can take many form

Loop Antennas

A simple antenna for RDF work is a small *loop antenna* tuned to resonan with a capacitor. Several factors must be considered in the design of an RDF loo The loop must be small compared to the wavelength—in a single-turn loop, the co ductor should be less than 0.08 wavelength long. For 28 MHz, this represents a lengt of less than 34 inches (10-inch diameter). Maximum response from the loop antenr is in the plane of the loop, with nulls exhibited at right angles to that plane.

To obtain the most accurate bearings, the loop must be balanced electrostati ally with respect to ground. Otherwise the loop will exhibit two modes of operatio One is the mode of the true loop, while the other is that of an essentially nondire tional vertical antenna of small dimensions, sometimes referred to as *antenna effec* The voltages introduced by the two modes are not in phase, and may add or subtrac depending on the direction from which the wave is coming.

Refer to Fig. 4-9. The theoretical true loop pattern is shown in Fig. 4-9A. Whe properly balanced, the loop exhibits two nulls that are 180° apart. Thus, a sing

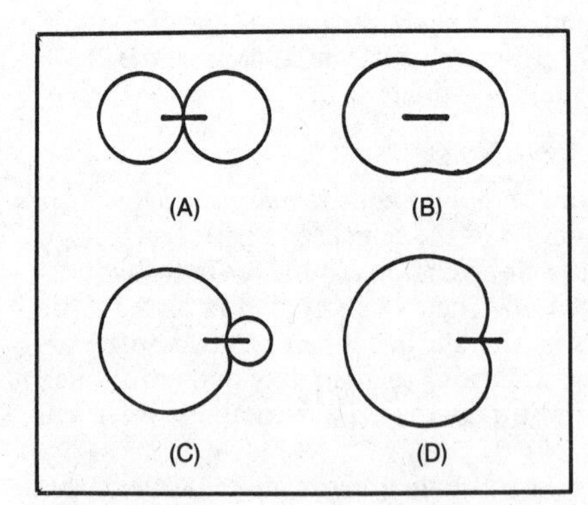

Fig.4-9—Small loop-antenna field patterns. The heavy lines show the plane of the loop. A is for an ideal loop, B is with appreciable antenna effect present, C represents a loop that has been detuned to shift the phasing and D is the optimum detuning to produce a cardioid radiation pattern.

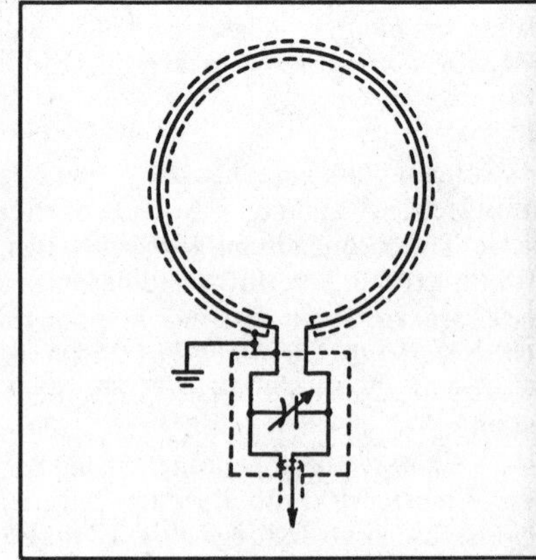

Fig. 4-10—Shielded loop for direction finding. The ends of the shielding turn are not connected. The shielding is effective against electric fields.

null reading with a small loop antenna will not indicate the exact direction toward the transmitter—only the line along which it lies. When the antenna effect is appreciable and the loop is tuned to resonance, the loop may exhibit little directivity (Fig. 4-9B). By detuning the loop to shift the phasing, a pattern similar to Fig. 4-9C may be obtained. This pattern does exhibit a pair of nulls, although they are not symmetrical. The nulls may not be as sharp as that obtained with a well-balanced loop, and may not be at right angles to the plane of the loop.

By suitable detuning, a unidirectional pattern may be approached (Fig. 4-9D). There is no complete null in the pattern, and the loop is adjusted for the best null. An electrostatic balance can be obtained by shielding the loop, as shown in Fig. 4-10. This eliminates the antenna effect, and the response of a well-constructed shielded loop is quite close to the ideal pattern of Fig. 4-9A.

For the lower-frequency amateur bands, single-turn loops are generally not satisfactory for RDF work. Therefore, multiturn loops, such as shown in Fig. 4-11, are generally used. This loop may also be shielded, and if the total conductor length remains below 0.08 wavelength, the pattern is that of Fig. 4-9A.

Ferrite Rod Antennas

The true loop antenna responds to the magnetic field of the radio wave, not to the electric field. The voltage delivered by the loop is proportional to the amount of magnetic flux passing through the coil, and to the number of turns in the coil. The action is much the same as in the secondary winding of a transformer. You could increase the output voltage of the loop by increasing the loop area or the number of turns in the loop. For a given size loop, the output voltage can be increased by increasing the flux density through the loop, and this is done with a ferrite core of high permeability. For additional increased output, the turns may be wound on two ferrite rods that are taped together, as shown in Fig. 4-12. Maximum response of the *ferrite loop* or "loopstick" antenna is broadside to the

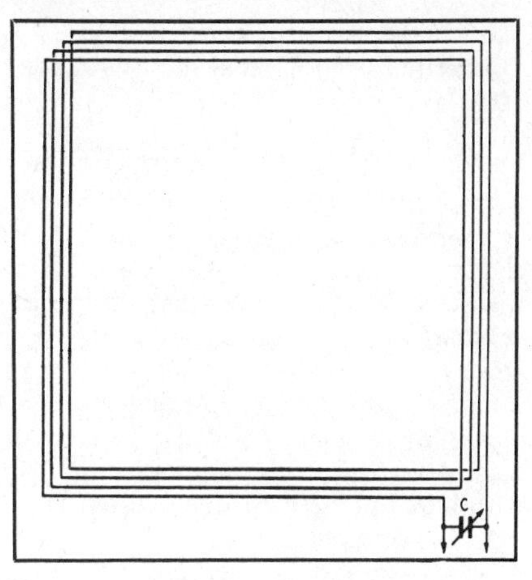

Fig 4-11—Small loop consisting of several turns of wire. The total conductor length is much less than a wavelength. Maximum response is in the plane of the loop.

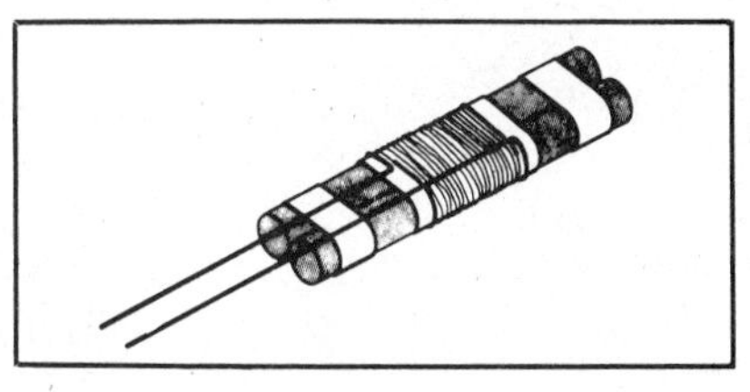

Fig 4-12—A ferrite rod or loopstick antenna.

Fig. 4-13—Field pattern for a ferrite rod antenna. The dark bar represents the rod on which the loop turns are wound.

axis of the rod, as shown in Fig. 4-13. A ½-inch-diameter, 7-inch-long rod of Q2 ferrite (μ_i = 125) is suitable as a loop core for use in the broadcast band through 10 MHz.

Sensing Antennas

Because there are two nulls that are 180° apart in the directional patterns of loops or loopsticks, an ambiguity exists as to which one indicates the true direction of the signal. If there is more than one receiving station, or if the single receiving station takes bearings from more than one position, the ambiguity may be resolved through *triangulation*. It may be more desirable for a pattern to have just one null, however, so there is no question about where the transmitter's true direction lies. A loop or a loopstick may be made to have a single null if a second antenna element, called a *sensing antenna*, is added. The second element must be omnidirectional, such as a short vertical. If the signals from the loop and the sensing antenna are combined with a 90° phase shift between the two, a *cardioid pattern* results.

The development of the cardioid pattern is shown in Fig. 4-14A. Fig. 4-14B shows a circuit for adding a sensing antenna to a loop or loopstick. For the best null in the composite pattern, the signals from the loop and the sensing antenna must be of equal amplitude. R1 is an internal adjustment, and is adjusted experimentally during setup to control the signal level from the sensing antenna. The null of the cardioid is 90° away from the nulls of the loop, so it is customary to first use the loop alone

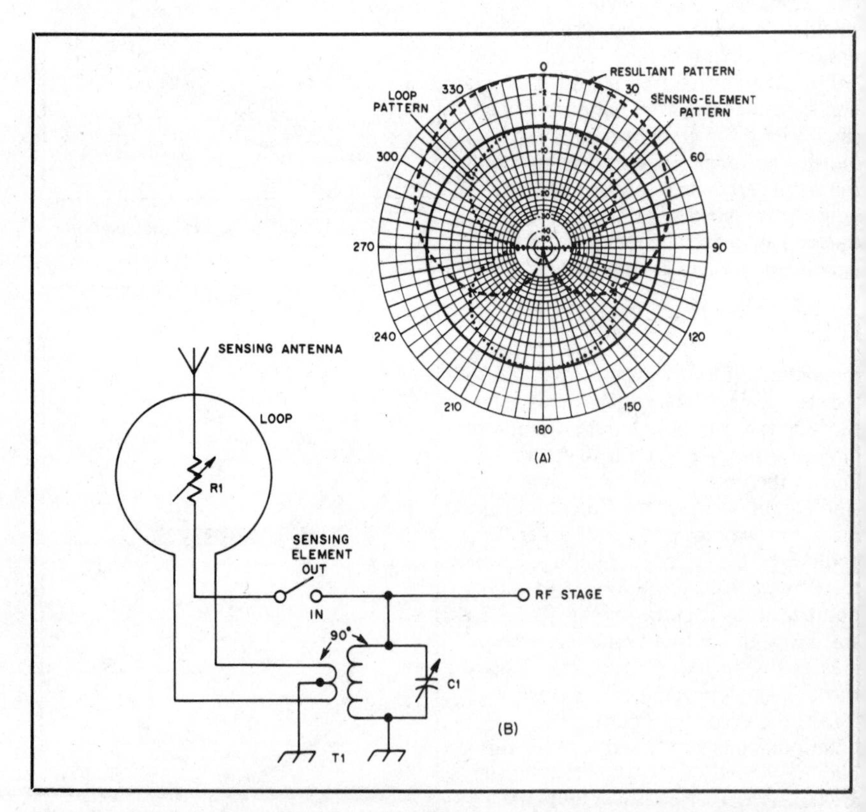

Fig. 4-14—At A, the directivity pattern of a loop antenna with sensing element. At B is a circuit for combining the signals from the two elements. C1 is adjusted for resonance with T1 at the operating frequency.

to obtain a precise bearing line, then switch in the sensing antenna to resolve the ambiguity.

Phased Arrays

There are two general classifications of phased arrays—broadside and end-fire. Broadside arrays are inherently bidirectional—there are always at least two nulls in the pattern. Broadside arrays are seldom used for amateur RDF applications. Depending on the spacing and phasing of the elements, end-fire patterns may exhibit a null off one end of the axis of the elements, while at the same time, the response is maximum off the other end of the axis, in the opposite direction from the null. A common arrangement uses two elements, spaced ¼ wavelength apart, and fed 90° out of phase. The result is a cardioid pattern, with the null in the direction of the leading element.

One of the most popular types of end-fire arrays was invented by F. Adcock, and patented in 1919. The *Adcock array* consists of two vertical elements, fed 180° apart, and mounted so the system can be rotated. Element spacing is not critical, and may be in the range 1/10 to 3/4 wavelength. The two elements must be of the same length, but need not be self-resonant. Elements shorter than resonant are commonly used. Because neither spacing nor length is critical, the Adcock array may be operated over more than one amateur band. The Adcock array is often used at HF for skywave RDF work. In this application, it is not considered a portable system.

The radiation pattern for the Adcock array is shown in Fig. 4-15A. The nulls are in directions broadside to the array axis, and become sharper with greater element spacing. With an element spacing greater than ¾ wavelength, however, the pattern begins to have additional nulls off the ends of the array axis. At a spacing of one wavelength, the pattern is that of Fig. 4-15B, and the array is unsuitable for RDF applications. The Adcock array, with its two nulls, has the same ambiguity as the loop and the loopstick. Adding a sensing element to the Adcock array has not met with much success because of mutual coupling between the array elements and the sensing element, among other things. Because Adcock arrays are usually used for

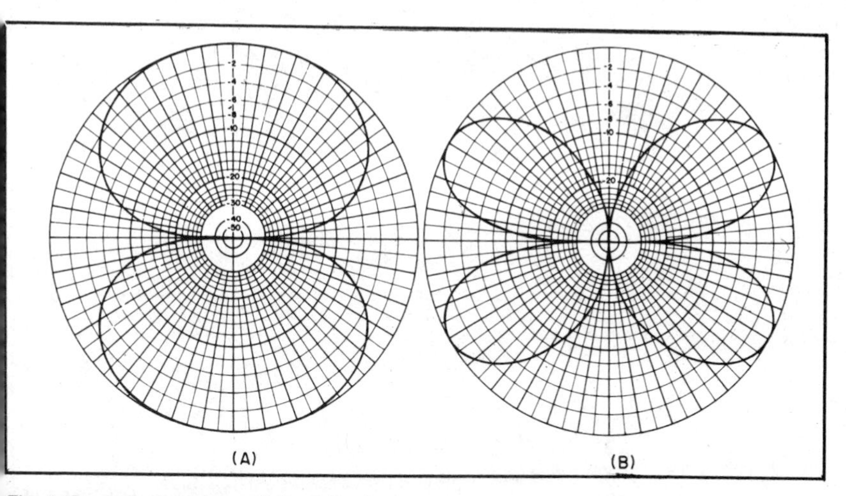

Fig. 4-15—At A, the pattern of the Adcock array with an element spacing of ½ wavelength. In these plots the elements are aligned with the vertical axis. As element spacing is increased beyond ¾ wavelength, additional nulls develop off the ends of the array, and at a spacing of 1 wavelength the pattern at B is produced.

fixed-station operation, as part of a group of stations in an RDF network, the ambiguity presents no serious problems.

Triangulation

If two, or more, RDF bearing measurements are made at locations that are separated by a significant distance, the bearing lines can be drawn from those positions as represented on a map. See Fig. 4-16. Notice that it is important that the two DF sites not be on the same straight line with the signal you are trying to find. The point where the lines cross (assuming the bearings are not the same nor 180° apart) will indicate a "fix" of the approximate transmitter location. The word "approximate" is used because there is always some uncertainty in the bearings obtained. Propagation effects may add to the uncertainty. In order to best indicate the probable location of the transmitter, the bearings from each position should be drawn as narrow sectors instead of single lines. Fig. 4-16 shows the effect of drawing bearings in sectors—the location of the transmitter is likely to be found in the area bounded by the intersection of the various sectors.

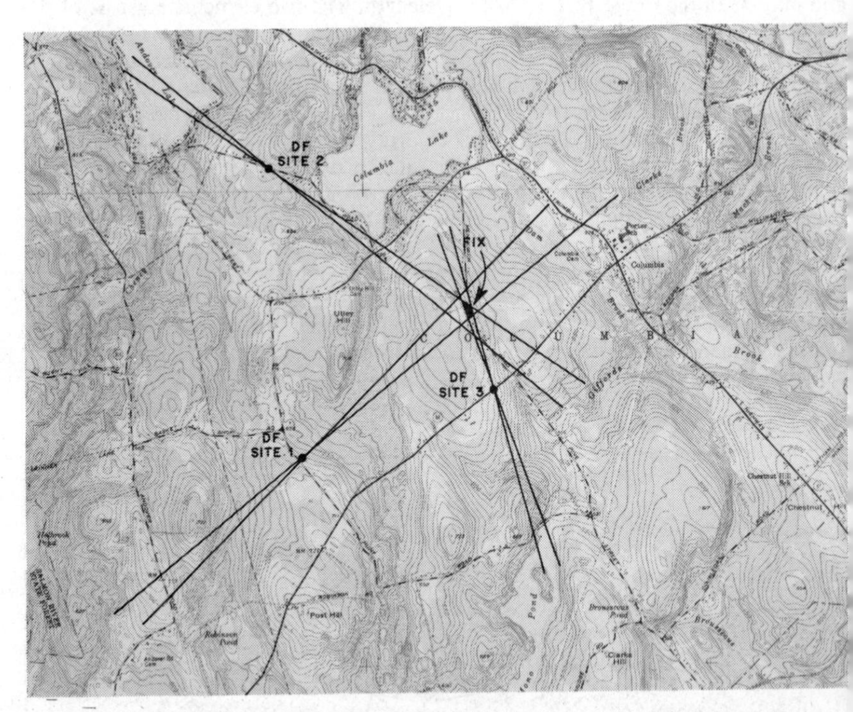

Fig. 4-16—Bearing sectors from three RDF positions drawn on a map. The method is known as triangulation. Note that sensing antennas are not required at any of the RDF sites; antennas with two null indications 180° apart are quite acceptable when several separate bearings can be taken.

Terrain Effects and Night Effect

Most amateur RDF activity is conducted with *ground-wave signals*. The best accuracy in determining a bearing to a signal source is when the propagation path is over homogeneous terrain, and when only the vertically polarized component of the ground wave is present. If a boundary exists, such as between land and water, the different

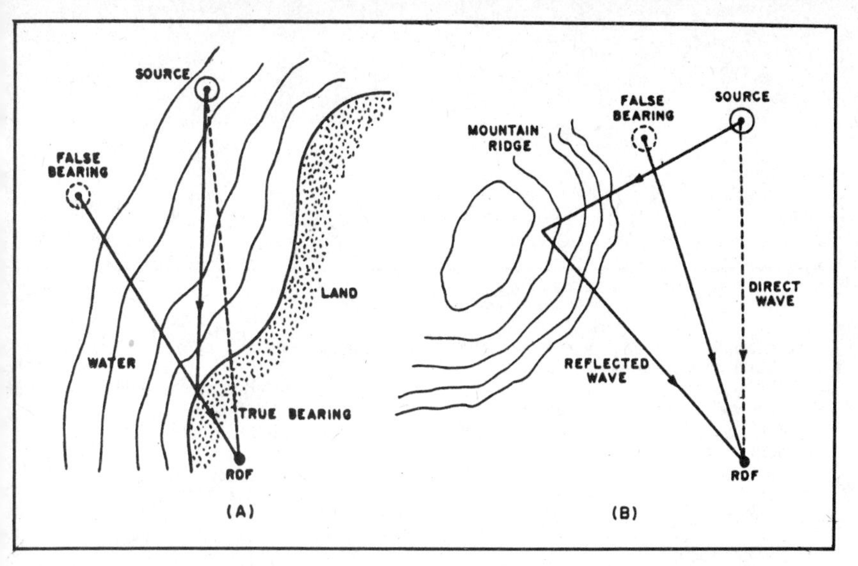

Fig. 4-17—RDF errors caused by refraction (A) and reflection (B). At A a false reading is obtained because the signal actually arrives from a direction that is different from that to the source. At B a direct signal from the source combines with a reflected signal from the mountain ridge. The two signals are averaged at the antenna, giving a false bearing somewhere between the two apparent sources.

conductivities of the two mediums under the ground wave can cause bending (refraction) of the wave front. In addition, reflection of RF energy from vertical objects, such as mountains or buildings, can add to the direct wave and cause RDF errors.

The effects of refraction and reflection are shown in Fig. 4-17. At A, the signal is actually arriving from a direction different than the true direction of the transmitter. This happens because the wave is refracted at the shoreline. Even the most sophisticated RDF equipment will not indicate the true bearing in this instance, as the equipment can only show the direction from which the signal is arriving.

In Fig. 4-17B, there are two apparent sources for the incoming signal—a direct wave from the source itself, and another wave that is reflected from the mountain ridge. In this case, the two signals add at the antenna of the RDF equipment. The uninitiated observer would probably obtain a false bearing in a direction somewhere between the directions to the two sources. The experienced RDF operator might notice that the null reading in this situation is not as sharp, or deep, as it usually is, but these indications would be subtle and easy to overlook.

Water towers, tall radio towers, and similar objects can also lead to false bearings. The effects of these objects become significant when they are large in terms of a wavelength. Local objects, such as buildings of concrete and steel construction, power lines, and the like, also tend to distort the field. It is important that the RDF antenna be in the clear, well away from surrounding objects.

Sky-wave propagation, where both vertically and horizontally polarized components are present, will also give unreliable results. A particular phenomenon, known as *night effect*, was noted with lower-frequency RDF systems years ago. During the daylight hours, D-layer absorption eliminates most of the sky wave and good bearings are obtainable. Once the D layer disappears after sundown, bearings shift, and in some cases a null is either very broad or doesn't exist at all.

[Before proceeding to Chapter 5, study FCC exam questions 4BD-4. Review this section if you have any difficulty with these questions.]

Key Words

Admittance—The reciprocal of impedance, often used to aid the solution of a parallel-circuit impedance calculation.

Back EMF—An opposing electromotive force (voltage) produced by a changing current in a coil. It can be equal to the applied EMF under some conditions.

Conductance—The reciprocal of resistance. This is the real part of a complex admittance.

Imaginary number—A value that sometimes comes up in solving a mathematical problem, equal to the square root of minus one. Since there is no real number that can be multiplied by itself to give a result of minus one, this quantity is imaginary.

Optical shaft encoder—A device consisting of two pairs of photoemitters and photodetectors, used to sense the rotation speed and direction of a knob or dial. Optical shaft encoders are often used with the tuning knob on a modern radio to provide a tuning signal for the microprocessor controlling the frequency synthesizer.

Optocoupler (optoisolator)—A device consisting of a photoemitter and a photodetector used to transfer a signal between circuits using widely varying operating voltages.

Photoconductive effect—A result of the photoelectric effect that shows up as an increase in the electric conductivity of a material. Many semiconductor materials exhibit a significant increase in conductance when electromagnetic radiation strikes them.

Photoelectric effect—An interaction between electromagnetic radiation and matter resulting in photons of radiation being absorbed and electrons being knocked loose from the atom by this energy.

Polar-coordinate system—A method of representing the position of a point on a plane by specifying the radial distance from an origin, and an angle measured counterclockwise from the right.

Rectangular-coordinate system—A method of representing the position of a point on a plane by specifying the distance from an origin in two perpendicular directions.

Smith Chart—A coordinate system developed by Phillip Smith to represent complex impedances on a graph. This chart makes it easy to perform calculations involving antenna and transmission-line impedances and SWR.

Susceptance—The reciprocal of reactance. This is the imaginary part of a complex admittance.

Time constant—The time required for a voltage across a capacitor or a current through an inductor to build up to 63.2% of its steady-state value, or to decay to 36.8% of the initial value. After a total of 5 time constants have elapsed, the voltage or current is considered to have reached its final value.

Chapter 5

Electrical Principles

While working your way up the ladder of Amateur Radio license classes, you have been learning some important electronics principles. You have studied both dc and ac circuit theory. To pass the Amateur Extra Class exam, you will have to know about some more complex topics. This chapter covers some principles of the photoelectric effect and photoconductivity of solid-state materials. It also explains the concept of RC and RL time constants. You will have to know about the Smith Chart, and the basic steps for using it to solve antenna and transmission-line impedance problems; those topics are also covered here. Finally, this chapter explains how to solve complex impedance problems, with the results expressed either as a pair of numbers on the complex number plane or as a magnitude and phase angle.

PHOTOELECTRIC EFFECT

In simple terms, the *photoelectric effect* refers to electrons being knocked loose from a material when light shines on the material. A thorough understanding of the photoelectric effect would require a course in the branch of physics known as quantum mechanics. Since that is beyond the scope of this book, we will simply describe some of the basic principles behind photoelectricity.

You are probably familiar with the basic structure of an atom, the building block for all matter. Fig. 5-1 is a simplified illustration of the parts of an atom. The nucleus contains protons (positively charged particles) and neutrons (with no electrical charge). The number of protons in the nucleus determines which type of element the atom will be. Carbon has 6 protons, oxygen has 8 and copper has 29, for example. The nucleus of the atom is surrounded by the same number of negatively charged electrons as there are protons in the nucleus. So an atom has zero net electrical charge.

The electrons surrounding the nucleus are found in specific energy levels, as shown in Fig. 5-1. The increasing energy levels are shown as larger and larger spheres surrounding the nucleus. While this picture is not strictly accurate, it will help you get the idea of the atomic structure. For an electron to move to a different energy level it must either gain or lose a certain amount of energy. One way that an electron can gain the required energy is by being struck by a photon of light (or other electromagnetic radiation). The electron absorbs the energy from the light photon, and jumps to a new energy level. We say the electron is excited, at this point. Since the electrons in an atom prefer to be at the lowest energy level possible, an excited electron will tend to radiate a photon of the required energy so that it can fall back into the lower level, if an opening exists at that level. (Only a certain number of electrons are allowed to exist at each energy level in an atom.)

This principle of electrons being excited to higher energy levels, then jumping back to lower levels and giving off a photon of light, is the operating principle of "neon" signs. Electrical energy, in the form of a spark, passes through a gas inside

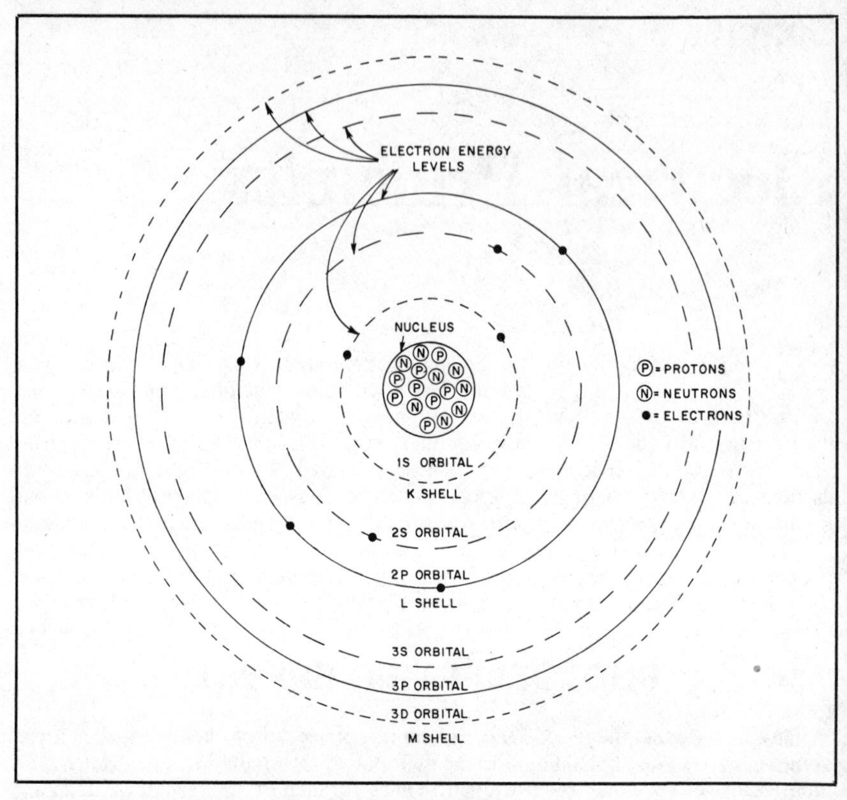

Fig. 5-1—A simple drawing to illustrate the structure of an atom. The nucleus is made up of protons and neutrons. The electrons surrounding the nucleus are found in increasing energy levels as you go out from the center. This drawing represents an oxygen atom.

a glass tube. Electrons are excited, then give off specific frequencies of light. Different gases inside the tubes produce different colors of light.

Each photon of an electromagnetic wave has energy given by:

$$E = hf \qquad \text{(Eq. 5-1)}$$

where

E is the photon energy in joules

f is the wave frequency in hertz

h is a constant, called Planck's Constant, equal to 6.62×10^{-34} joule seconds.

So if an electromagnetic wave (usually in the range of visible light frequencies) with the right amount of energy strikes the surface of a material, electrons will be excited into higher energy levels. If the light photon has sufficient energy, then it may be able to knock an electron completely free of the atom. We will then have a free electron and a positive ion left from what was an atom before the photon collided with it.

If a voltage is applied between two metal surfaces, and one of the surfaces is illuminated with light of the proper frequency (energy), then a spark can be made to jump across the gap at a lower voltage than if the surface were not illuminated. This discovery was made by Heinrich Hertz in 1886. It wasn't until 1905 that Albert Einstein was able to offer a mathematical explanation of the photoelectric effect.

The Photoconductive Effect

With our simple model of the atom in mind, it is easy to see that an electric current through a wire or other material depends on electrons being pulled or knocked free of one atom and moving along to another. The rate of electrons moving past a certain point in the wire specifies the current. Every material presents some opposition to this flow of electrons, and that opposition is referred to as the resistivity of the material. If you include the length and cross-sectional area of a specific object or piece of wire, then you know the resistance of the object:

$$R = \frac{\rho \ell}{A} \qquad \text{(Eq. 5-2)}$$

where
- ρ is the resistivity of the material
- ℓ is the length of the object
- A is the cross-sectional area of the object
- R is the resistance.

Conductivity is the reciprocal of resistivity, and conductance is the reciprocal of resistance:

$$\sigma = \frac{1}{\rho} \text{ and} \qquad \text{(Eq. 5-3)}$$

$$G = \frac{1}{R} \qquad \text{(Eq. 5-4)}$$

We learned earlier that with the photoelectric effect, electrons can be knocked loose from atoms when light strikes the surface of the material. With this principle in mind, you can see that those free electrons will make it easier for a current to flow through the material. But even if electrons are not knocked free of the atom, excited electrons in the higher energy-level regions are more easily passed from one atom to another. So it is easier to make a current flow when some of the electrons associated with an atom are excited. The conductivity of the material is increased, and the resistivity is decreased. So the total conductance of a piece of wire may increase and the resistance decrease when light shines on the surface.

The *photoconductive effect* is more pronounced, and more important, with regard to semiconductor materials than with ordinary conductors. With a piece of copper wire, for example, the conductance is normally high, so any slight increase because of light striking the wire surface will be almost unnoticeable. The conductivity of semiconductor materials such as germanium, silicon, cadmium sulphide, cadmium selenide, gallium arsenide, lead sulphide and others is low when they are not illuminated, but the increase in conductivity is significant when light shines on their surfaces. Each material will show the biggest change in conductivity over a different range of light frequencies. For example, lead sulphide responds best to frequencies in the infrared region, while cadmium sulphide and cadmium selenide are both commonly used in visible light detectors, such as are found in cameras.

Of course, it is important to realize that most semiconductor devices are sealed in plastic or metal cases, so no light will reach the semiconductor junction. Light will not affect the conductivity, and hence the operating characteristics, of a transistor or diode. But if the case is made with a window to allow light to pass through and reach the junction, then the device characteristics will depend on how much light is shining on it. Such specially made devices have a number of important applications in Amateur Radio electronics.

A phototransistor is a special device designed to allow light to reach the transistor junctions. Light, then, acts as the control element for the transistor. In fact, in some phototransistors, the base lead is not even brought out of the package. In others, a base lead is provided, so you can control the output signal in the absence of light. In general, the gain of the transistor is directly proportional to the amount of light shining on the transistor. A phototransistor can be used as a photodetector, or photomultiplier.

Electroluminescence in Semiconductors

If a semiconductor diode junction is forward biased, majority charge carriers from both the P- and N-type material will cross the junction, where they become minority carriers. As these minority carriers move away from the junction, they meet and combine with the opposite-type carriers. In this process, a photon of electromagnetic radiation will be produced. If the radiation frequency is in the range of visible light, and if the diode is constructed to allow the light to escape, we will have a light-emitting diode (LED)! The material used to dope the semiconductor to form the P- and N-type materials will determine the light color emitted. Some materials will emit light in the infrared range, rather than the visible range.

Optocouplers

An *optocoupler*, or *optoisolator*, is an LED and a phototransistor in a common IC package. Because they use light instead of a direct electrical connection, optoisolators provide one of the safest ways to interface circuits using widely differing voltages. Signals from a high-voltage circuit can be fed into a low-voltage circuit without fear of damage to the low-voltage devices. The LEDs in most optocouplers are infrared emitters, although some operate in the visible-light portion of the electromagnetic spectrum.

Fig. 5-2A shows the schematic diagram of a typical optocoupler. In this example, the phototransistor base lead is brought outside the package. As shown at B, a Darlington phototransistor can be used to improve the current transfer ratio of the device.

In an IC optoisolator, the light is transmitted from the LED to the phototransistor detector by means of a plastic light pipe or small gap between the two sections. It is also possible to make an optoisolator by using discrete components. A separate LED or infrared emitter and matching phototransistor detector can be separated by some small distance to use a reflective path or other external gap. In this case, changing the path length or blocking the light will change the transistor output. This can be used to detect an object passing between the detector and light source, for example.

Other applications for a separate emitter and detector might be in a punched-paper-tape or computer-card reader, where you want to detect the position of holes to read a code. Such a system could also be used to send digital pulses over considerable distances using fiber optics. Amateur-Radio applications for this type of

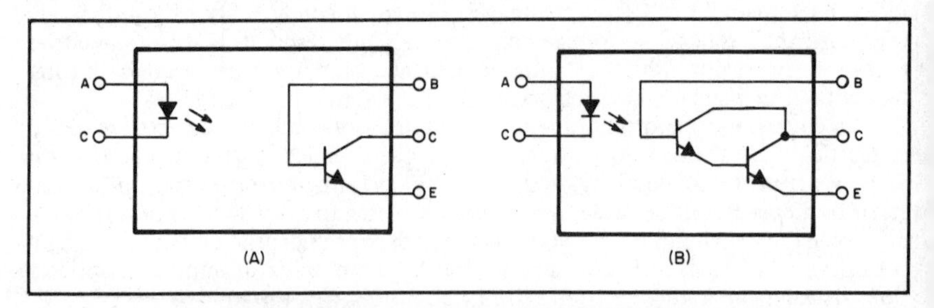

Fig. 5-2—Optocouplers, consisting of an LED and a phototransistor detector. The one shown at A uses a single-transistor detector, while the unit shown at B uses a Darlington phototransistor for improved transfer ratio.

system might include the use of light pipes to connect various pieces of equipment in the presence of strong RF fields. Optical coupling can minimize interference between digital and analog equipment, such as computers and radios.

The Optical Shaft Encoder

An *optical shaft encoder* usually consists of two pair of emitters and detectors. A plastic disc with alternating clear and black radial bands rotates through a gap between the emitters and detectors. See Fig. 5-3. By using two emitters and two detectors, a microprocessor can detect which direction and how fast the wheel is turning. Modern transceivers use a system like this to control the frequency of a synthesized VFO. To the operator, the tuning knob has the feel of an ordinary VFO, but there is no tuning capacitor or other mechanical linkage connected to the knob and light-chopping wheel.

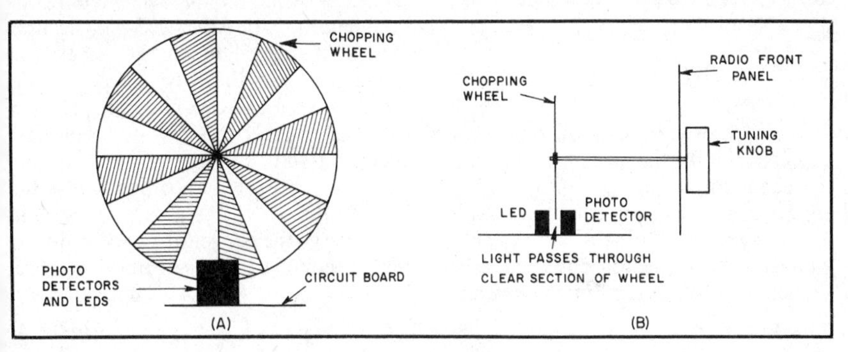

Fig. 5-3—Illustration of the operation of an optical shaft encoder, often used as a tuning mechanism on modern transceivers.

A Solid-State Relay

An optoisolator and a triac, either packaged as one unit or as discrete components, form the solid-state equivalent of a mechanical relay. A solid-state relay can replace electromechanical units in many applications. Some of the advantages include freedom from contact bounce, no arcing, no mechanical wear and no noisy, clicking relay sounds. Some solid-state relays are rated to switch 10 A at 117 V using CMOS control signals! Fig. 5-4 illustrates how a solid-state relay can be connected in a circuit.

[Now turn to Chapter 10 and study FCC examination questions with numbers that begin 4BE-1. Review this section as needed.]

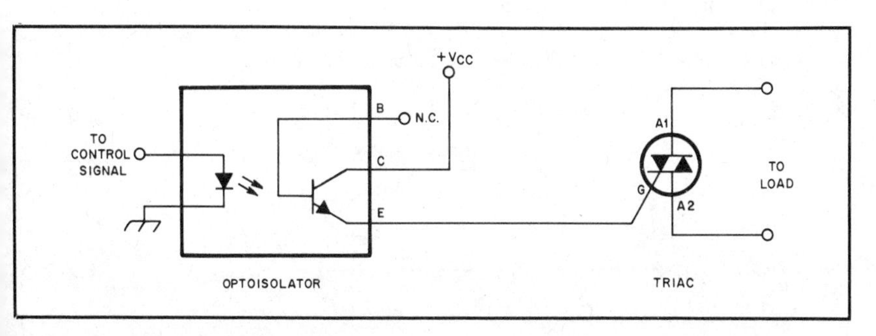

Fig. 5-4—An optoisolator and a triac can be connected to make the solid-state equivalent of a mechanical relay for 117-V ac household current.

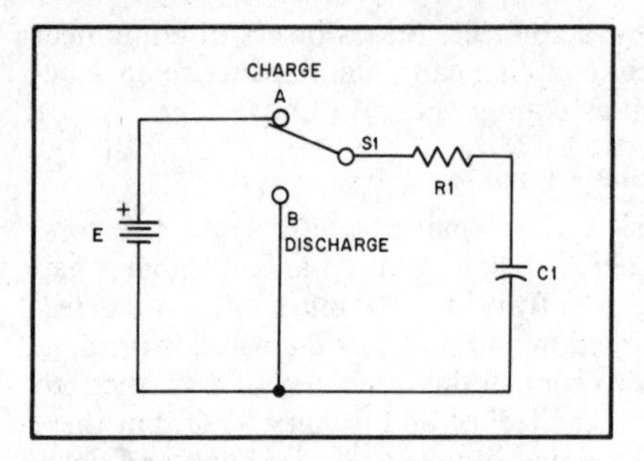

Fig. 5-5—A simple series circuit to illustrate the principle of an RC time constant for charging and discharging a capacitor through a resistor.

TIME CONSTANT FOR AN RC CIRCUIT

If you connect a voltage source directly to a capacitor, the capacitor will be charged to the full voltage almost instantly. If the circuit contains resistance, however, the current will be limited, and it will take some time to charge the capacitor. The higher the resistance value, the longer it will take to charge the capacitor. You probably remember, from studying for your Advanced class license, that as the capacitor charges, energy is being stored in the form of an electric field between the capacitor plates.

Fig. 5-5 shows a circuit that can be used to alternately charge and discharge a capacitor. With the switch in position A, a current will flow through R to charge the capacitor to the battery voltage. When the switch is moved to position B the capacitor will give the energy stored in it back to the circuit, as a current through R. The amount of time it takes to charge or discharge the capacitor depends on the value of the capacitor and the resistor. The product of the resistance and the capacitance is called a *time constant*:

$$\tau = RC \qquad \text{(Eq. 5-5)}$$

where

τ is the Greek letter tau, used to represent the time constant
R is the total circuit resistance in ohms
C is the capacitance in farads

The simple circuit of Fig. 5-5 is a series circuit. It is possible to have a circuit with several resistors and capacitors connected either in series or parallel. If the components are wired in series, simply combine all of the resistors into one equivalent value, and all of the capacitors into one equivalent value. Then calculate the time constant using Eq. 5-5, as before. If the components are connected in parallel, there is an added complication when the circuit is charging, but for a discharging circuit you can still calculate a time constant. Again combine all of the resistors and all of the capacitors into equivalent values, and calculate the time constant using Eq. 5-5. If you have forgotten how to combine resistors and capacitors in series and parallel, review the appropriate sections of *The ARRL Technician/General Class License Manual* or *The ARRL Handbook*. The required equations are:

$$R_T \text{ (series)} = R1 + R2 + R3 + \ldots + Rn \qquad \text{(Eq. 5-6)}$$

$$R_T \text{ (parallel)} = \frac{1}{\frac{1}{R1} + \frac{1}{R2} + \frac{1}{R3} + \ldots + \frac{1}{Rn}} \qquad \text{(Eq. 5-7)}$$

$$C_T \text{ (series)} = \frac{1}{\frac{1}{C1} + \frac{1}{C2} + \frac{1}{C3} + \ldots + \frac{1}{Cn}} \qquad \text{(Eq. 5-8)}$$

$$C_T \text{ (parallel)} = C1 + C2 + C3 + \ldots + Cn \qquad \text{(Eq. 5-9)}$$

Let's look at an example of calculating the time constant for a circuit like the one shown at Fig. 5-5. We will pick values of 1 μF and 1 MΩ for C and R. To calculate the time constant, τ, we simply multiply the R and C values, in ohms and farads.

$\tau = RC = 1 \times 10^6 \text{ ohms} \times 1 \times 10^{-6} \text{ farads} = 1 \text{ second}$

You can calculate the time constant for any RC circuit in this manner. If you have a 1500-pF capacitor and a 220-kΩ resistor, the time constant is:

$\tau = RC = 220 \times 10^3 \text{ ohms} \times 1500 \times 10^{-12} \text{ farads} = 0.00033 \text{ second}$

Exponential Charge/Discharge Curve for Capacitors

The capacitor charge and discharge follows a pattern known as an exponential curve. Fig. 5-6 illustrates the charge and discharge curves, where the time axis is in terms of τ, and the vertical axis is expressed as a percentage of the battery voltage. These graphs hold true for any RC circuit, as long as you know the time constant and the maximum voltage the capacitor is charged to.

We can write equations to calculate the voltage on the capacitor at any instant of time, based on the exponential charge or discharge. For a charging capacitor:

$$V_{(t)} = E\left(1 - e^{\frac{-t}{\tau}}\right) \quad \text{(Eq. 5-10)}$$

where

$V_{(t)}$ is the charge on the capacitor at time t
E is the maximum charge on the capacitor, or the battery voltage
t is the time in seconds that have elapsed since the capacitor began charging or discharging
e is the base for natural logarithms, 2.718
τ is the time constant for the circuit, in seconds

If the capacitor is discharging, we have to write a slightly different equation:

$$V_{(t)} = E\left(e^{\frac{-t}{\tau}}\right) \quad \text{(Eq. 5-11)}$$

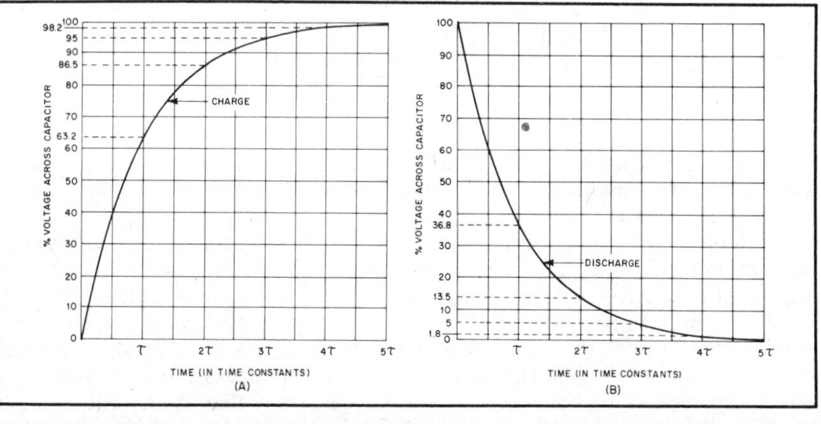

Fig. 5-6—The graph at A shows how the voltage across a capacitor rises, with time, when charged through a resistor. The curve at B shows the way in which the voltage decreases across the capacitor terminals while discharging through the same value of resistance. From a practical standpoint, a capacitor may be considered as charged (or discharged) after a time equal to 5τ.

These exponential equations can be solved fairly easily with an inexpensive calculator that is able to work with natural logarithms (a key labeled LN or LN X). In that case you could calculate the value for $e^{-t/\tau}$ as the inverse ln of $-t/\tau$. Actually, you do not have to know how to solve these equations if you are familiar with the results at a few important points.

As shown on the graphs of Fig. 5-6, it is common practice to think of time in terms of a multiple of the circuit time constant when relating to the capacitor charge or discharge. If we select times equal to zero (starting time), 1 time constant (1τ), 2 time constants (2τ), and so on, then the exponential term in Eqs. 5-10 and 5-11 becomes simply e^0, e^{-1}, e^{-2}, e^{-3} and so forth. Then we can solve Eq. 5-6 for those values of time. For simplicity, let's pick a value of E = 100 V. Then our answers will be in the form of a percentage of any battery voltage you are working with.

$$V_{(t)} = E\left(1 - e^{\frac{-t}{\tau}}\right)$$

$$\begin{aligned}
V_0 &= 100\text{ V }(1 - e^{-0}) = 100\text{ V }(1 - 1) = 0\text{ V, or }0\% \\
V_{(1\tau)} &= 100\text{ V }(1 - e^{-1}) = 100\text{ V }(1 - 0.368) = 63.2\text{ V, or }63.2\% \\
V_{(2\tau)} &= 100\text{ V }(1 - e^{-2}) = 100\text{ V }(1 - 0.135) = 86.5\text{ V, or }86.5\% \\
V_{(3\tau)} &= 100\text{ V }(1 - e^{-3}) = 100\text{ V }(1 - 0.050) = 95.0\text{ V, or }95\% \\
V_{(4\tau)} &= 100\text{ V }(1 - e^{-4}) = 100\text{ V }(1 - 0.018) = 98.2\text{ V, or }98.2\% \\
V_{(5\tau)} &= 100\text{ V }(1 - e^{-5}) = 100\text{ V }(1 - 0.007) = 99.3\text{ V, or }99.3\%
\end{aligned}$$

After a time equal to five time constants has passed, the capacitor is charged to 99.3% of the battery voltage. This is fully charged, for all practical purposes.

You should have noticed that the equation used to calculate the capacitor voltage while it is discharging is slightly different from the one for charging. The exponential term is not subtracted from 1 in Eq. 5-11, as it is in Eq. 5-10. At times equal to multiples of the circuit time constant, the solutions to Eq. 5-11 are very similar to those for Eq. 5-10.

$$V_{(t)} = E\left(e^{\frac{-t}{\tau}}\right)$$

At $t = 0$, $e^{-0} = 1$, so $V_{(0)} = 100$ V, or 100%

$$\begin{aligned}
t &= 1\tau,\ e^{-1} = 0.368,\text{ so } V_{(1\tau)} = 36.8\text{ V, or }36.8\% \\
t &= 2\tau,\ e^{-2} = 0.135,\text{ so } V_{(2\tau)} = 13.5\text{ V, or }13.5\% \\
t &= 3\tau,\ e^{-3} = 0.050,\text{ so } V_{(3\tau)} = 5\text{ V, or }5\% \\
t &= 4\tau,\ e^{-4} = 0.018,\text{ so } V_{(4\tau)} = 1.8\text{ V, or }1.8\% \\
t &= 5\tau,\ e^{-5} = 0.007,\text{ so } V_{(5\tau)} = 0.7\text{ V, or }0.7\%
\end{aligned}$$

Here we see that after a time equal to five time constants has passed, the capacitor has discharged to less than 1% of its initial value. This is fully discharged, for all practical purposes.

Another way to think of these results is that the discharge values are the complements of the charging values. Subtract either set of percentages from 100 and you will get the other set. You may also notice another relationship between the discharging values. If you take 36.8%, or 0.368 as the value for 1 time constant, then after 2 time constants, it is $0.368^2 = 0.135$, 3 time constants is $0.368^3 = 0.050$, 4 time constants is $0.368^4 = 0.018$ and after 5 time constants the value is $0.368^5 = 0.007$. You can change these values to percentages, or just remember that you have to multiply the decimal fraction times the battery voltage. If you subtract these decimal values from 1, you will get the values for the charging equation.

In many cases, you will want to know how long it will take a capacitor to charge or discharge to some particular voltage. Probably the easiest way to handle such problems is to first calculate what percentage of the maximum voltage you are charging or discharging to. Then compare that value to the percentages listed for either charging or discharging the capacitor. Most of the time you will be able to approximate the time as some number of time constants.

Suppose you have a 100-μF capacitor and a 470-kilohm resistor wired in parallel with a battery. The capacitor is charged to 50 V, and then the battery is removed. How long will it take for the capacitor to discharge to 2.5 V? First, let's calculate the percentage decrease in voltage:

$$\frac{2.5 \text{ V}}{50 \text{ V}} = 0.05 = 5\%$$

You should recognize this as the value for the discharge voltage after 3 time constants. We can calculate the time constant for our circuit using Eq. 5-5:

$$\tau = RC = 470 \times 10^3 \text{ ohms} \times 100 \times 10^{-6} \text{ farads} = 47 \text{ seconds}$$

It will take 3τ to discharge the capacitor to 2.5 V, or 3×47 s = 141 s.

TIME CONSTANT FOR AN RL CIRCUIT

A similar situation exists when resistance and inductance are connected in series. Fig. 5-7 shows a circuit for storing a magnetic field in an inductor. When S1 is closed, a current will tend to flow immediately. The instantaneous transition from no current to the value that would flow in the circuit with no resistance, represents a very large change in current, and a *back EMF* is developed by the inductance. This back EMF is proportional to the rate of change of the current, and is of a polarity opposite to that of the applied voltage. The result is that the initial current is very small.

The back EMF depends on the rate of change of the current, so it decreases as the current stops increasing so fast. As the current builds up to its final value, as given by Ohm's Law (I = E/R), the back EMF decreases toward zero.

Fig. 5-8 shows how the current through an inductor increases as time passes. At any given instant, the back EMF will be equal to the difference between the voltage drop across R1 and the battery voltage. You can see that initially, with no current, when the switch is closed the back EMF is equal to the full battery voltage. Theoretically, the back EMF will never quite disappear, and so the current never quite reaches the value predicted by Ohm's Law when you ignore the inductance. In practice, the current is essentially equal to the final value after 5 time constants. The curve looks just like the one we found for a charging capacitor.

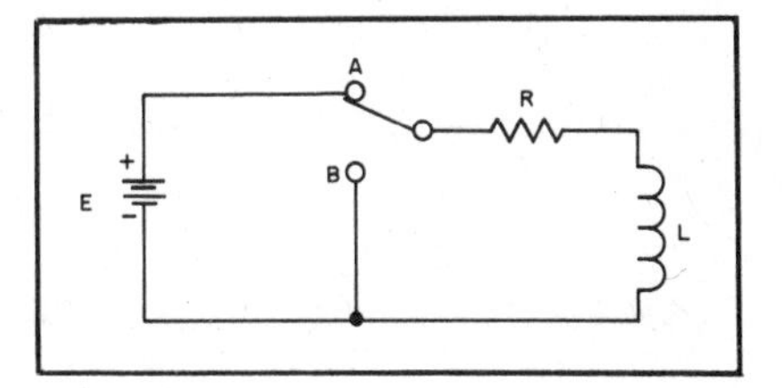

Fig. 5-7—A simple series circuit to illustrate the principle of an RL time constant. The circuit is not practical unless you can consider the switch to close at B before it opens at A.

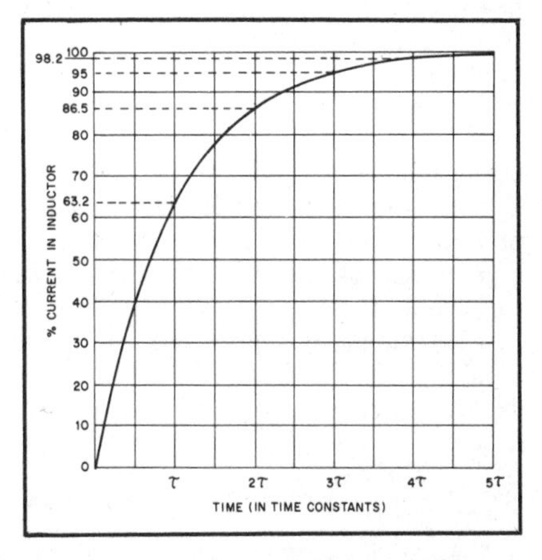

Fig. 5-8—Graph showing the current build up in an RL circuit. Notice that the curve is identical to the case of a charging capacitor.

You can probably guess that this is another exponential curve, and that we can write an equation similar to Eq. 5-10 to calculate the current in the circuit.

$$I_{(t)} = \frac{E}{R}\left(1 - e^{\frac{-t}{\tau}}\right) \qquad \text{(Eq. 5-12)}$$

where

$I_{(t)}$ is the current in amperes at time t
E is the power-supply potential in volts
R is the circuit resistance in ohms
t is the time in seconds after the switch is closed

Here again, we can define a time constant that depends on the circuit components, as we did with the RC circuit. For an RL circuit, the time constant is given by:

$$\tau = \frac{L}{R} \qquad \text{(Eq. 5-13)}$$

where τ is the greek letter tau, used to represent a time constant for both RC and RL circuits.

If we choose values of time equal to multiples of the circuit time constant, as we did for the RC circuit, then we will find that the current will build up to its maximum value in the same fashion as the voltage does when a capacitor is being charged. This time let's pick a value of 100 A for the maximum current, so that our results will again come out as a percentage of the maximum current for any RL circuit.

At $t = 0$, $e^{-0} = 1$, so $I_{(0)} = 100$ A $(1 - 1) = 0$ A, or 0%
$t = 1\tau$, $e^{-1} = 0.368$, so $I_{(1\tau)} = 100$ A $(1 - 0.368) = 63.2$ A, or 63.2%
$t = 2\tau$, $e^{-2} = 0.135$, so $I_{(2\tau)} = 100$ A $(1 - 0.135) = 86.5$ A, or 86.5%
$t = 3\tau$, $e^{-3} = 0.050$, so $I_{(3\tau)} = 100$ A $(1 - 0.050) = 95.0$ A, or 95%
$t = 4\tau$, $e^{-4} = 0.018$, so $I_{(4\tau)} = 100$ A $(1 - 0.018) = 98.2$ A, or 98.2%
$t = 5\tau$, $e^{-5} = 0.007$, so $I_{(5\tau)} = 100$ A $(1 - 0.007) = 99.3$ A, or 99.3%

Exponential Decay in an RL Circuit

The electrical energy stored in an inductor as a magnetic field cannot be returned to the circuit in the same way as you discharge a capacitor. The magnetic field disappears as soon as the current stops. Opening S1 does not result in an inductor with energy stored in it, to be released at some later time. The energy stored in that magnetic field returns to the circuit instantly when S1 is opened. This rapid discharge of the field causes a very large voltage to be induced in the coil. This induced voltage, or back EMF, is normally many times larger than the applied voltage, because the induced voltage is proportional to the speed with which the field changes.

The most common result of opening the switch in such a circuit is that a spark, or arc, forms at the switch contacts. If the inductance is large and the current is high, a great deal of energy is released in a very short time. It is not at all unusual for the switch contacts to burn or melt under such circumstances.

Looking at Fig. 5-7, if we could move the switch to position B without actually breaking the circuit, then the current would decay exponentially, just as with a capacitor that is discharging. We can write an equation to describe this exponential decay:

$$I_{(t)} = \frac{E}{R}\left(e^{\frac{-t}{\tau}}\right) \qquad \text{(Eq. 5-14)}$$

Solving for times equal to multiples of one time constant, we find the same percentages as we did for a discharging capacitor.

At $t = 0, e^{-0} = 1$, so $I_{(0)} = 100$ A, or 100%
$t = 1\tau, e^{-1} = 0.368$, so $I_{(1\tau)} = 36.8$ A, or 36.8%
$t = 2\tau, e^{-2} = 0.135$, so $I_{(2\tau)} = 13.5$ A, or 13.5%
$t = 3\tau, e^{-3} = 0.050$, so $I_{(3\tau)} = 5$ A, or 5%
$t = 4\tau, e^{-4} = 0.018$, so $I_{(4\tau)} = 1.8$ A, or 1.8%
$t = 5\tau, e^{-5} = 0.007$, so $I_{(5\tau)} = 0.7$ A, or 0.7%

After one time constant the current will lose 63.2% of its steady-state value. (The current will decay to 36.8% of the starting value.) The graph of Fig. 5-9 shows the current-decay pattern of an RL circuit to be identical to the voltage-decay pattern of a capacitor. You should be careful about applying the terms charge and discharge to an inductive circuit, however, because an inductor stores energy in a magnetic field rather than in an electric field as a capacitor does.

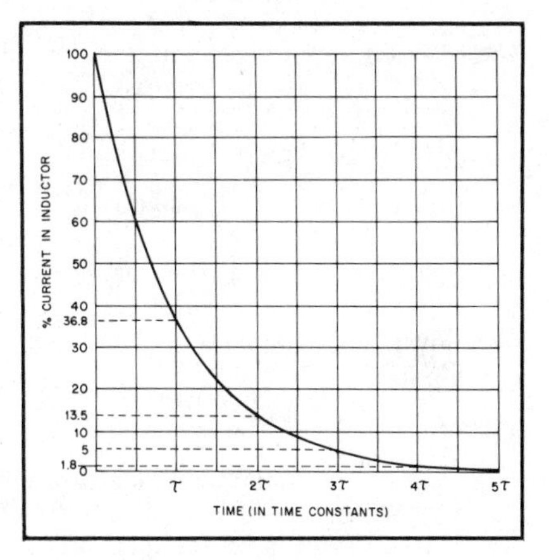

Fig. 5-9—Graph showing the current decay in an RL circuit. This curve is like the one for a discharging capacitor.

Time constants play an important part in numerous devices, such as electronic keys, timing and control circuits and the shaping of waveforms. Circuit time constants are also important in such applications as automatic gain control and noise limiters.

Let's try a couple of problems to be sure you understand the principles of time constants. If the resistor in Fig. 5-7 is 330 Ω, and the inductor has a value of 15 mH, what is the time constant? Use Eq. 5-13, being sure to use the fundamental units of ohms and henrys:

$$\tau = \frac{L}{R} = \frac{15 \times 10^{-3} \text{ henrys}}{330 \text{ ohms}} = 4.55 \times 10^{-5} \text{ seconds} = 45.5\ \mu\text{s}$$

If the current in the circuit is building up, then after 45.5 μs it will be at 63.2% of its maximum value. How long will it take for the current to be within 1% of its maximum value? Since we know it will take 5 time constants, or 5τ to reach the maximum value, we simply multiply the time constant by five:

$$5\tau = 5 \times 45.5\ \mu\text{s} = 227.5 \text{ microseconds, or } 0.0002275 \text{ second.}$$

The maximum current in this circuit is 4 A. If the current is building up from zero, how long will it take for a current of 3.5 A to be flowing? This represents 3.5 A/4 A = 87.5% of the maximum. This is just longer than 2 time constants, so in 2 × 45.5 μs = 91 μs the current will reach 3.5 A.

Another circuit has a 220-μF capacitor, a 470-kΩ resistor, a 100-μF capacitor and a 330-kΩ resistor all connected in series. A 50-V power supply is connected to the circuit until the capacitors are charged and current stops flowing. How long will it take for the voltage across this circuit to drop from 50 V to 2.5 V after the power supply has been replaced by a short circuit? First we must find the total equivalent resistance and capacitance. Since the resistors are in series, we can use Eq. 5-6:

$$R_T = 470 \times 10^3\ \Omega + 330 \times 10^3\ \Omega = 800 \times 10^3\ \Omega$$

$$R_T = 800\ \text{k}\Omega$$

To find the equivalent capacitance from the series combination, use Eq. 5-8:

$$C_T = \frac{1}{\dfrac{1}{220 \times 10^{-6}\ \text{F}} + \dfrac{1}{100 \times 10^{-6}\ \text{F}}}$$

$$C_T = \frac{1}{4.55 \times 10^3 + 1 \times 10^4} = \frac{1}{14.55 \times 10^3}$$

$$C_T = 6.87 \times 10^{-5}\ \text{F} = 68.7\ \mu\text{F}$$

Then use Eq. 5-5 to calculate the time constant:

$$\tau = RC = 800 \times 10^3\ \Omega \times 68.7 \times 10^{-6}\ \text{F}$$

$$\tau = 5.50 \times 10^1\ \text{sec} = 55.0\ \text{sec}$$

We want the circuit to discharge to a voltage that is

$$\frac{2.5}{50} \times 100\% = 5\%\ \text{of the starting voltage.}$$

We know that it will take a total of 3 time constants to discharge to 5% of the starting voltage. Then:

$$\text{time} = 3\tau = 3 \times 55.0\ \text{sec} = 165\ \text{sec}$$

[Now turn to Chapter 10 and study FCC examination questions with numbers that begin 4BE-2 and 4BE-3. Be sure to work out all of the problems there, and review this section if you have any difficulty.]

THE SMITH CHART

Named after its inventor, Phillip H. Smith, this chart was originally described in the January 1939 issue of *Electronics*. The *Smith Chart* is an invaluable tool for calculating impedances along transmission lines and many types of matching networks, both for antennas and other electronic circuitry.

At first glance it may seem formidable, but the Smith Chart is really nothing more than a specialized type of graph, with curved, rather than all straight lines. This coordinate system consists simply of two families of circles—the resistance family and the reactance family.

Fig. 5-10 illustrates how the resistance circles are drawn on the chart. These circles are centered on the resistance axis, which is the only straight line on an unused chart. The resistance circles are tangent to the outer circle at the bottom of the chart. Each circle is assigned a resistance value, which is indicated at the point where the circle crosses the resistance axis. All points along any one circle have the same resistance value.

Values assigned to these circles vary from zero at the top of the chart to infinity at the bottom, and actually represent a ratio with respect to the impedance value assigned to the center point on the chart, indicated by a 1.0 on Fig. 5-10. This center point is called prime center. If prime center is assigned a value of 100 ohms, then 200 ohms of resistance is represented by the 2.0 circle, 50 ohms by the 0.5 circle, 20 ohms by the 0.2 circle and so on. If a value of 50 is assigned to prime center, then the 2.0 circle represents 100 ohms, the 0.5 circle 25 ohms and the 0.2 circle 10 ohms. In each case you can see that the value on the chart is determined by dividing the actual resistance by the number assigned to prime center. This process is called

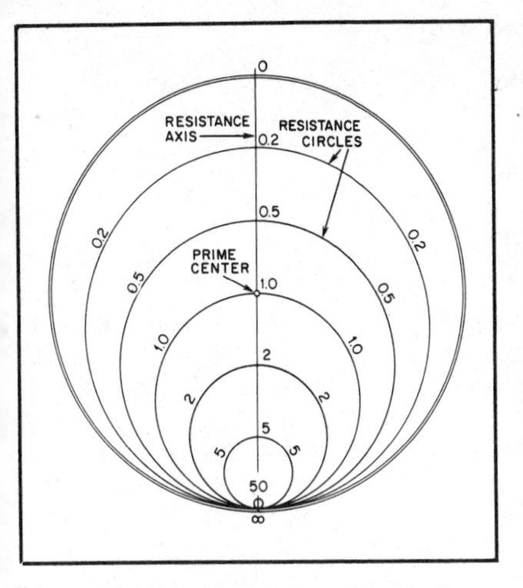

Fig. 5-10—Resistance circles of the Smith Chart coordinate system.

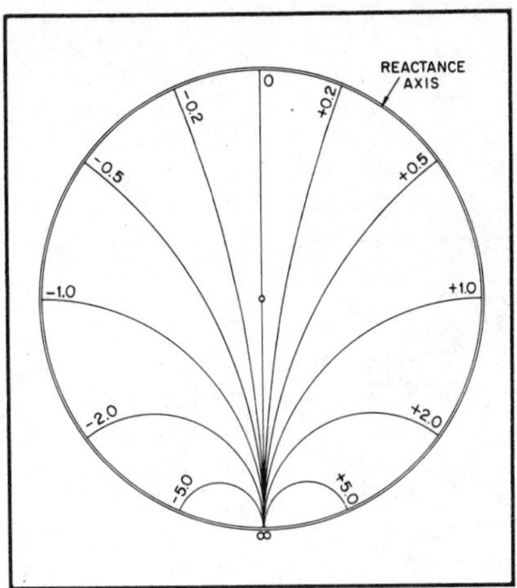

Fig. 5-11—Reactance circles (only segments are actually shown) of the Smith Chart coordinate system.

normalizing. Conversely, values from the chart can be converted back to actual resistance values by multiplying the chart value times the value assigned to prime center. Normalization permits the Smith Chart to be used for any impedance value.

Now let's consider just the reactance circles, as shown on Fig. 5-11. These circles appear as curved lines on the chart because only segments of the complete circles are drawn. These circles are tangent to the resistance axis, which itself is a member of the reactance family (it has a radius of infinity). The centers of the reactance circles are displaced to the right or left of the resistance axis, along a line that is perpendicular to the resistance axis at the bottom of the chart. The large outer circle bounding the coordinate portion of the chart is called the reactance axis.

Each reactance-circle segment is assigned a value of reactance, indicated near the point where the circle touches the reactance axis. All points along any one segment have the same reactance value. As with the resistance circles, the values assigned to each reactance circle are normalized with respect to the value assigned to prime center. Values to the right of the resistance axis are considered to be positive (inductive reactance) and those to the left of the resistance axis are considered to be negative (capacitive reactance).

When the resistance family and the reactance family of circles are combined, the coordinate system of the Smith Chart results, as shown in Fig. 5-12. Complex series impedances can be plotted on this coordinate system. More circles are normally added, so there will be less interpolation between chart circles for any value you have to plot.

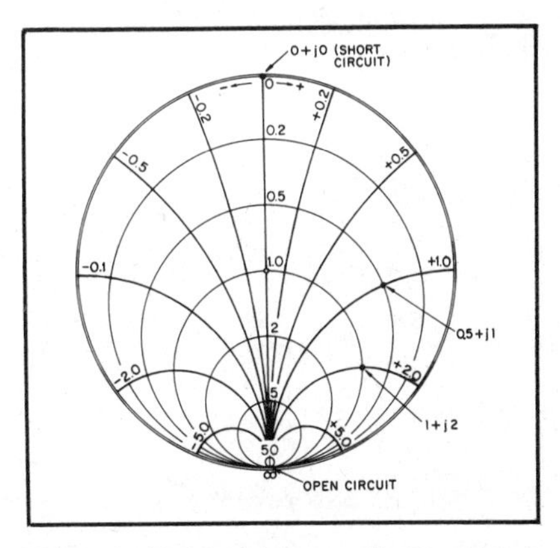

Fig. 5-12—The complete coordinate system of the Smith Chart. For simplicity, only a few divisions are shown for the resistance and reactance values.

Impedance Plotting

Suppose we have an impedance consisting of 50 ohms resistance and 100 ohms inductive reactance ($Z = 50 + j100$ ohms). If a value of 100 ohms is assigned to prime center, you would normalize this impedance by dividing each component by 100. The normalized impedance would then be:

$$\frac{50}{100} + \frac{j100}{100} \text{ ohms} = 0.5 + j1.0 \text{ ohms}$$

This impedance would be plotted on the Smith Chart at the intersection of the 0.5 resistance circle and the +1.0 reactance circle, as indicated on the chart shown in Fig. 5-12. If a value of 50 ohms had been assigned to prime center, as for 50-ohm coaxial line, the same impedance would be plotted at the intersection of the 1.0 resistance circle and the +2.0 reactance circle (also indicated on the chart shown in Fig. 5-12.

From these examples, you can see that the same impedance may be plotted at different points on the chart, depending on the value assigned to prime center. It is customary, when solving transmission-line problems, to assign a value to prime center that is equal to the characteristic impedance, or Z_0, of the line being used.

Short and Open Circuits

On the subject of plotting impedances, two special cases deserve consideration. These are short circuits and open circuits. A true short circuit has zero resistance and zero reactance, or an impedance of $0 + j0$ ohms. This impedance is plotted at the top of the Smith Chart, at the intersection of the resistance and reactance axes. An open circuit has infinite resistance, and would therefore be plotted at the bottom of the chart, at the intersection of the resistance and reactance axes. These two points will bé the same regardless of the value assigned to prime center. They are special cases sometimes used to calculate matching stubs.

Standing Wave Ratio Circles

Members of a third family of circles, which are not printed on the chart but which are added during the problem-solving process, are standing-wave-ratio (SWR) circles. This family is centered on prime center, and appears as concentric circles inside the reactance axis. During calculations, one or more of these circles may be added with a drawing compass. Each circle represents an SWR value, and every point on a given circle may be determined directly from the chart coordinate system by reading the resistance value where the SWR circle crosses the resistance axis, below prime center. (The reading where the circle crosses the resistance axis above prime center indicates the inverse ratio.) Fig. 5-13 illustrates a Smith Chart with SWR circles added.

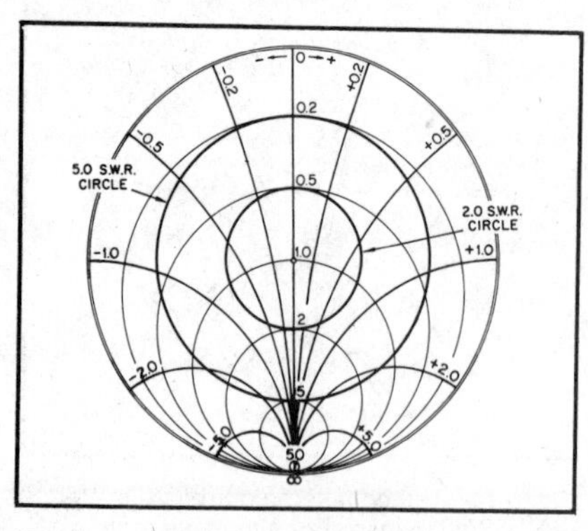

Fig. 5-13—A Smith Chart with SWR circles added.

Consider the situation where a load mismatch in a length of transmission line causes an SWR of 3 to 1. If we temporarily disregard line losses, we may state that the SWR remains constant throughout the entire length of this line. This is represented on the Smith Chart by drawing a 3:1 constant SWR circle (a circle with a radius of 3 on the resistance axis), as shown in Fig. 5-14. The chart design

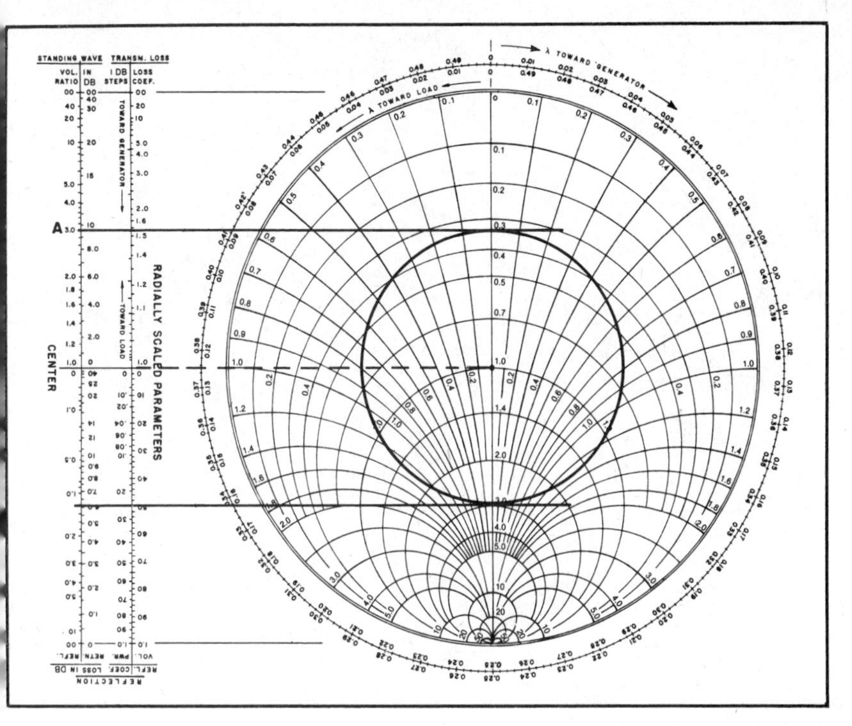

Fig. 5-14—An example of impedance calculations using a complete Smith Chart. See text for an explanation of the steps involved.

is such that an impedance encountered anywhere along the length of this mismatched line will fall on the SWR circle, and may be read directly from the coordinates merely by moving around the SWR circle by an amount corresponding to the length of the line involved.

This brings us to using the wavelength scales, which appear around the perimeter of the Smith Chart circle, as shown in Fig. 5-14. These scales are calibrated in terms of electrical wavelength along a transmission line. One scale, running counterclockwise, starts with the generator, or input end of the line at the top of the chart, and progresses toward the load end of the line. The other scale starts with the load end of the line at the top of the scale, and proceeds toward the generator in a clockwise direction. The complete circle represents one half wavelength. Progressing once around the perimeter of these scales corresponds to progressing along a transmission line for a half wavelength. Because impedances will repeat every half wavelength along a piece of line, the chart may be used for any line length by subtracting an integer (whole number) multiple of half wavelengths from the total length of the line.

A set of external scales are also shown along the edge of the chart in Fig. 5-14. These scales can be thought of as replacements for the resistance axis. For example, there is a scale labled Standing Wave Voltage Ratio. This scale starts with an SWR of 1 at prime center, and increases toward the top of the scale. You could measure the radius for your 3:1 SWR circle along this scale, then transfer it to the resistance axis and draw the circle. The advantage of using this external scale is that finer calibration marks can be included than is practical on the resistance scale. Circle radii for SWR values less than 2:1, for example, would have to be estimated on the resistance scale, but could be measured more accurately on the SWR scale.

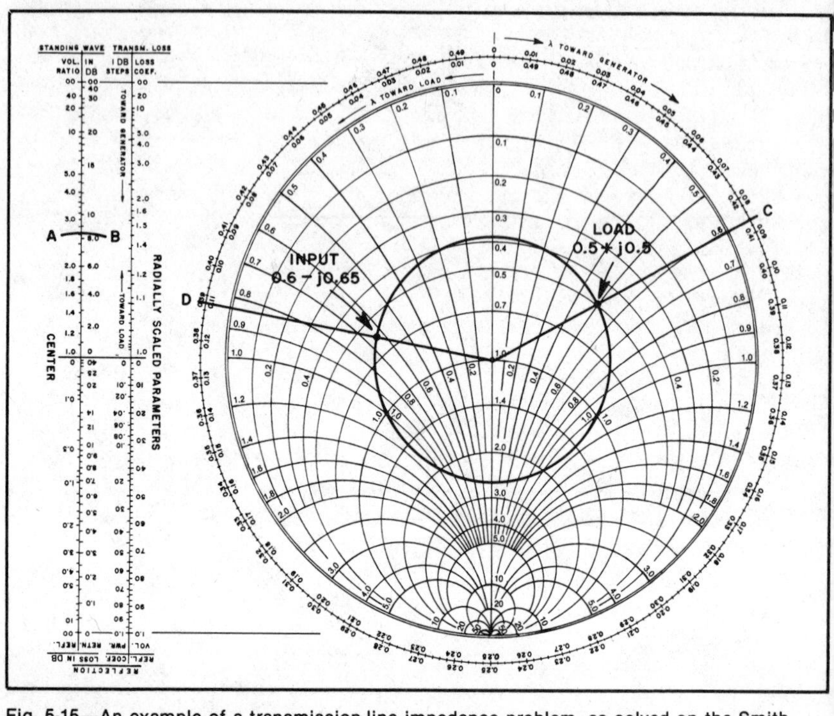

Fig. 5-15—An example of a transmission-line impedance problem, as solved on the Smith Chart. See text for an explanation of the steps involved.

Solving Problems with the Smith Chart

Suppose we have a transmission line with a characteristic impedance of 50 ohms and an electrical length of 0.3 wavelength. Also suppose we terminate this line with an impedance having a resistive component of 25 ohms and an inductive reactance of 25 ohms ($Z = 25 + j25$ ohms), and we want to determine the input impedance to the line. Because the line is not terminated in its characteristic impedance, we know that standing waves will be present on the line, and therefore, that the input impedance to the line will not be 50 ohms.

Proceed as follows: First, normalize the load impedance by dividing both the resistive and reactive components by 50 (the characteristic impedance, Z_0, of the line being used). The normalized impedance in this case is $0.5 + j0.5$ ohms. Plot this impedance on a blank Smith Chart at the intersection of the 0.5 resistance and the +0.5 reactance circles, as shown in Fig. 5-15. Now draw a constant SWR circle that passes through this point. The radius of this circle may then be transferred to the external scales by using your drawing compass. From the SW-VR scale, you can see that our transmission line is operating with an SWR of 2.6:1.

Next, with a straightedge, draw a radial line from prime center through the plotted impedance to intersect the wavelength scale marked TOWARD-GENERATOR (the outermost calibration), and read the value where the line meets this scale. Fig. 5-15 shows this value to be 0.088 wavelength (point C). To find the input impedance of our line, we simply find the point on the SWR circle that is 0.3 wavelength toward the generator from the plotted load impedance. This is accomplished by adding 0.3 (the length of the line in wavelengths) to the reference, or starting point, 0.088; $0.3 + 0.088 = 0.388$.

Now locate 0.388 on the TOWARD-GENERATOR scale (point D on Fig. 5-15), and draw a second radial line from this point to prime center. The intersection of this new radial line with the SWR circle represents the line input impedance, in this case 0.6 − j0.65 ohms. To find the actual line input impedance, multiply both of these values by 50—the value assigned to prime center. This calculation gives a result of 30 − j32.5 ohms, or 30-ohms resistance and 32.5-ohms capacitive reactance. This is the impedance that the transmitter must match if our system represents an antenna and transmission line. It is also the impedance that would be measured with an impedance bridge, if the measurement were taken at the input end of the line.

Most Smith Charts include other scales that will give us even more information about the impedance of our system. For example, there are scales to read phase angle, reflection coefficient, additional signal loss caused by the SWR, percent of the incident voltage, and power that is reflected from the load end. So this really is a versatile and powerful tool for impedance calculations.

Admittance Coordinates

Quite often it is desirable to convert impedance information to *admittance* data—resistance and reactance to *conductance* and *susceptance*. The reciprocals of impedance, reactance and resistance are given the special names admittance (Y), susceptance (B) and conductance (G), respectively. The units associated with admittance, susceptance and conductance are called siemens (S). We can write equations to express the reciprocal relationships: $Y = 1 / Z$, $B = 1 / X$, and $G = 1 / R$. Working with admittances greatly simplifies determining the resultant when two complex impedances are connected in parallel, as in calculating stub matching systems. The conductance values can then be added directly, as can the susceptance values, to arrive at the overall admittance for the parallel combination. This admittance may then be converted back to impedance data.

This conversion can be made very simply on the Smith Chart. The equivalent admittance of a plotted impedance value lies diametrically opposite the impedance point on the chart. Draw a straight line from the impedance point through prime center to the opposite side of the SWR circle. Normalized admittance values can be converted to actual values by dividing by the prime center value. Of course admittance coordinates may be converted to impedance coordinates just as easily—by locating the point on the Smith Chart that is diametrically opposite the admittance, on the same SWR circle.

Summary of Smith Chart Use

We can briefly summarize the major steps used in solving any impedance problem on the Smith Chart:

- First select a value for prime center that your known impedance values can be normalized to. If it is a transmission-line problem, select the characteristic impedance of the line. Otherwise, select a value that will result in normalized resistance and reactance values in the range of 0.5 to 5 if possible. Now you are ready to plot the normalized values and draw a constant SWR curve.
- Next, draw a line through prime center and the plotted impedance point. Extend the line through both sides of the SWR circle, or to cross the transmission-line wavelength scale, depending on what kind of problem you are solving.
- Third, you should read values from any other scales pertinent to your problem, such as finding the attenuation or loss in the line by drawing a second SWR circle.
- Finally, you calculate the resulting impedance by reversing the normalization process. (Multiply by the prime center value.)

Those are the basic steps involved in solving any Smith Chart problem. Each different type of problem may require some adjustment to that procedure, in terms of which scales to read, or what additional information you can determine from the chart.

[At this point, you should turn to Chapter 10 and study FCC examination questions with numbers that begin 4BE-4. Review this section as needed.]

IMPEDANCE CALCULATIONS

When a circuit contains both resistance and reactance, the combined effect of the two is called impedance, symbolized by the letter Z. Impedance is a more general term than either resistance or reactance. The term is often used with circuits that have only resistance or reactance.

The reactance and resistance comprising an impedance may be connected in series or in parallel, as shown in Fig. 5-16. In these circuits, the reactance is shown as a box, to indicate that it could be either inductive or capacitive. In the series circuit, shown at A, the current is the same through both elements, but with different voltages appearing across the resistance and reactance. In the parallel circuit shown at B, the same voltage is applied to both elements, but different currents may flow in the two branches.

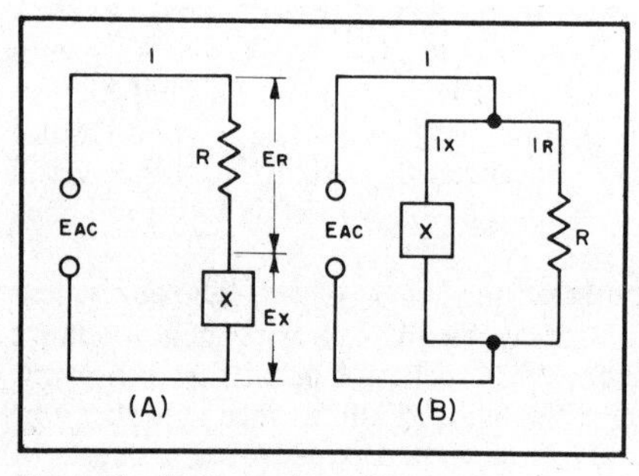

Fig. 5-16—Series and parallel circuits containing resistance and reactance.

You should remember that the current is in phase with the applied voltage through a resistance, but it is 90° out of phase with the voltage in a reactance. You probably learned the mnemonic ELI the ICE man when you studied for your Advanced class exam, so you will have no trouble remembering that the voltage leads the current in an inductor and the current leads the voltage in a capacitor.

You can see, then, that the phase relationship between current and voltage for the whole circuit can be anything between zero and 90°, depending on the relative amounts of resistance and reactance in the circuit.

It is important to realize that if there is more than one resistor in the circuit, you must combine them to get one equivalent resistance value. Likewise, if there is more than one reactive element, they must be combined to one equivalent reactance. If there are several inductors and several capacitors, combine all the like elements, then subtract the capacitance value from the inductance.

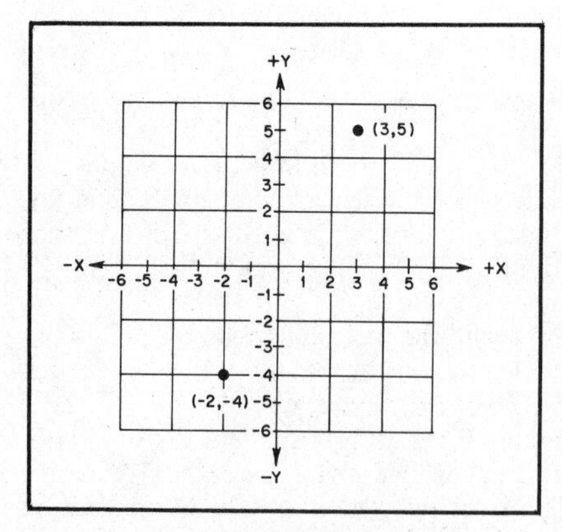

Fig. 5-17—A rectangular-coordinate system uses axes that are at right angles to each other. Any point on the plane can be expressed in terms of an X and a Y coordinate value.

Coordinate Systems

We commonly use coordinate systems to help us visualize or picture problems that we cannot see directly. For example, we can't actually see the electrons flowing in a circuit, or look at the voltage or impedance associated with the circuit. But we can draw pictures that will represent these quantities, and use those pictures to d our solution to the problems. There re a number of commonly used coor-inate systems. You learned about one stem in the last section. Two other com-only used coordinate systems are the *ctangular-coordinate system* shown in ig. 5-17 (sometimes called Cartesian oordinates) and the *polar-coordinate stem* shown in Fig. 5-18. Both of these stems are two dimensional, but they can e extended to three dimensional systems asily. If you add a third axis, usually beled the Z axis (no relation to im-edance), coming out of the page at right ngles to the other two on the rectangular stem, you have a three-dimensional sys-m. If you rotate the polar-coordinate rcle around the center point so it escribes a sphere, you have a spherical oordinate system, which is handy for presenting satellite orbits.

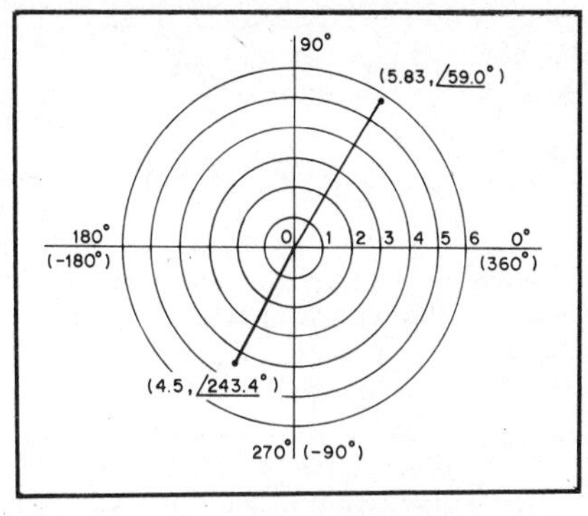

Fig. 5-18—A polar-coordinate system uses a radius, as measured from the center to the point and an angle measured counter-clockwise from the right side. Any point on the plane can be expressed in terms of a radius and an angle.

When using rectangular coordinates, a point anywhere on the plane formed by e X and Y axes can be described by a pair of numbers, (X,Y). For example a point at is 3 units from the origin along the X axis and 5 units from the origin along e Y axis could be written as (3,5). A point that is 2 units to the left of the origin long the X axis and 4 units below the origin along the Y axis would be described s (−2, −4).

In the polar-coordinate system, points on the plane are also described by a number air. In this case we use a length, or radius, measured from the origin and an angle, easured counterclockwise from the 0° line. So the two points described in the last aragraph could also be written as (5.83, /59.0°) and (4.5, /243.3°). If you measured is angle clockwise from the 0° line, then it would be (4.5, /−116.6°)

Complex Numbers

Sometimes during the solution of a mathematical problem, you come up with term that includes the square root of minus one. $\sqrt{-1}$ is a mathematical curiosity ecause it is impossible to find a number that you can multiply by itself and get a egative value ($-1 \times -1 = +1$). Since that number does not really exist, it is called n *imaginary number*. Mathematicians use the symbol i to represent $\sqrt{-1}$, but in lectronics we use j instead, to avoid confusion with the symbol for current. Com-leting such a problem involves the use of complex mathematics. (This name has othing to do with how hard—or easy—it is to understand and use the mathematics!)

In electronics, it is common to use both the rectangular- and the polar-coordinate ystems when dealing with impedance problems. In the rigorous mathematical solu-ion of general impedance problems, the $\sqrt{-1}$ term comes up. So these problems ead us to using the techniques of complex-number algebra. This also helps make clear that the point being described is an impedance that consists of a resistance nd a reactance component. It is quite simple to learn how to calculate impedances sing this method. We say that the X axis, called the real axis, represents the voltage r current associated with the resistances in our circuit. Then the Y axis, or the im-ginary axis, represents the voltage or current associated with the reactances.

Series Circuits

When the resistance and reactance are in series, then the two values can be combined in a relatively straightforward manner. The current is the same in all parts c the circuit ($I_R = I_X$), and the voltage is different across each part. We can write a equation for the impedance in the form:

$$Z = \frac{E}{I} = \frac{E_R + E_X}{I} \qquad \text{(Eq. 5-15)}$$

You will notice that this is really just Ohm's Law, written for impedance instead c resistance, as we are used to seeing it. This equation also shows that we can conside the voltage and current associated with the resistive and reactive elements separately

Since we are really interested in the impedance, and not the actual voltage c current in most cases, it is convenient to choose a current of 1 A, so that the voltag and impedance have the same magnitude. If the reactance is inductive, the voltag will lead the current by 90°, so if we use the voltage across the resistor as a referenc then the voltage across the reactance must be drawn in the +j direction on our grapl If the reactance is capacitive, the voltage will lag the current by 90°. This time th voltage across the reactance is drawn in the −j direction.

It is a common practice to eliminate the voltage calculation, and just plot th resistance and reactance values on the graph directly. It would be helpful to remembe however, that the reason we label inductive reactance as +j and capacitive reactanc as −j is because of the leading and lagging current-voltage relationship described her

To specify an impedance on the rectangular-coordinate complex-number plan you only need to know the resistance and the reactance value. If an inductance c capacitance is specified instead of a reactance, then you will first have to calculat the reactance. When you studied for the Element 3 exam, you learned equations t calculate inductive and capacitive reactance, given the inductance or capacitance an the applied signal frequency:

$$X_L = 2\pi f L \qquad \text{(Eq. 5-16)}$$

$$X_C = \frac{1}{2\pi f C} \qquad \text{(Eq. 5-17)}$$

where

X_L is inductive reactance in ohms
X_C is capacitive reactance in ohms
f is the applied signal frequency in hertz
L is the inductance in henrys
C is the capacitance in farads

Suppose you have a 500-ohm resistor in series with a 100-pF capacitor. Wha is the impedance of this combination when it is connected to a 2-MHz signal generator Calculate the capacitive reactance using Eq. 5-17. X_C = 796 ohms. Because th voltage across a capacitor lags the current, we place the X_C value on the −j axis The impedance value for this circuit, then, is 500 − j796 ohms. If we replace th capacitor with a 25-μH inductor, what is the impedance of this new circuit? First calculate the inductive reactance using Eq. 5-16. X_L = 314 ohms. So the new im pedance is 500 + j314 ohms. Fig. 5-19 shows how these two impedance values ca be plotted on a rectangular-coordinate plane.

Sometimes you may want to express the impedance in a polar-coordinate form There are a number of methods that you could use to perform that conversion. Prob ably the easiest way is if you have an inexpensive calculator that is able to do th conversion for you! Enter the two rectangular-coordinate values, push a few othe buttons, and read the polar-coordinate equivalent value. With this type of calculato you can also convert from polar to rectangular coordinates.

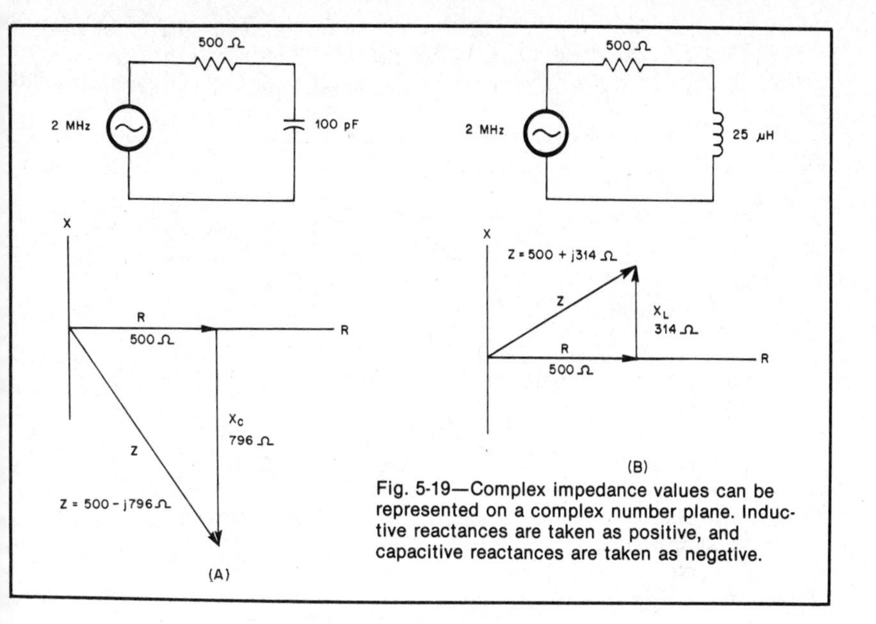

Fig. 5-19—Complex impedance values can be represented on a complex number plane. Inductive reactances are taken as positive, and capacitive reactances are taken as negative.

Lacking such a device, you will have to resort to some basic trigonometry. To help you visualize the solution, make a drawing on a piece of graph paper, or at least draw a set of axes, and label them R and j. Then mark off a rough scale and plot the impedance point using the resistance and reactance values. (Draw one line starting at the origin and going to the right to represent the resistance. Then draw another line starting at the end of that one and going either up or down depending on whether you have an inductive or capacitive reactance, to represent the reactance.) If you draw a line from the origin to the point at the end of the reactance line, you will form a right triangle, as shown on Fig. 5-19. The Pythagorean Theorem equation is one way to calculate the impedance in polar coordinates:

$$|Z| = \sqrt{R^2 + X^2} \qquad \text{(Eq. 5-18)}$$

We put vertical lines around the Z to indicate that we are calculating the total magnitude of the impedance.

To determine the phase angle between the current and voltage you will have to use the tangent function.

$$\tan \theta = \frac{\text{side opposite the angle}}{\text{side adjacent to the angle}} \qquad \text{(Eq. 5-19)}$$

Let's convert our two examples above to polar-coordinate form. First we will convert the 500 − j796 ohms impedance.

$$|Z| = \sqrt{(500 \text{ ohms})^2 + (-j796 \text{ ohms})^2}$$

$$|Z| = \sqrt{88.4 \times 10^4 \text{ ohm}^2} = 940 \text{ ohms}$$

Then find the phase angle:

$$\tan \theta = \frac{-j796 \text{ ohms}}{500 \text{ ohms}} = -1.59$$

$$\tan^{-1}(-1.59) = -57.9°$$

So we can also express this impedance in the form; 940 ohms at a phase angle of $-57.9°$. The negative phase angle implies that the voltage across the circuit lags the current through it, which is what we would expect from a circuit with capacitive reactance.

We can convert the 500 + j314 ohms impedance in the same way.

$$|Z| = \sqrt{(500 \text{ ohms})^2 + (+j314 \text{ ohms})^2}$$

$$|Z| = \sqrt{34.9 \times 10^4 \text{ ohms}^2} = 590 \text{ ohms}$$

and

$$\tan \theta = \frac{+j314 \text{ ohms}}{500 \text{ ohms}} = 0.628$$

$$\tan^{-1}(0.628) = 32.1°$$

$$500 + j314 \text{ ohms} = 590 \text{ ohms at a phase angle of } 32.1°.$$

Parallel Circuits

When a resistance and a reactance are connected in parallel, it is the applied voltage that is common to all parts of the circuit ($E_R = E_X$ this time), and the current through each part will be different. We can still use Ohm's Law to calculate impedance, however:

$$Z = \frac{E}{I} = \frac{E}{I_R + I_X} \qquad \text{(Eq. 5-15A)}$$

Keeping in mind our mnemonic, ELI the ICE man, you can see that the current through an inductor lags the voltage, and that the current through a capacitor leads the voltage. Again, the current and voltage in a resistor are in phase, so we can use the resistive element as the reference point for a drawing to calculate the impedance and phase angle associated with the circuit. We can choose a convenient value for the voltage applied to the circuit, if all we want to know is the impedance and phase angle.

To find the current through each element, we can consider the resistance and reactance elements separately.

$$I_R = \frac{E_R}{R} = E_R \times G \qquad \text{(Eq. 5-20)}$$

and

$$I_X = \frac{E_X}{X} = E_X \times B \qquad \text{(Eq. 5-21)}$$

If we choose $E = 1$ V, for convenience, the current turns out to be numerically equivalent to the reciprocal of the resistance or reactance. In the section describing the Smith Chart, we defined these quantities as conductance (G) and susceptance (B). For convenience, we often skip the current calculation, and just take the reciprocal of the resistance and reactance.

Let's look at an example to see how we can work through a parallel impedance problem. We will use the same 100-pF capacitor, 500-ohm resistor and 2-MHz signal generator that we used before, but this time they are connected in parallel. See Fig. 5-20. The capacitive reactance is the same as it was in the series circuit, $X_C = 796$ ohms. As indicated above, to calculate the impedance of a parallel circuit, first find the conductance and susceptance:

$$G = \frac{1}{R} = \frac{1}{500 \text{ ohms}} = 0.00200 \text{ S}$$

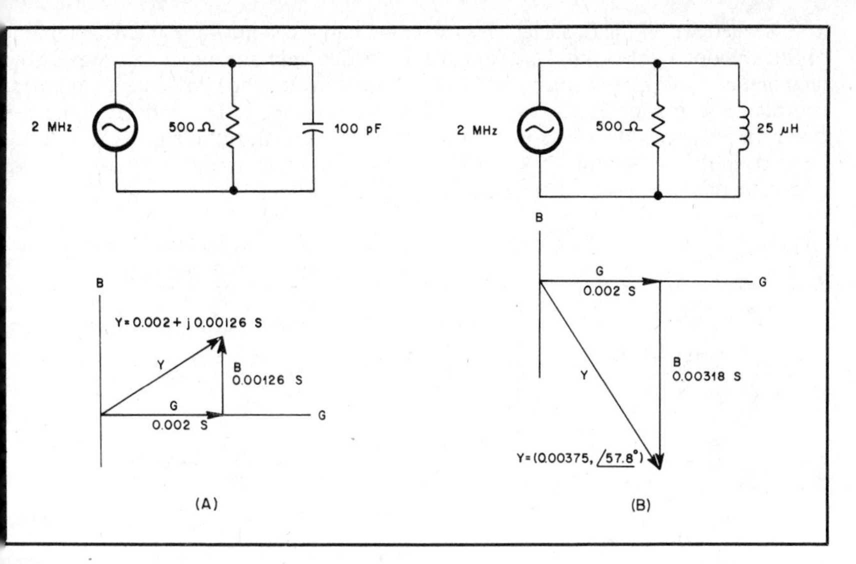

Fig. 5-20—With a parallel circuit, it is common to use admittance, conductance and susceptance values to calculate the circuit impedance.

and

$$B = \frac{1}{X} = \frac{1}{-j796 \text{ ohms}} = +j0.00126 \text{ S}$$

Don't be confused by the change of sign in front of the j here. That comes about because of a rule of algebra for complex numbers. When you have a j operator in the denominator, you must eliminate it before performing the division. That is accomplished using a technique called complex conjugates. If you multiply a complex number by the same number, but with an opposite-sign j operator, you get a $-j^2$ term. But since $j = \sqrt{-1}$, $j^2 = \sqrt{-1}^{\,2} = -1$, and $-(-1) = +1$! The only catch is that there is another rule of algebra that says if you multiply the denominator of a fraction by some number, you must multiply the numerator by the same value. Here are the steps involved:

$$B = \frac{1}{X} = \frac{1}{-j796 \text{ ohms}} \times \frac{(+j796 \text{ ohms})}{(+j796 \text{ ohms})}$$

$$B = \frac{+j796 \text{ ohms}}{-j^2\,(796 \text{ ohms})^2} = \frac{+j}{-(-1)(796 \text{ ohms})}$$

$$B = +j0.00126 \text{ S}$$

Of course, you can see that the only real effect of this process is that the sign in front of the j operator has to change when you convert from reactance to susceptance. If you have a complex number that includes a real and an imaginary part, the process becomes just a little more involved.

Now we will turn to Eq. 5-15A, and write our impedance equation, choosing a value of 1 for the voltage:

$$Z = \frac{E}{I_R + I_X} = \frac{1}{2.00 \times 10^{-3} + j1.26 \times 10^{-3}}$$

This is where we hit the first snag. We can either multiply numerator and denominator of this fraction by the complex conjugate of the denominator, or we can convert the rectangular-coordinate form to polar-coordinate form. When you change to polar-coordinate form there is no j operator left in the expression. To divide by a number that is in polar-coordinate form, you just divide by the magnitude and subtract the angle from the numerator angle. (In this case the numerator angle is 0°.) Let's take a look at the complex conjugate method first.

$$Z = \frac{1}{2.00 \times 10^{-3} + j1.26 \times 10^{-3}} \; \frac{(2.00 \times 10^{-3} - j1.26 \times 10^{-3})}{(2.00 \times 10^{-3} - j1.26 \times 10^{-3})}$$

$$Z = \frac{(2.00 \times 10^{-3} - j1.26 \times 10^{-3}}{(2.00 \times 10^{-3})^2 - (j^2)(1.26 \times 10^{-3})^2}$$

$$Z = \frac{(2.00 \times 10^{-3} - j1.26 \times 10^{-3})}{4.00 \times 10^{-6} - (-1)(1.59 \times 10^{-6})} = \frac{(2.00 \times 10^{-3} - j1.26 \times 10^{-3})}{5.59 \times 10^{-6}}$$

$$Z = 3.58 \times 10^2 - j2.25 \times 10^2 \text{ ohms} = 358 - j225 \text{ ohms}$$

If you want this answer specified in polar-coordinate form, you should have no difficulty doing the conversion. $Z = 423$ ohms at $-32.2°$.

Suppose you replace the 100-pF capacitor with a 25-μH inductor again, as we did in our example of a series impedance problem. The reactance of the inductor will still be $+j314$ ohms. To calculate the impedance we can find the susceptance of this inductor, $B = -j3.18 \times 10^{-3}$ S. (Don't forget about changing the sign of the j operator when you take the reciprocal!) Again we will assume a value of 1 V for our applied voltage. Then we can write the equation to find impedance, using Eq. 5-15A:

$$Z = \frac{E}{I_R + I_X} = \frac{1}{2.00 \times 10^{-3} - j3.18 \times 10^{-3}}$$

This time let's use the polar-coordinate concept and convert the denominator to polar-coordinate form. The numerator is especially easy. Since it is a value of 1, and there is no imaginary component, the angle will be zero; 1 V, $\underline{/0}°$. To calculate the magnitude of the denominator we use the Pythagorean Theorem (Eq. 5-18).

$$|I| = \sqrt{(2.00 \times 10^{-3} \text{ A})^2 + (-j3.18 \times 10^{-3} \text{ A})^2}$$

$$|I| = \sqrt{(4.00 \times 10^{-6} \text{ A}^2) + (-j^2)(10.1 \times 10^{-6} \text{ A}^2)}$$

$$|I| = \sqrt{14.1 \times 10^{-6} \text{ A}^2}$$

$$|I| = 3.75 \times 10^{-3} \text{ A}$$

Notice that since we are really just combining the conductance and susceptance values here, the result is a value for the admittance, so $Y = 3.75 \times 10^{-3}$ S. To calculate the phase angle, you will have to use the tangent function from trigonometry:

$$\tan \theta = \frac{\text{side opposite}}{\text{side adjacent}} = \frac{-3.18 \times 10^{-3}}{2.00 \times 10^{-3}} = -1.59$$

Then take the inverse tangent function to find the angle:

$$\tan^{-1}(-1.59) = -57.8°$$

n polar-coordinate form, we can write the admittance as

$\cdot.75 \times 10^{-3}$ S, at $-57.8°$

Now substitute this value back into our impedance equation:

$$Z = \frac{1, \angle 0°}{3.75 \times 10^{-3}, \angle -57.8°} = 267 \text{ ohms}, 57.8°$$

f you want the impedance specified in rectangular-coordinate form, you can use the ine and cosine functions from trigonometry.

$$\text{Sine } \theta = \frac{\text{side opposite}}{\text{hypotenuse}} = \frac{\text{reactance}}{\text{impedance}} = \frac{X_L}{Z}$$

$$X = Z \sin \theta \qquad \text{(Eq. 5-22)}$$

$$X_L = \sin(57.8°) \times 267 \text{ ohms} = 0.846 \times 267 \text{ ohms}$$

$$X_L = 226 \text{ ohms}$$

$$\text{Cosine } \theta \frac{\text{side adjacent}}{\text{hypotenuse}} = \frac{\text{resistance}}{\text{impedance}} = \frac{R}{Z}$$

$$R = Z \cos \theta \qquad \text{(Eq. 5-23)}$$

$$R = \cos(57.8°) \times 267 \text{ ohms} = 0.533 \times 267 \text{ ohms}$$

$$R = 142 \text{ ohms}$$

So $Z = 142 + j226$ ohms

For some extra practice you should also solve the first example using polar co-ordinates. That will also verify that both methods give the same results.

There are many other ways to manipulate the mathematics of complex impedances or series and parallel circuits. Some books choose to present special equations that allow you to transform a parallel circuit into an equivalent series circuit, and then have you write the impedance of the equivalent circuit. That technique requires you o memorize several different equations, and then to be sure you know when to ap-ply each equation. *The ARRL Handbook* includes such "circuit transformation" equations. The method described in this license manual requires only that you know ome basic algebra of complex numbers, and the electronics principles that you are already familiar with, such as Ohm's Law. The FCC syllabus for the Extra Class icense includes algebraic operations using complex numbers as a topic that you may be tested on. To solve the actual impedance problems on your exam you can use any method that you choose. If you follow the step-by-step procedure described here, however, you won't go wrong.

[Before proceeding to Chapter 6, turn to Chapter 10 and study FCC examina-ion questions with numbers that begin 4BE-5 and 4BE-6. Review this section as needed.]

Key Words

AND gate—A logic circuit whose output is 1 only when both of its inputs are 1.

CMOS—Complementary metal-oxide semiconductor digital integrated circuits. CMOS is composed of both N channel and P channel MOS devices on the same chip.

Depletion mode—Type of operation in a JFET or MOSFET where current flow is reduced by reverse bias on the gate.

Digital IC—An integrated circuit whose output is either on (1) or off (0).

Drain—The point at which the charge carriers exit an FET. Corresponds to the plate of a vacuum tube.

Enhancement mode—Type of operation in a MOSFET in which current flow is increased by forward bias on the gate.

Field-effect transistor (FET)—A voltage-controlled semiconductor device. Output current can be varied by varying the input voltage. The input impedance of an FET is very high.

Gate—Control terminal of an FET. Corresponds to the grid of a vacuum tube.

Integrated circuit—A device composed of many bipolar or field-effect transistors manufactured on the same chip, or wafer, of silicon.

Inverter—A logic circuit with one input and one output. The output is 1 when the input is 0, and the output is 0 when the input is 1.

Ion trap—A magnet installed in a cathode-ray tube to prevent negative ions from burning a brown spot on the center of the CRT.

Junction field-effect transistor (JFET)—A field-effect transistor created by diffusing a gate of one type of semiconductor material into a channel of the opposite type of semiconductor material.

Linear IC—An integrated circuit whose output voltage is a linear (straight line) representation of its input voltage.

Metal-oxide semiconductor FET (MOSFET)—A field-effect transistor that has its gate insulated from the channel material. Also called an IGFET, or insulated gate FET.

NAND (NOT AND) gate—A logic circuit whose output is 0 only when both inputs are 1.

Noninverting buffer—A logic circuit with one input and one output, and whose output level is the same as the input level.

NOR (NOT OR) gate—A logic circuit whose output is 0 if either input is 1.

Operational amplifier (op amp)—A linear IC that can amplify dc as well as ac. Op amps have very high input impedance, very low output impedance and very high gain.

OR gate—A logic circuit whose ouput is 1 when either input is 1.

Phase-locked loop (PLL)—A servo loop consisting of a phase detector, low-pass filter, dc amplifier and voltage-controlled oscillator.

Source—The point at which the charge carriers enter an FET. Corresponds to the cathode of a vacuum tube.

Truth table—A chart showing the outputs for all possible inputs to a digital circuit.

Transistor-transistor logic (TTL)—Digital integrated circuits composed of bipolar transistors, possibly as discrete components, but usually part of a single IC. Power supply voltage should be 5 V.

Vidicon tube—A type of photosensitive vacuum tube widely used in TV cameras.

Chapter 6

Circuit Components

Before you can understand the operation of most complex electronic circuits, you must know some basic information about the parts that make up those circuits. This chapter presents the information about circuit components that you need to know in order to pass your Amateur Extra Class license exam. You will find descriptions of several types of field-effect transistors (FETs); linear and digital integrated circuits (ICs) including phase-locked loops, operational amplifiers and basic logic gates; and vidicon tubes. These components are combined with other devices to build practical electronic circuits, some of which are described in Chapter 7.

As you study the characteristics of the components described in this chapter, be sure to turn to the FCC Element 4B questions in Chapter 10 when you are directed to do so. That will show you where you need to do some extra studying. If you thoroughly understand how these components work, you should have no problem learning how they can be connected to make a circuit perform a specific task.

FIELD-EFFECT TRANSISTORS (FETs)

Field-effect transistors are given that name because the current through them is controlled by a varying electric field created by the voltage applied to the *gate* lead. By contrast, in a bipolar transistor, output current is controlled by the current applied to the base.

There are two types of field-effect transistors in use today: the *junction FET (JFET)* and the *metal-oxide semiconductor FET (MOSFET)*. Like bipolar transistors, the JFET has no insulation between its elements. The MOSFET has a thin layer of oxide between the gate or gates and the drain/source junction.

The basic characteristic of both FET types is a very high input impedance—typically 1 megohm or greater. This is considerably higher than the input impedance of a bipolar transistor. Although some FETs have but one gate, others have two. Single-gate FETs can be equated to triode vacuum tubes. The gate represents the grid, the *drain* is similar to the anode, and the *source* is like the cathode.

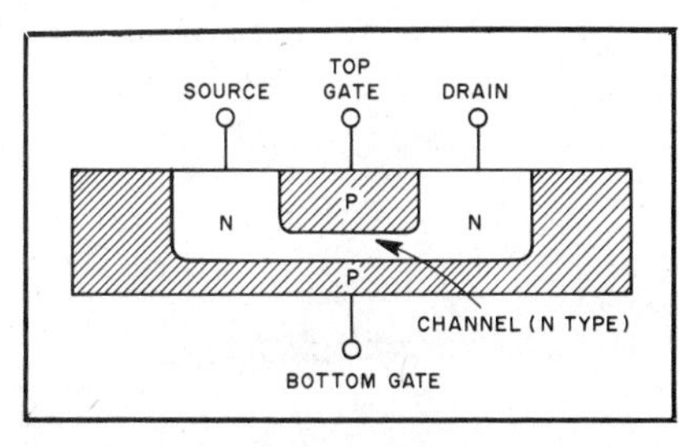

Fig. 6-1—Construction of a junction field-effect transistor (JFET).

JFETs

The basic JFET construction is shown in Fig. 6-1. The JFET can be thought of

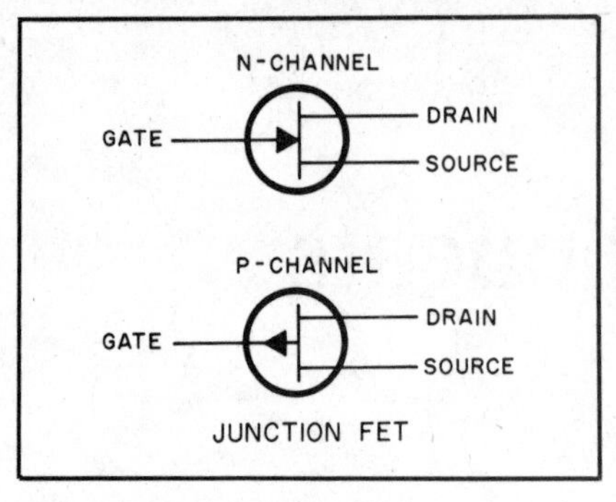

Fig. 6-2—Schematic symbols for N-channel and P-channel JFETs.

simply as a bar of silicon semiconductor material that acts like a resistor. The terminal into which the charge carriers enter is called the source. The opposite terminal is called the drain. There are two types of JFET (N channel and P channel), so named for the type of material used to form the drain/source channel. The schematic symbols for the two JFET types are illustrated in Fig. 6-2.

Two gate regions, made of the opposite type of semiconductor material, are diffused into the JFET channel. When a reverse voltage is applied to the gates, an electric field is set up perpendicular to the current through the channel. The electric field interferes with the normal electron flow through the channel. As the gate voltage is varied, this electric field varies, and that causes a variation in source-to-drain current. Thus, the FET is an example of a voltage-controlled current source, similar to a triode vacuum tube. The gate terminal is always reverse biased, so the JFET has a very high input impedance, much like a vacuum tube with its high grid input impedance.

MOSFETs

The construction of a MOSFET, sometimes called an insulated gate field-effect transistor (IGFET), and its schematic symbol are illustrated in Fig. 6-3. In the MOSFET, the gate is insulated from the source/drain channel by a thin dielectric layer. Since there is very little current through this dielectric, the input impedance

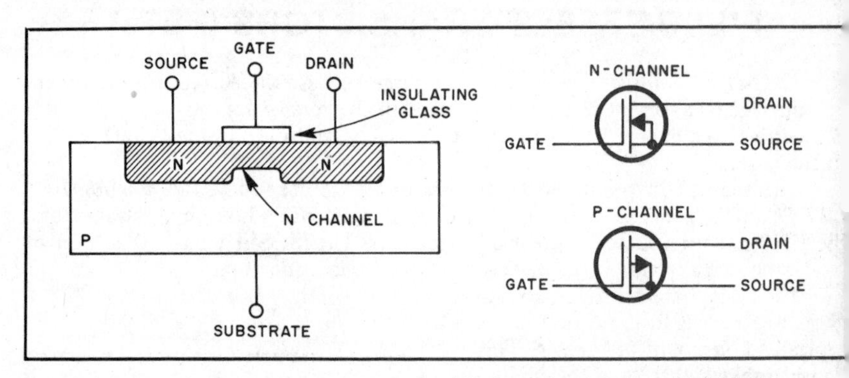

Fig. 6-3—Construction and schematic symbols of N-channel and P-channel MOSFETs.

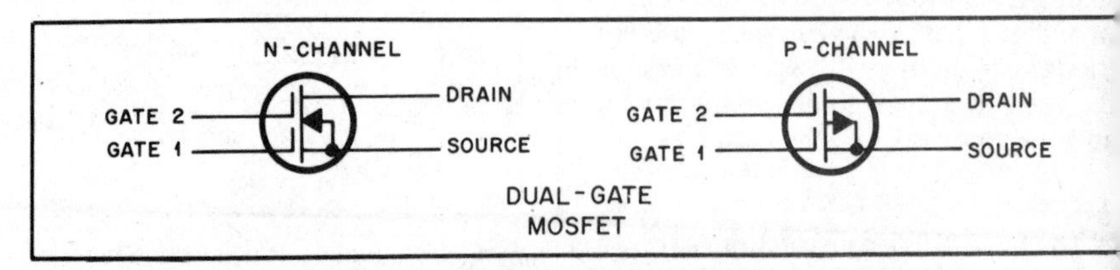

Fig. 6-4—Schematic symbols for N-channel and P-channel, dual-gate MOSFETs.

s even higher than in the JFET, typically 10 megohms or greater.

Some MOSFETs are made with two gates rather than one. This type of FET s widely used as an RF amplifier or mixer in receivers and converters. The schematic symbols for N-channel and P-channnel dual-gate MOSFETs are shown in Fig. 6-4.

Nearly all the MOSFETs manufactured today have built-in gate-protective Zener diodes. Without this provision the gate insulation can be perforated easily by small static charges on the user's hand or by the application of excessive voltages to the device. The protective diodes are connected between the gate (or gates) and the source lead of the FET.

Enhancement- and Depletion-Mode FETs

Field-effect transistors are classified into two main groupings for applications in circuits—*enhancement mode* and *depletion mode*. A depletion-mode device corresponds to Fig. 6-1, where a channel exists with no bias applied. The gate of a depletion-mode device is reverse biased in operation. When the reverse bias is applied to the gate, the channel is depleted of charge carriers, and current decreases.

Enhancement-mode devices are those specifically constructed so they have no channel when there is no voltage on the gate. They become useful only when a gate voltage is applied, which causes a channel to be formed. When the gate of an enhancement-mode device is forward biased, current begins to flow through the source/drain channel. The more forward bias on the gate, the more current through the channel. JFETs cannot be used as enhancement-mode devices, because if the gate is forward biased it will conduct, like a forward-biased diode. MOSFETs have their gates insulated from the channel region, so they may be used as enhancement-mode devices, since both polarities may be applied to the gate without the gate becoming forward biased and conducting. Some MOSFETs are designed to be used with no bias on the gate. In this type of operation, the MOSFET operates in the enhancement mode when the gate is forward biased, and in the depletion mode when the gate is reverse biased.

To sum up, a depletion-mode FET is one that has a channel constructed; thus, a current will flow through the channel with zero gate voltage applied. Enhancement-mode FETs are those that have no channel, so there is no current when there is zero gate voltage applied.

Designing with FETs is much like designing with vacuum tubes. They are used in RF amplifiers and oscillators, and they are ideal for use in voltmeters, where their high input impedance will not load down the circuit being tested.

FETs are made in the same types of packages as bipolar transistors. Some different case styles are shown in Fig. 6-5.

[Now study FCC exam questions 4BF-1.1 through 4BF-1.13 and 4BF-1.16. Review this section if you have any difficulty with these questions.]

Fig. 6-5—FETs are packaged in cases much like those used for bipolar transistors.

INTEGRATED CIRCUITS

Integrated circuits comprise many transistors on a single wafer, or chip, of silicon. Integrated circuits may be made up of bipolar or field-effect transistors, and may be linear (smoothly varying output) or digital (on/off output) in operation. Most ICs today are packaged in a plastic dual-in-line package, or DIP. Some are packaged

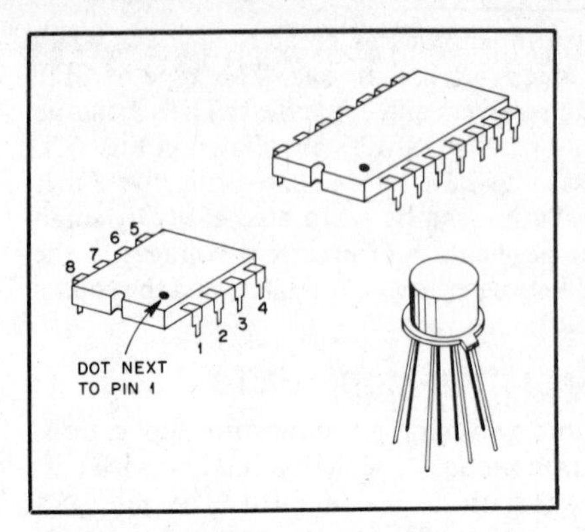

Fig. 6-6—Integrated circuits may be packaged in 8-pin and 14-pin (or more) dual-in-line packages (DIP) or a TO-5 case similar to a transistor, with many leads.

in a metal can similar to a transistor, with many leads. The DIP and metal-can packages are illustrated in Fig. 6-6.

Linear Integrated Circuits

Linear ICs are so-called because in their usual operating mode, the output voltage is a linear function of the input voltage. This does not mean they cannot be operated in a nonlinear mode, such as for a class-C amplifier. The bias will determine the operating mode, class A through class C. You will need to understand two types of linear ICs for the Extra Class exam: operational amplifiers (op amps) and phase-locked loops (PLL).

Operational Amplifiers

The *operational amplifier (op amp)* is a high-gain amplifier that will amplify dc signals as well as ac signals. An op amp is characterized by its high input impedance and low output impedance. The first op amps were designed for use in analog computers, where they performed such mathematical operations as multiplying numbers and extracting square roots; hence the name operational amplifier. The op amp is perhaps the most versatile IC in the amateur and commercial field today.

The most obvious use for operational amplifiers is as a low-distortion audio amplifier, but it has many other uses as well. Op amps can be made into oscillators to generate sine, square and even sawtooth waves. Used with negative feedback, their high input impedance and linear characteristics make them ideal for use as instrumentation amplifiers. It would take an entire book to illustrate the many uses to which op amps can be put.

A theoretically perfect (ideal) op amp would have the following characteristics: infinite input impedance, zero output impedance, infinite voltage gain, flat frequency response within its frequency range and zero output when the input is zero. These criteria can be approached in a practical situation, but not realized entirely.

Operational amplifiers have two inputs, an inverting input and a noninverting

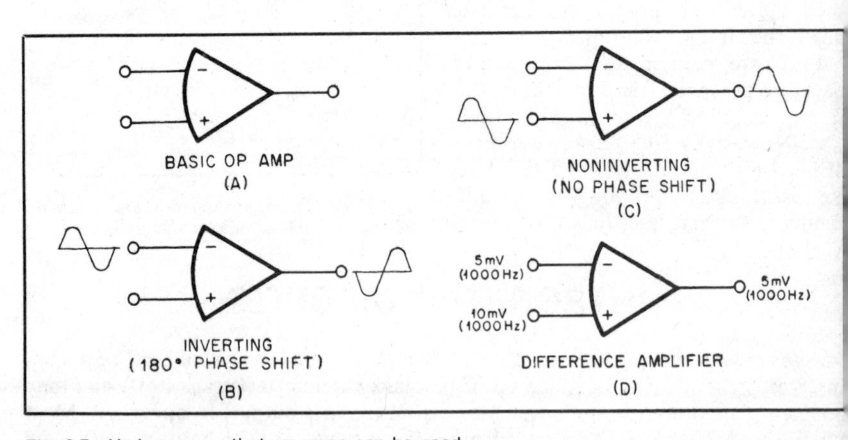

Fig. 6-7—Various ways that op amps can be used.

nput. They can be used in a variety of configurations. Three of the more common op amp configurations are the inverting amplifier, noninverting amplifier and difference amplifier. These modes are compared in Fig. 6-7, which also shows the phase relationship of the input and output for each configuration.

The gain of a practical op amp without feedback (open-loop gain) is often as high as 500,000. Op amps are rarely used as amplifiers in the open-loop configuration, however. Usually, some of the output is fed back to the inverting input, where it acts to reduce the stage gain. The more negative feedback, the more stable the amplifier circuit will be.

The gain of the circuit with negative feedback is called the closed-loop gain. The higher the open-loop gain, the more negative feedback that can be used and still provide enough closed-loop gain. If the open-loop gain is many times greater than the closed-loop gain, we find that the stage gain is determined by external components rather than by the gain of the op amp itself.

Voltage gain and voltage output for the inverting op-amp circuit can be determined easily. Fig. 6-8 ilustrates the basic circuit arrangement. R1 and Rf determine the circuit gain.

$$V_{out} = \frac{Rf}{R1} V_{in} \qquad \text{(Eq. 6-1)}$$

$$V_{gain} = \frac{V_{out}}{V_{in}} \qquad \text{(Eq. 6-2)}$$

Then substituting Eq. 6-1 into Eq. 6-2, we get:

$$V_{gain} = \frac{\frac{Rf}{R1} V_{in}}{V_{in}} = \frac{Rf}{R1} \qquad \text{(Eq. 6-3)}$$

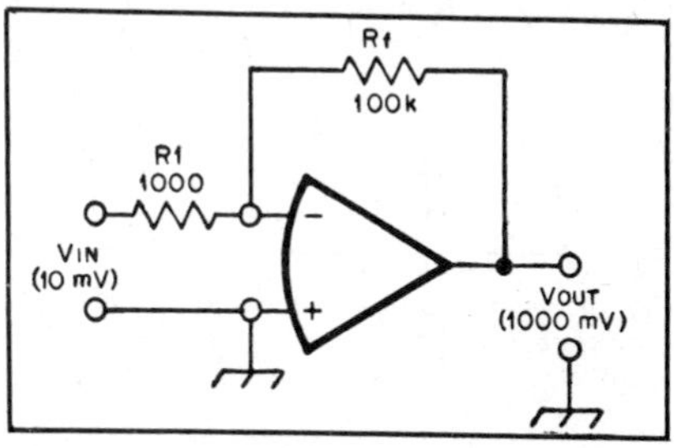

Fig. 6-8—Inverting op-amp configuration. Resistance values determine the gain.

We can also express this gain in decibels if we remember that V_{gain} is a voltage ratio, and that we have to multiply the log of a voltage ratio by 20 when converting to decibels.

$$\text{Gain (dB)} = 20 \log(V_{gain}) \qquad \text{(Eq. 6-4)}$$

These equations illustrate that the circuit gain is determined by the resistors rather than by the op-amp characteristics and the power-supply voltage. In the inverting configuration, the op amp can act as a summing amplifier. For example, two or more separate audio signals may be brought to the input; the circuit then acts as an active audio mixer.

Fig. 6-9 shows a noninverting op-amp circuit. The gain of this circuit can be derived in a manner similar to the one used with the inverting circuit.

$$V_{out} = \frac{R1 + Rf}{R1} V_{in} \qquad \text{(Eq. 6-5)}$$

Again substituting for V_{out} in Eq. 6-2 we have:

$$V_{gain} = \frac{V_{out}}{V_{in}} = \frac{\frac{R1 + Rf}{R1} V_{in}}{V_{in}}$$

$$V_{gain} = \frac{R1 + Rf}{R1} \qquad \text{(Eq. 6-6)}$$

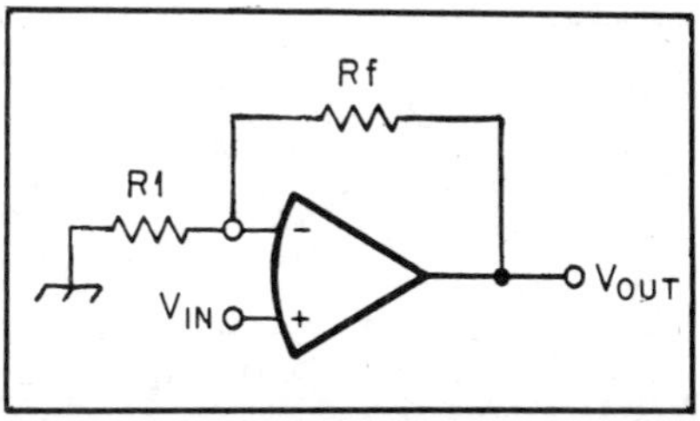

Fig. 6-9—A noninverting op-amp configuration. The resistance values still determine the gain, but the relationship is a bit different than for the inverting configuration.

In both the noninverting and the inverting configuration, op amps can be used in many audio applications, such as low-level audio amplifiers and microphone preamps. When an op-amp circuit is adjusted so the gain of the circuit is one, the output voltage follows the input exactly. In this configuration, the device is called a voltage follower. The voltage follower is the ultimate impedance transformer. With the high input impedance and low output impedance of the op amp, the voltage follower can be used to match a low-power signal to subsequent stages in a circuit. If the voltage follower is set up in the inverting configuration, it can be used as a 180° phase-shift network as well as an impedance-matching circuit. A voltage-follower circuit is shown in Fig. 6-10.

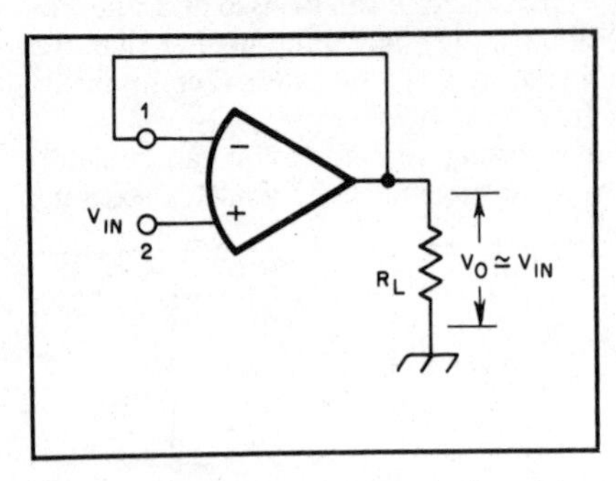

Fig. 6-10—The voltage follower (unity gain) configuration. Notice that 100% negative feedback is used.

Fig. 6-11—An op amp configured as a difference amplifier.

Fig. 6-11 illustrates a difference amplifier. A reference voltage can be applied to one of the op amp inputs while the remaining input is fed a changing dc level. The difference in the two input voltages will appear at the op-amp output. The output voltage of this circuit is given by:

$$V_{out} = \left[\frac{R1 + Rf}{R2 + R3}\right] \left[\frac{R3}{R1} V2\right] - \left[\frac{Rf}{R1} V1\right] \quad \text{(Eq. 6-7)}$$

Phase-Locked Loops

Phase-locked loop (PLL) circuits have recently come to have many applications in Amateur Radio: as FM demodulators, frequency synthesizers, FSK demodulators and a host of other applications. A basic phase-locked loop circuit is really an electronic servo loop, consisting of a phase detector, a loop filter, a dc amplifier and a voltage-controlled oscillator. Fig. 6-12 shows a block diagram of a phase-locked-loop circuit.

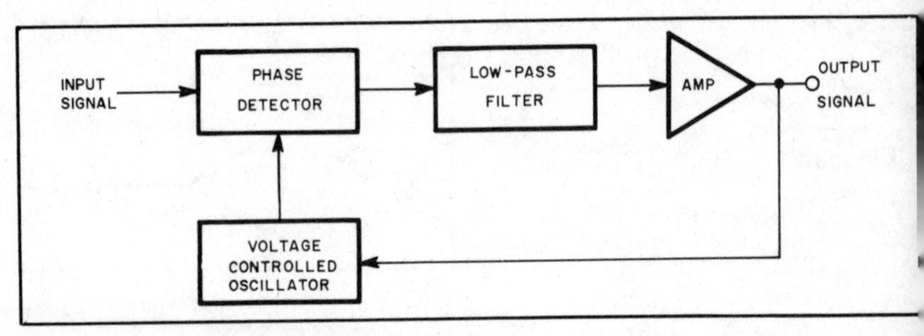

Fig. 6-12—Block diagram of a phase-locked-loop system, which could be used to demodulate an FM signal.

The signal from the voltage-controlled oscillator (VCO) and the input signal are fed to the phase detector, which produces an error voltage corresponding to the frequency difference between the VCO and the input signals. This error voltage is filtered, amplified and sent back to the VCO so the oscillator adjusts to the same frequency as the input signal. When the input signal to the phase detector and the output from the VCO approach the same frequency, the error voltage coming from the phase detector decreases toward zero, and we say that the VCO is locked onto the incoming-signal frequency. Any changes in the phase of the input signal, indicating a change in its frequency, are sensed at the phase detector and the error voltage readjusts the VCO so that it remains locked to that signal.

One important use of a phase-locked-loop circuit is as an FM demodulator or detector. If the incoming signal is frequency modulated, then the error voltage coming out of the dc amplifier is a copy of the audio variations used to modulate the FM transmitter. Thus, taking the output from the dc amplifier allows the phased-locked loop to be used directly as a demodulator or detector in an FM receiver.

Many frequency synthesizers used in amateur transceivers involve phase-locked-loop ICs. Fig. 6-13 illustrates the parts of such a circuit. The frequency divider block allows a wide range of output frequencies to be generated using a single, stable reference oscillator. In most circuits the division ratio is controlled by electronic means. To change the output frequency you vary the division ratio. Again, we are using the principle of an electronic servo, or negative feedback system. The average frequency error is reduced to zero.

Notice that if the VCO is not stable when it is outside of the phase-locked loop, it will still change frequency when it is in the loop. The average frequency is correct, but there is a frequency variation that produces a phase modulation of the output signal. This "phase noise" is directly audible in some receivers, and the noise is also present on the transmitted signal of some rigs that use a PLL synthesizer.

The phase detector, filter and amplifier functions shown in the block diagrams of Figs. 6-12 and 6-13 can be contained in a single IC. A few external components can be used to set the VCO frequency and loop filter characteristics. PLL integrated circuits are also used for signal conditioning, AM demodulation, FSK demodulation and a host of other applications.

[Now study FCC exam questions that begin 4BF-2. Review this section as needed.]

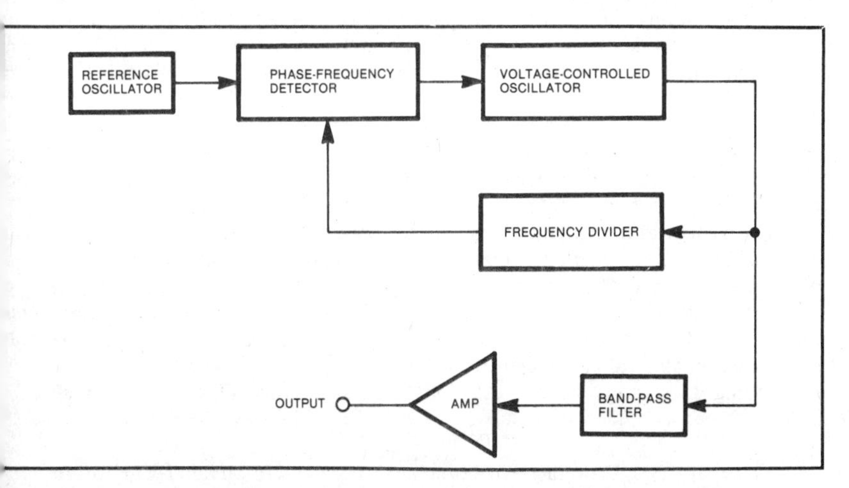

Fig. 6-13—An indirect frequency synthesizer, which uses a phase-locked loop and a variable-ratio frequency divider.

Digital Electronics Basics

Digital electronics is an important aspect of Amateur Radio. Everything from simple digital circuits to sophisticated microcomputer systems are used in modern Amateur Radio systems. The applications include digital communications, code conversion signal processing, station control, frequency synthesis, amateur satellite telemetry message handling, word processing and other information-handling operations.

The fundamental principle of digital electronics is that a device can have only a finite number of states. In binary digital systems, there are two discrete states, represented in base-2 arithmetic by the numerals 0 and 1. The binary states described as 0 and 1 may represent an off and on condition or a space and mark in a communications transmission such as CW or RTTY. Fig. 6-14 illustrates a typical binary signal.

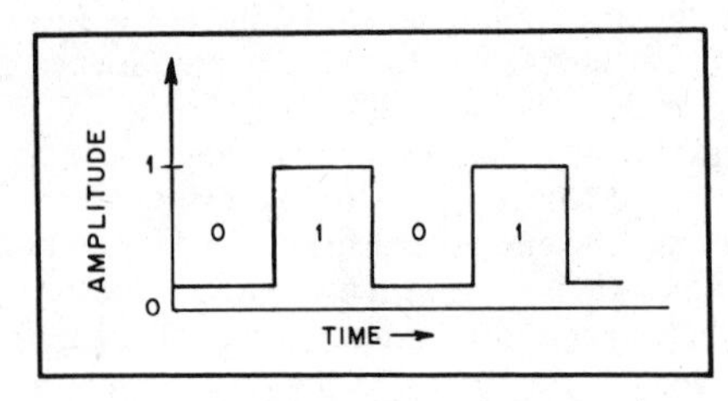

Fig. 6-14—A typical binary signal, which may have either of two signal levels. The signal shown is a square wave, but a binary signal may spend unequal times at each level, depending on how the signal is intended to be used.

The simplest digital devices are switches and relays. Computers built before 1950 were almost entirely made with mechanical relays. Low speed and rapid wear were the main objections to such mechanical devices. The next generation of digital instruments used electron tubes as the switching elements. Physical size and power consumption then limited the complexity of possible circuits. Modern semiconductor technology allows digital systems with tremendous complexity to be built at a small fraction of the cost of previous methods. A *digital IC*, then, is an integrated circuit chip that generates, detects or in some way processes digital signals.

In electronic digital logic systems, binary information is transmitted as voltage levels. For example, 0 V may represent a binary 0, and +5 V equals a binary 1. Because it is not always possible to achieve those exact voltages in practical circuits, digital circuits consider the signal to be a 1 or 0 if the voltage comes within certain bounds.

Combinational Logic—Boolean Algebra

In binary digital logic circuits, a combination of inputs results in a specific output or combination of outputs. Except during switching transitions, the state of the output is determined by the simultaneous state(s) of the input channel(s). A combinational logic function has one and only one output state corresponding to each combination of input states. Combinational logic networks have no feedback loops.

Networks made up of many combinational logic elements may perform arithmetic or logical manipulations. Regardless of the purpose, these operations are usually expressed in arithmetic terms. Digital networks add, subtract, multiply and divide, but normally do it in binary form using the numerals 0 and 1.

Binary digital circuit functions may be represented by equations using Boolean algebra. The symbols and laws of Boolean algebra are somewhat different from those of ordinary algebra. The symbol for each logical function is shown here in the descriptions of the individual logical elements. The logical function of a particular element may be described by listing all possible combinations of input and output values in a *truth table*, also called a state table. Standard logic circuit symbols and their corresponding Boolean expressions and truth tables are shown in Figs. 6-15 through 6-22.

One-Input Elements

There are two logic elements that have only one input and one output: the *noninverting buffer* and the *inverter* or NOT circuit (Figs. 6-15 and 6-16). The noninverting buffer simply passes the same state (0 or 1) from its input to its output

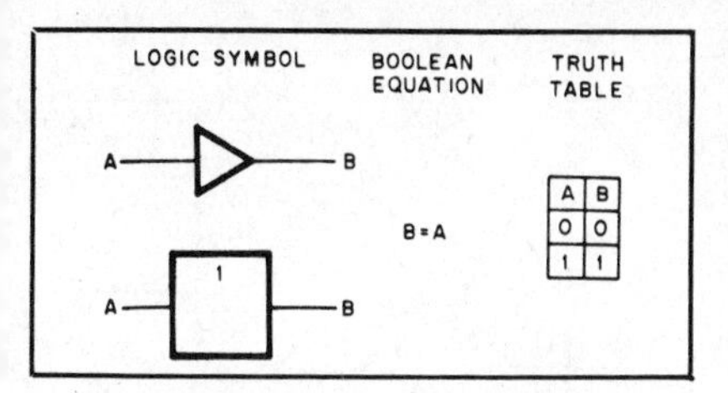

Fig. 6-15—Schematic symbols for a non-inverting buffer are shown. The distinctive (triangular) shape is used by ARRL and in most U.S. publications. The square symbol is used in some other countries. The Boolean equation for the buffer and a truth table for the operation are also given.

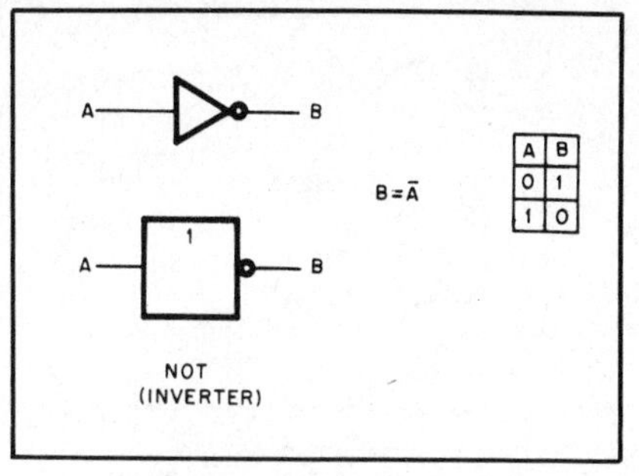

Fig. 6-16—Schematic symbols for an inverter (NOT) are shown. The Boolean equation and truth table for the operation are also given.

In an inverter or NOT circuit, a 1 at the input produces a 0 at the output, and vice versa. NOT indicates inversion, negation or complementation. The Boolean algebra notation for NOT is a bar over the variable or expression.

The AND Operation

A gate is usually defined as a combinational logic element with two or more inputs and one output state dependent on the state of the inputs. (Emitter-coupled logic—ECL—devices violate this definition by having two outputs.) Gates perform simple logical operations and can be combined to form complex switching functions. So as we talk about the logical operations used in Boolean algebra, you should keep in mind that each function is implemented by using a gate with the same name. So an *AND gate* implements the AND operation.

The AND operation results in a 1 only when all operands are 1. That is, if the inputs are called A and B, the output is 1 only if A and B are both 1. In Boolean notation, the logical operator AND is usually represented by a dot between the variables, centered on the line (•). The AND function may also be signified by no space between the variables. Both forms are shown in Fig. 6-17, along with the schematic symbol for an AND gate.

The OR Operation

The OR operation results in a 1 at the output if any or all inputs are 1. In Boolean notation, the + symbol is used to indicate the OR function. The *OR gate* shown in Fig. 6-18 is somtimes called an INCLUSIVE OR.

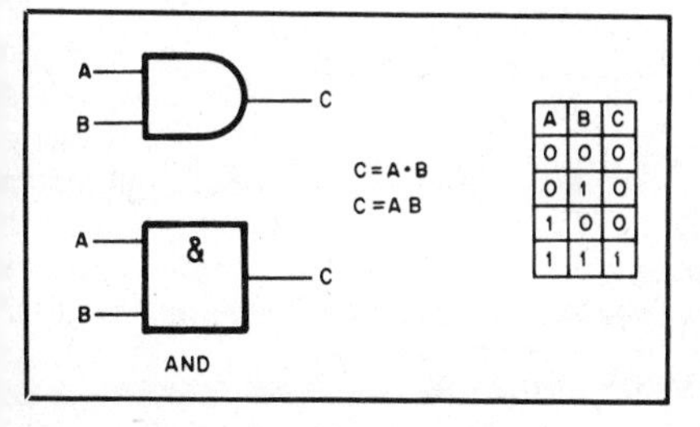

Fig. 6-17—Schematic symbols for a two-input AND gate. The Boolean equation and truth table for the operation are also given.

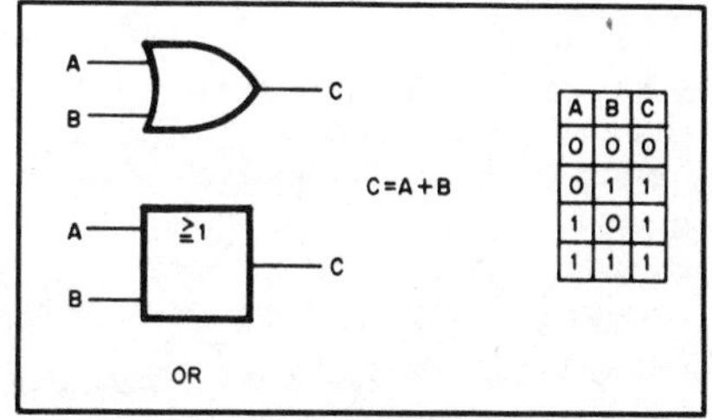

Fig. 6-18—Schematic symbols for a two-input OR gate. The Boolean equation and truth table for the operation are also given.

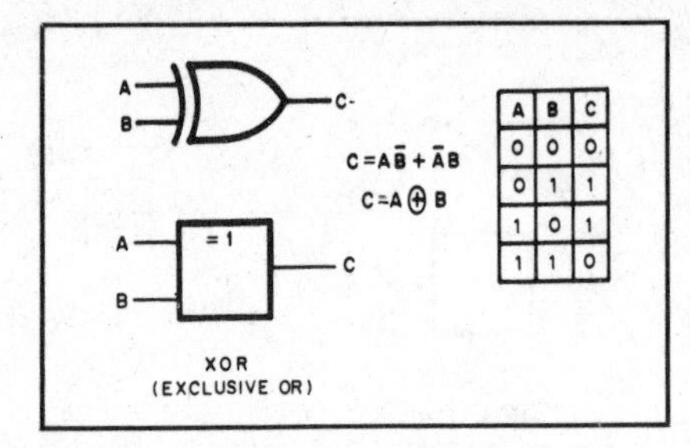

A	B	C
0	0	0
0	1	1
1	0	1
1	1	0

Fig. 6-19—Schematic symbols for a two-input EXCLUSIVE OR (XOR) gate. The Boolean equation and truth table for the operation are also given.

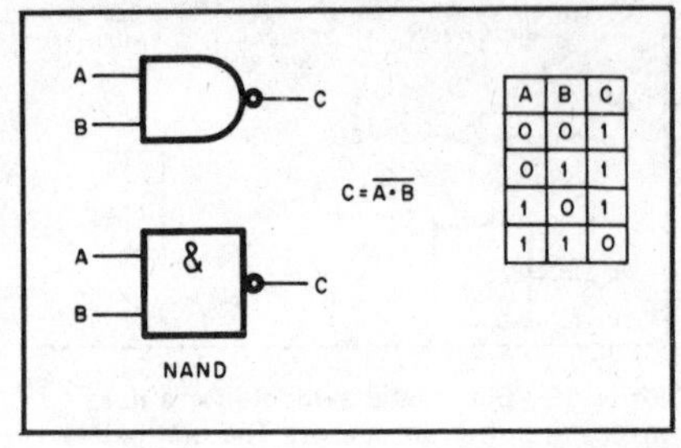

A	B	C
0	0	1
0	1	1
1	0	1
1	1	0

Fig. 6-20—Schematic symbols for a two-input NAND gate. The Boolean equation and truth table for the operation are also given.

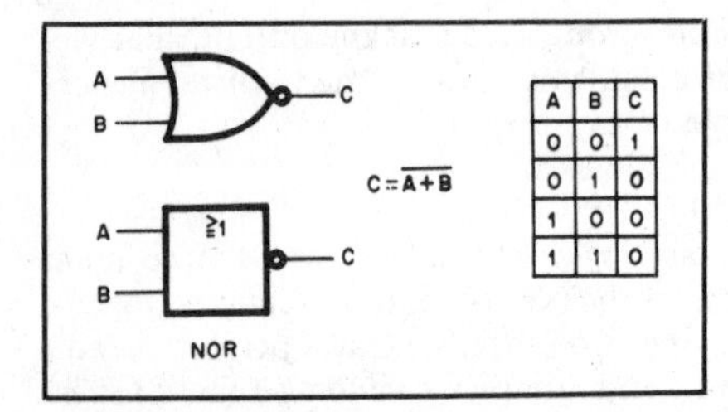

A	B	C
0	0	1
0	1	0
1	0	0
1	1	0

Fig. 6-21—Schematic symbols for a two-input NOR gate. The Boolean equation and truth table for the operation are also given.

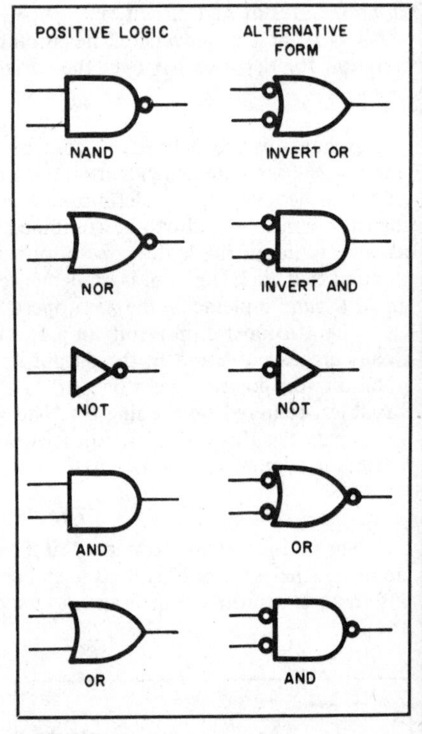

Fig. 6-22—Positive and negative true-logic symbols for the common functions.

The EXCLUSIVE OR Operation

The EXCLUSIVE OR (XOR) operation results in an output of 1 if only one of the inputs is 1, but if both inputs are 1, then the output is 0. The Boolean expression ⊕ represents the EXCLUSIVE OR function. Fig. 6-19 shows the schematic symbol for an EXCLUSIVE OR gate and its truth table.

The NAND Operation

The NAND operation means NOT AND. A NAND gate (Fig. 6-20) is an AND gate with an inverted output. A *NAND gate* produces a 0 at its output only when all inputs are 1. In Boolean notation, NAND is usually represented by a dot between the variables and a bar over the combination, as shown in Fig. 6-20.

The NOR Operation

The NOR operation means NOT OR. A *NOR gate* (Fig. 6-21) produces a 0 output if any or all of its inputs are 1. In Boolean notation, the variables have a + symbol between them and a bar over the entire expression to indicate the NOR function.

Logic Polarity

Logic systems can be designed to use two types of polarity. If the highest voltage level (high) represents a binary 1, and the lowest level (low) represents a 0, the logic is positive. If the opposite representation is used, the logic is negative. In the gate descriptions discussed so far, positive logic was assumed.

Positive and negative logic symbols are compared in Fig. 6-22. Small circles (state indicators) on the input side of a gate signify negative logic. The use of negative logic sometimes simplifies the Boolean algebra associated with logic networks. Consider a circuit having two inputs and one output, and suppose you desire a high output only when both inputs are low. A search through the truth tables shows the NOR gate has the proper characteristics. The way the problem is posed (the words only and both) suggests the AND (or NAND) function, however. A negative-logic NAND is functionally equivalent to a positive-logic NOR gate. The NAND symbol better expresses the circuit function in the application just described. Fig. 6-23 traces the evolution

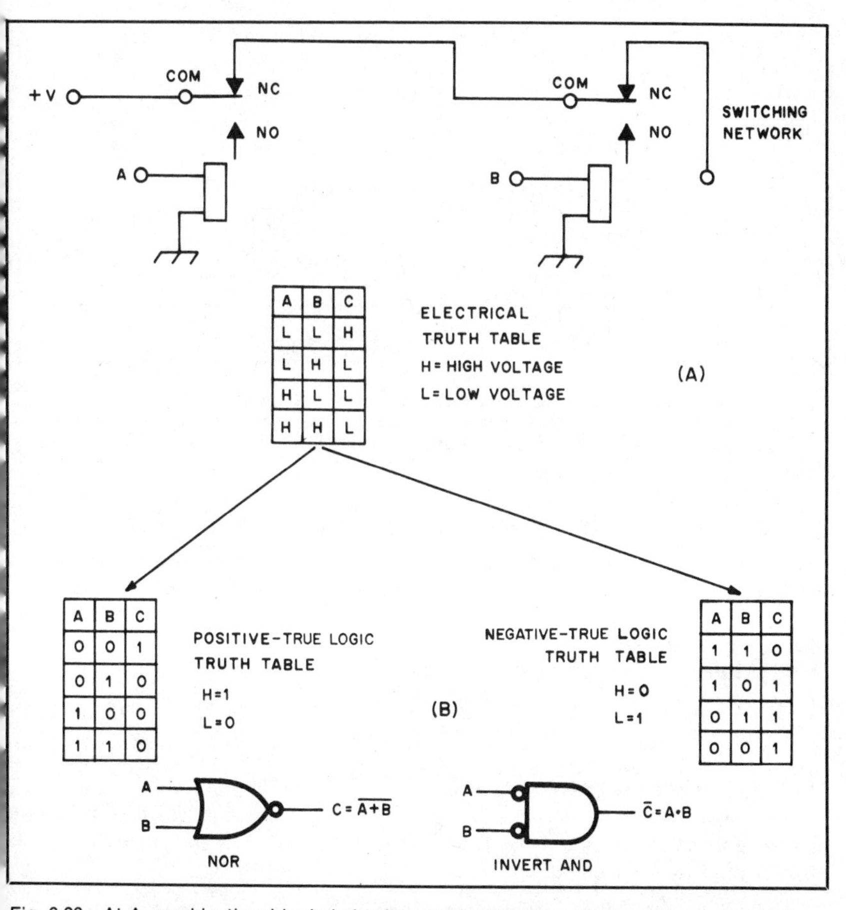

A	B	C
L	L	H
L	H	L
H	L	L
H	H	L

A	B	C
0	0	1
0	1	0
1	0	0
1	1	0

A	B	C
1	1	0
1	0	1
0	1	1
0	0	1

Fig. 6-23—At A, combinational logic is implemented with relays, shown with an electrical truth table. Assigning values of 1 and 0 to the electrical states as shown in B leads to two schematic symbols, one for positive-true logic and one for negative-true logic. The two symbols are electrically equivalent; depending on the application, one symbol may represent the logical operation being performed better than the other.

of an electromechanical switching circuit into a NOR or NAND gate, depending on the logic convention chosen.

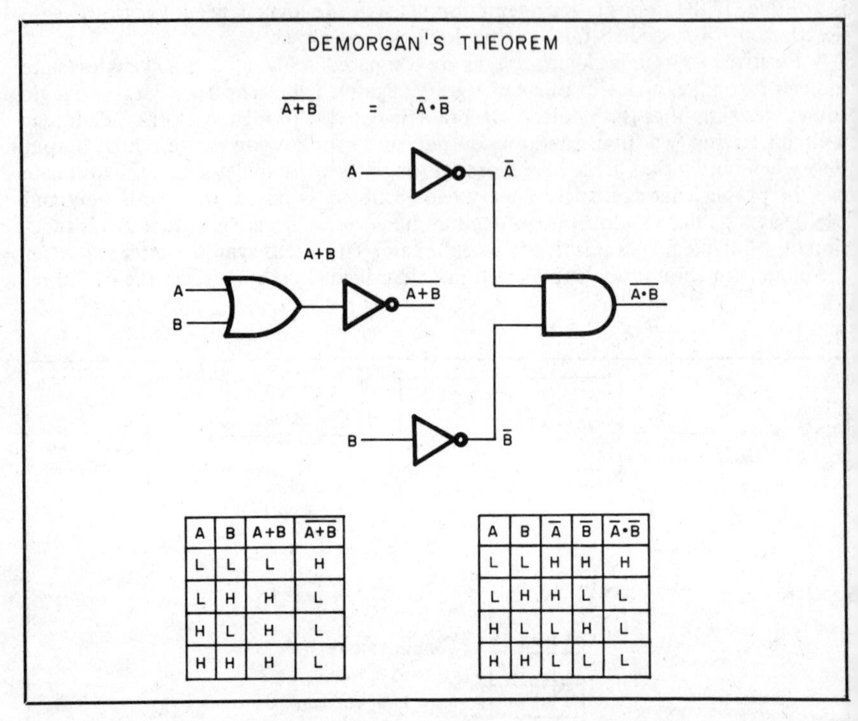

A	B	A+B	$\overline{A+B}$
L	L	L	H
L	H	H	L
H	L	H	L
H	H	H	L

A	B	$\overline{A}$	$\overline{B}$	$\overline{A} \cdot \overline{B}$
L	L	H	H	H
L	H	H	L	L
H	L	L	H	L
H	H	L	L	L

Fig. 6-24—One of the fundamental theorems of combinational logic. The block diagrams and their corresponding expanded truth tables verify DeMorgan's Theorem, which can also be stated as $\overline{A} \cdot \overline{B} = \overline{A + B}$. The diagrams also show the relationship between the AND, OR and NOT functions.

DeMorgan's Theorem

Algebraic identities, such as A × B = B × A, help solve complicated mathematics problems. There are several identities in Boolean algebra that are useful for solving complex logic problems. One of them, DeMorgan's Theorem, deserves special attention.

DeMorgan's Theorem is stated in the form of two identities:

$$\overline{A \cdot B} = \overline{A} + \overline{B} \qquad \text{(Eq. 6-8)}$$

$$\overline{A + B} = \overline{A} \cdot \overline{B} \qquad \text{(Eq. 6-9)}$$

The truth tables shown in Fig. 6-24 prove these identities. The important consequence of DeMorgan's Theorem is that any logical function can be implemented with AND gates and inverters or OR gates and inverters, as shown by the schematic diagrams given in Fig. 6-24.

Logic Families

There are several types of digital-logic integrated circuits. You should be familiar with the characteristics of two types for the Extra Class exam: *TTL* (transistor-transistor logic) and *CMOS* (Complementary metal-oxide semiconductor).

Transistor-Transistor-Logic Characteristics

Transistor-transistor logic (TTL) is a bipolar logic family, so called because the gates are made up of bipolar transistors. Discrete-component logic circuits could be built using TTL, but modern IC technology builds complete circuits containing many transistors on a single semiconductor wafer. Most TTL ICs are identified by 7400/5400 series numbers. For example, the 7490 is a decade counter IC. This IC, as its name implies, can count to 10 (one decade). Internally, the decade counter has 10 distinct states. Some decade counters have a separate output pin for each of the 10 states, while others have only one output connected to the last bit of the counter. Decade counters produce one output pulse for every 10 input pulses.

Other examples of 7400 series ICs are the 7404 hex inverter, and the 7476 dual flip-flop. The 7404 contains six separate inverters, each with one input and one output, in a single 14-pin DIP. A diagram of the 7404 is shown in Fig. 6-25. The 7476 is a dual J-K flip-flop IC. The flip-flop circuit is sometimes known as a latch, because the flipflop can be set, or latched, either high or low to store one bit of information. Chapter 7 will give more details about flip-flops and how they work.

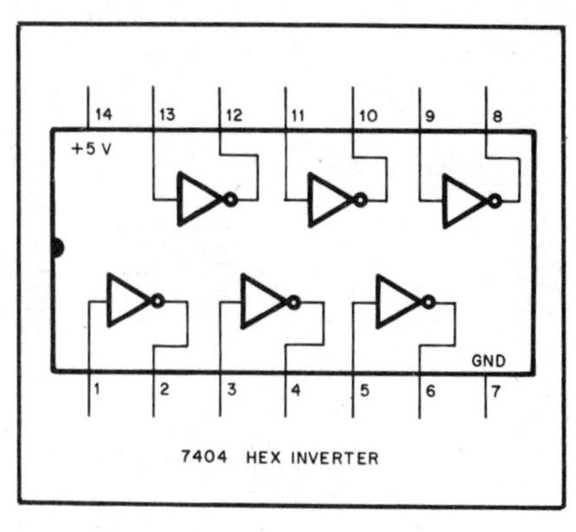

Fig. 6-25—Schematic representation of a 7404 hex inverter.

TTL ICs require a +5-V power supply. The supply voltage can vary between 4.7 and 5.3, but 5 is optimum. There are also limits on the input-signal voltages. To ensure proper logic operation, a HI, or 1, input must be at least 2 V and a LO, or 0, input must be no greater than 0.8 V. To prevent permanent damage to a TTL IC, HI inputs must be no greater than 5.5 V, and LO inputs no more negative than −0.6 V. TTL HI outputs will fall somewhere between 2.4 V and 5.0 V, depending on the individual chip. TTL LO outputs will range from 0 V to 0.4 V. The ranges of input and output levels are shown in Fig. 6-26. Note that the guaranteed output levels fall conveniently within the input limits. This ensures reliable operation when TTL ICs are interconnected. TTL inputs that are left open, or allowed to "float," will assume a HI or 1 state, but operation of the gate may be unreliable. If an input should be HI, it is better to tie the input to the positive supply through a pull-up resistor.

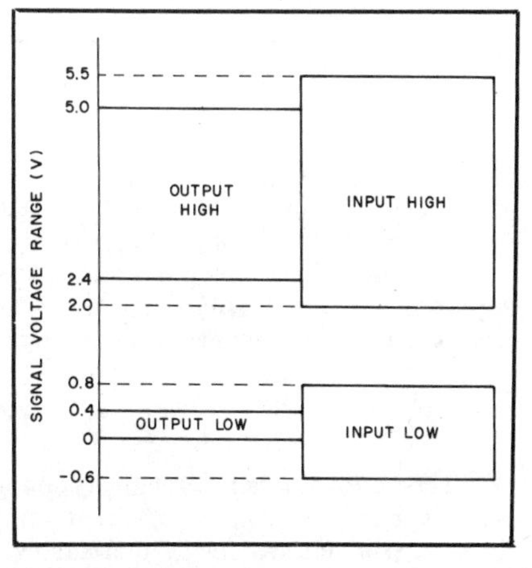

Fig. 6-26—TTL-device input and output signal-voltage ranges.

When a TTL gate changes state, the amount of current that it draws changes rapidly. These current changes, called switching transients, appear on the power-supply line and can cause false triggering of other devices. Good engineering practice suggests being generous with decoupling (or bypass) capacitors. Typical decoupling schemes for the +5-V lines include: a 20-μF electrolytic capacitor at the input to the circuit board, a 6.8-μF electrolytic capacitor in parallel with a 0.1- to 1.5-μF tantalum capacitor near every 1 to 8 devices and a 0.01- to 0.5-μF ceramic capacitor for every 1 or 2 adjacent devices.

CMOS Logic Characteristics

Complementary metal-oxide semiconductor (CMOS) gates are composed of N-channel and P-channel field-effect transistors (FETs) combined on the same substrate. The major advantages of using the CMOS family are low current consumption and high noise immunity. CMOS ICs are identified by 4000 series numbers. For example, a 4001 IC is a quad, two-input NOR gate. The 4001 contains four separate NOR gates, each with two inputs and one output.

CMOS ICs will operate over a much larger power-supply range than TTL ICs. The power-supply voltage can vary from 3 V to as much as 18 V. CMOS output voltages depend on the power supply voltage. A HI output is generally within 0.1 V of the positive supply connection, and a LO output is within 0.1 V of the negative supply connection (ground in most applications). For example, if you are operating CMOS gates from a +9-volt battery, a logic 1 output will be somewhere between 8.9 and 9 volts, and a logic 0 output will be between 0 and 0.1 volts. The switching threshold for CMOS inputs is approximately half the supply voltage. Fig. 6-27 shows these input and output voltage characteristics. The wide range of input voltages gives the CMOS family great immunity to noise, since noise spikes will generally not cause a transition in the input state.

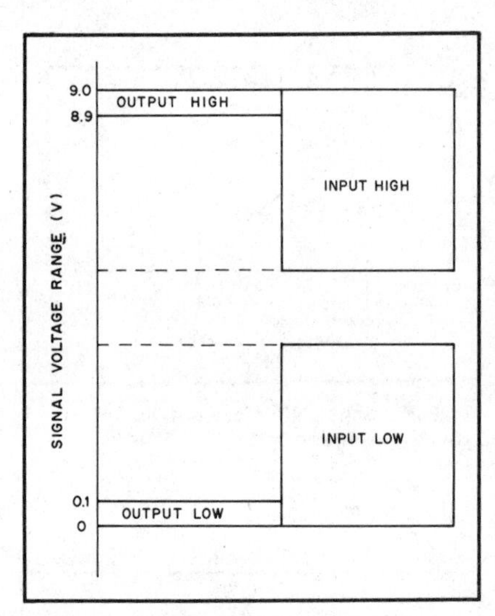

Fig. 6-27—CMOS-device input and output signal-voltage ranges.

CMOS ICs require special handling because of the thin layer of insulation between the gate and substrate of the MOS transistors. Even small static charges can cause this insulation to be punctured, destroying the gate. CMOS ICs should be stored with their pins pressed into special conductive foam, they should be installed in a socket or a soldering iron with a grounded tip should be used to install them on a circuit board. Wear a grounded wrist strap when handling CMOS ICs.

[Now turn to Chapter 10 and study FCC exam questions 4BF-1.14, 4BF-1.15, questions that begin 4BF-3 and 4BF-4 and questions 4BG-1.6 through 4BG-1.10. Review this section as needed.]

THE VIDICON TUBE

The *vidicon tube* is a relatively simple, inexpensive TV-camera pickup tube. It is the type employed in closed-circuit applications in banks and factories because of its small size and low cost. Amateurs use vidicon tubes almost exclusively for both fast- and slow-scan television. Fig. 6-28 shows the physical construction of a vidicon tube.

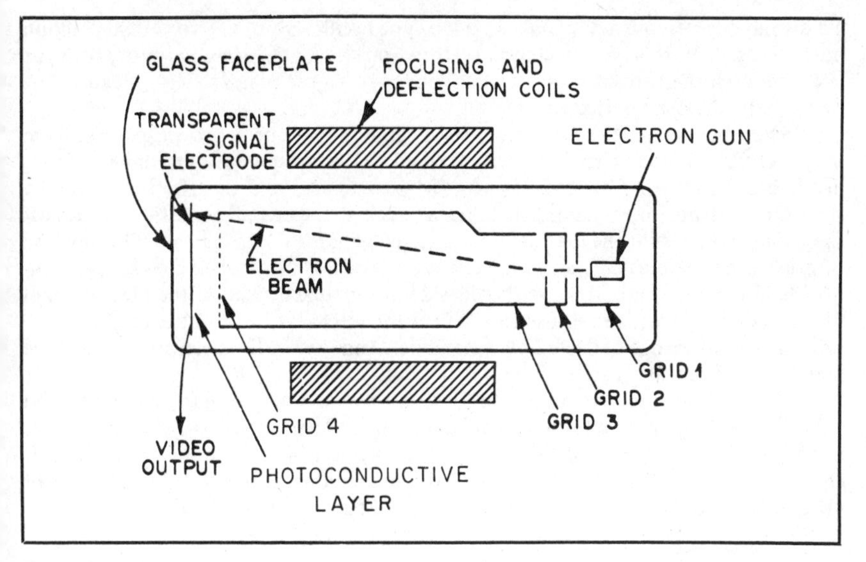

Fig. 6-28—A diagram showing the construction of a vidicon tube.

The photoconductive layer and signal electrode can best be thought of as an array of leaky capacitors. As the electron beam scans the photoconductor, it charges each minicapacitor to the cathode voltage (usually about −20 V with respect to the signal electrode.) Horizontal and vertical deflection of the electron beam in a vidicon is accomplished with magnetic fields generated by coils on the outside of the tube.

As soon as each capacitor is charged, it starts discharging through its leakage resistance, the rate of discharge depending on the amount of light reaching it. On the next scan of the area, the electron beam will redeposit enough electrons to recharge the capacitor to cathode potential. In the instant that it does this, a net current flows through the cathode/signal electrode circuit with an amplitude proportional to the amount of capacitor discharge. Since this discharge depends on the amount of light hitting that portion of the screen, the beam current, as the beam sweeps by that area, is proportional to the light intensity. The output beam current is very low (a fraction of a microamp), and the output impedance of the vidicon is very high, so the video preamp must be designed with care to minimize hum and noise problems.

CATHODE-RAY TUBE

A cathode-ray tube (CRT) is a display device that converts an electrical signal into a visible image. The CRTs used in TV sets and oscilloscopes differ quite a bit, but both operate on the same basic principles. An electron beam is produced in the tube and directed toward a phosphorescent material on the inside face of the tube. When the electrons in the beam strike this "phosphor," the phosphor begins to glow. By scanning the electron beam across the tube face, an image is produced. The relative brightness of the image depends on the number of electrons striking the surface at that point.

Horizontal and vertical deflection of the electron beam is produced with magnetic fields generated by external coils in a TV picture tube. While sufficient for television purposes, this electromagnetic deflection is not suitable for measurement purposes.

To display high-frequency signals on a lab-type oscilloscope, electrostatic deflection must be used. In this case, a charge is stored on the plates of a capacitor, with the CRT between the capacitor plates. As the charge on the plates varies, it causes the electron beam to be deflected.

When magnetic deflection is used for a cathode-ray tube, a brown spot can form at the center of the screen. This spot is formed by the constant bombardment of ions in the beam current. The ions have much greater mass and therefore less velocity than the electrons in the beam. It takes a much stronger magnetic field to deflect these ions. As a result, the ions are not deflected with the electron beam. One method of preventing the burned spot at the center of a cathode ray tube is to cause the entire beam from the cathode to strike the side of the tube instead of the face. A small magnet, or *ion trap*, is then used to deflect the electrons back to the center of the tube. The ions are not deflected by the magnet, and they strike electrodes in the tube and are collected as waste current.

[To check your understanding of these last two sections, turn to Chapter 10 and study FCC examination questions that begin 4BF-5. By now, you should have a basic understanding of how all of the circuit components work that we described in this chapter. If you had trouble with any of the related questions on the FCC Element 4B question pool, review those sections before proceeding to Chapter 7.]

Key Words

Astable (free-running) multivibrator—A circuit that alternates between two unstable states. This circuit could be considered as an oscillator that produces square waves.

Asynchronous flip-flop—A circuit that changes output state depending on the data inputs, but not requiring a clock signal.

Blocking—A receiver condition in which reception of a desired weak signal is prevented because of a nearby, unwanted strong signal.

Bypass capacitor—A capacitor for providing an alternating-current path of comparatively low impedance around some circuit element.

Counter (divider, divide-by-n counter)—A circuit that is able to change from one state to the next each time it receives an input signal. It produces an output signal every time a predetermined number of input signals have been received.

Coupling capacitor—A capacitor for providing a relatively low-impedance path from one stage to the next for ac signals above some frequency, while providing a high impedance to lower-frequency ac and dc signals. Also called a blocking capacitor for this reason.

Crystal-controlled marker generator—An oscillator circuit that uses a quartz crystal to set the frequency, and which has an output rich in harmonics that can be used to determine band edges on a receiver. An output every 100 kHz or less is normally produced.

Drift—As related to op amps, the change of offset voltage with temperature changes, typically a few microvolts per degree Celsius.

Dynamic (edge-triggered) input—An input, such as for a flip-flop, that is sampled only when the rising edge of a clock pulse is detected.

Dynamic range—A measure of receiver performance. The difference, in decibels, between the minimum usable signal level and the maximum signal that does not produce distortion in the audio output.

Flicker noise—As related to op amps, a wideband noise with an amplitude that decreases as frequency increases.

Flip-flop (bistable multivibrator)—A circuit that has two stable output states, and which can change from one state to the other when the proper input signals are detected.

Intercept point—As related to receiver performance, it is the input (or output) signal level at which the desired output power and the distortion product power have the same value.

Minimum discernible signal (MDS)—The smallest input-signal level that can just be detected by a receiver.

Monostable multivibrator (one shot)—A circuit that has one stable state. It can be forced into an unstable state for a time determined by external components, but it will revert to the stable state after that time.

Noise figure—A ratio of the noise output power to the noise input power when the input termination is at a standard temperature of 290 K. It is a measure of the noise generated in the receiver circuitry.

Offset voltage—As related to op amps, the differential amplifier output voltage when the inputs are shorted. It can also be measured as the voltage between the amplifier input terminals in closed-loop configuration.

Popcorn noise—As related to op amps, a low-frequency pulsing noise that decreases in amplitude as the temperature increases.

Prescaler—A divider circuit used to increase the useful range of a frequency counter.

Selectivity—A measure of the ability of a receiver to distinguish between a desired signal and an undesired one at some different frequency. Selectivity can be applied to the RF, IF and AF stages.

Sensitivity—A measure of the minimum input-signal level that will produce a certain output from a receiver.

Sequential logic—A type of circuit element that has at least one output and one or more input channels, and in which the output state depends on the previous input states. A flip-flop is one sequential logic element.

Slew rate—As related to op amps, a measure of how fast the output voltage can change with changing input signals.

Static (level-triggered) input—An input, such as for a flip-flop, which causes the output to change when it changes state.

Synchronous flip-flop—A circuit that changes output state depending on the data inputs, but that will change output state only when it detects the proper clock signal.

Chapter 7

Practical Circuits

Now that you have studied some advanced-level electronics principles, and have learned about the basic properties of some modern solid-state components, you are ready to learn how to apply those ideas to practical Amateur Radio circuits. This chapter will lead you through examples and explanations to help you gain experience with those circuits.

Digital frequency dividers and flip-flop circuits, audio operational amplifier circuits, FET amplifier circuits and some high-performance receiver characteristics are all explained in this chapter. At various places throughout your study you will be directed to turn to Chapter 10 to use the FCC questions as a review exercise.

You should keep in mind that there have been entire books written about every topic covered in this chapter. So if you do not understand some of the circuits from our brief discussion, or if you feel we have left out some of the details, you may want to consult some other reference books. *The ARRL Handbook* is a good starting point, but even that won't tell you everything there is to know about each topic. Our discussion in this chapter should help you understand the circuits well enough to pass your Extra Class exam, however.

DIGITAL LOGIC CIRCUITS

Flip-Flops

The output state of a *sequential-logic* circuit is a function of its present inputs and past output states. The dependence on previous output states implies a capability of, and requirement for, some type of memory.

A *flip-flop* (also known as a *bistable multivibrator*) is a binary sequential-logic element with two stable states: the set state (1 state) and the reset state (0 state). Thus, a flip-flop can store one bit of information. The schematic symbol for a flip-flop is a rectangle containing the letters FF. (These letters may be omitted if no ambiguity results.)

Flip-flop inputs and outputs are normally identified by a single letter, as outlined in Tables 7-1 and 7-2. A letter followed by a subscripted letter (such as D_C), means that input is dependent on the input of the subscripted letter (input D_C is dependent on input C). Note that simultaneous 1 states for the R and S inputs are not allowed. There are normally two output lines, which are complements of each other, designated Q and $\overline{Q}$ (Q NOT). If Q = 1 then $\overline{Q}$ = 0 and vice versa.

Synchronous and Asynchronous Flip-Flops

The terms synchronous and asynchronous are used to characterize a flip-flop

Table 7-1
Flip-Flop Input Designations

Input	*Action*	*Restriction*
R (Reset)	1 resets the flip-flop. A return to 0 causes no further action.	Simultaneous 1 states for R and S inputs are not allowed.
S (Set)	1 sets the flip-flop. A return to 0 causes no further action.	
R_D (Direct Reset)	1 causes flip-flop to reset regardless of other inputs.	
S_D (Direct Set)	1 causes flip-flop to set regardless of other inputs.	
J	Similar to S input.	Simultaneous 1 states for J and K inputs cause the flip-flop to change states.
K	Similar to the R input.	
G (Gating)	1 causes the flip-flop to assume the state of G's associated input.	
C (Control)	1 causes the flip-flop to assume the state of the D input.	A return to 0 produces no further action.
T (Toggle)	1 causes the flip-flop to change states.	A return to 0 produces no further action.
D (Data)	A D input is always dependent on another input, usually C. C = 1, D = 1 causes the flip-flop to set. C = 1, D = 0 causes the flip-flop to reset.	A return of C to 0 causes the flip-flop to remain in the existing state (set or reset).
H1 (Hold for 1 state)	A 1 input will prevent the flip-flop from being reset after it has been set.	The signal has no effect on the flip-flop if it is in the reset state.
H0 (Hold for 1 state)	A 1 input will prevent the flip-flop from being set after it has been reset.	The signal has no effect on the flip-flop if it is in the set state.

Table 7-2
Flip-Flop Output Designations

Output	*Action*	*Restrictions*
Q (Set)	Normal output	Only two output states are possible:
$\overline{Q}$ (Reset)	Inverted output	$Q = 1, \overline{Q} = 0$; and $Q = 0, \overline{Q} = 1$.

Notes
1) $\overline{Q}$ is the complement of Q.
2) The normal output is normally marked Q or unmarked.
3) The inverted output is normally marked $\overline{Q}$. If so, if there is a 1 state at Q, there will be a 0 state at $\overline{Q}$.
4) Alternatively the inverted output may have a (negative) polarity indicated (a small right triangle on the outside of the flip-flop rectangle at the inverted output line). For lines with polarity indicators, be aware that a 1 state in negative logic is the same as a 0 state in positive logic. This is the convention followed by the International Electrotechnical Commission.

or individual inputs to an IC. In *synchronous* (also called clocked, clock driven or gated) *flip-flops*, the output follows the input only at prescribed times determined by the clock input. *Asynchronous flip-flops* are sometimes called unclocked or data driven because the output is determined solely by the inputs. Asynchronous inputs are those that can affect the output state independently of the clock. Synchronous inputs affect the output state only on command of the clock.

Dynamic vs. Static Inputs

Dynamic (edge-triggered) inputs are sampled only when the clock changes state. This type of input is indicated on logic symbols by a small isosceles triangle (called a dynamic indicator) inside the rectangle where the input line enters. Unless there is a negation indicator (a small circle outside the rectangle), the 0-to-1 transition is recognized. This is called positive-edge triggering. The negation indicator means that the input is negative-edge triggered, and is responsive to 1-to-0 transitions.

Static (level-triggered) inputs are recognizable by the absence of the dynamic indicator on the logic symbol. Input states (1 or 0) cause the flip-flop to act.

Many different types of flip-flops exist. These include the clocked and unclocked R-S, D, T, J-K and master/slave (M/S) types. These names come from the type of input lines that the flip-flop has. See Tables 7-1 and 7-2 for a summary of the operation of these lines.

R-S Flip-Flop

One simple circuit for storing a bit of information is the R-S (or S-R) flip-flop. The inputs for this circuit are set (S) and reset (R). Fig. 7-1 shows the schematic symbol for the flip-flop along with a truth table to help you determine the outputs for given input conditions. Two implementations of this circuit using discrete digital logic gates are also shown. When S = 0 and R = 0 the output will stay the same as it was at the last input pulse. This is indicated by a Q in the truth table

If S = 1 and R = 0, the Q output will change to 1. If S = 0 and R = 1, the

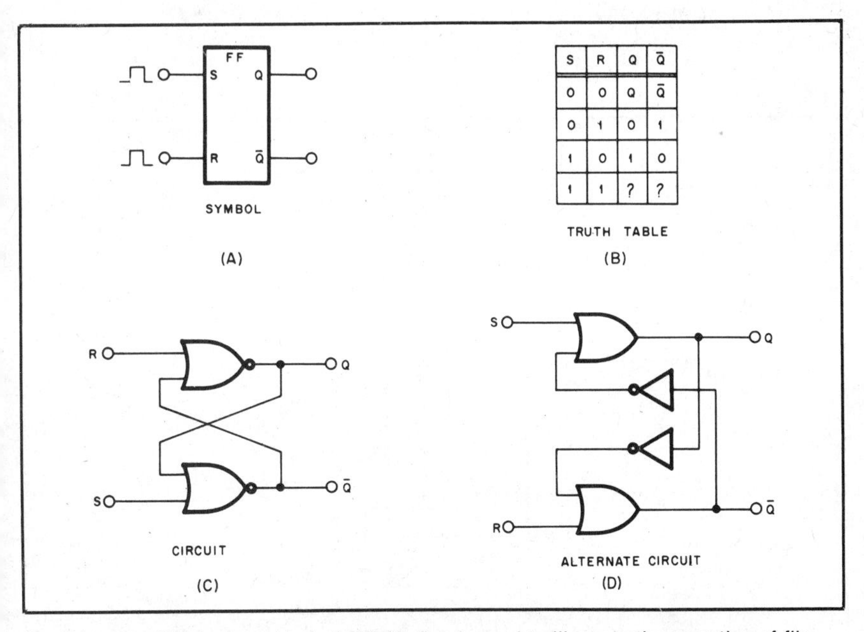

Fig. 7-1—A positive-logic, unclocked R-S flip-flop is used to illustrate the operation of flip-flops in general. Where Q and $\bar{Q}$ are shown in the truth table, the previous output states are retained. A question mark (?) indicates an invalid state, and you cannot be sure what the output will be. C shows an R-S flip-flop made from two NOR gates. The circuit shown at D is another implementation using two OR gates and two inverters.

Q output will change to 0. If both inputs became 1 simultaneously, the output states would be indeterminate, meaning there is no way to predict how they may change. A clocked R-S flip-flop also has a clock input, in which case no change in the output state can occur until a clock pulse is received.

There are other types of flip-flop circuits, each with different input and clocking arrangements. They too can be implemented using discrete digital logic elements, although all types of flip-flops are available as single IC packages. Some ICs include several flip-flop circuits in a single package.

One-Shot Multivibrator

A *one-shot* or *monostable multivibrator* has one stable state and an unstable (or quasi-stable) state that exists for a time determined by RC circuit components connected to the one-shot. When the time constant has expired, the one-shot reverts to its stable state until retriggered. A monostable multivibrator, then, is a flip-flop circuit.

In Fig. 7-2, a 555 timer IC is shown connected as a one-shot multivibrator. The action is started by a negative-going trigger pulse applied between the trigger input and ground. The trigger pulse causes the output (Q) to go positive and capacitor C to charge to two-thirds of V_{cc} through resistor R. At the end of the trigger pulse, the capacitor is quickly discharged to ground. The output remains at logic 1 for a time determined by:

$$T = 1.1\ RC \qquad \text{(Eq. 7-1)}$$

where

R = resistance in ohms
C = capacitance in farads
T = time in seconds

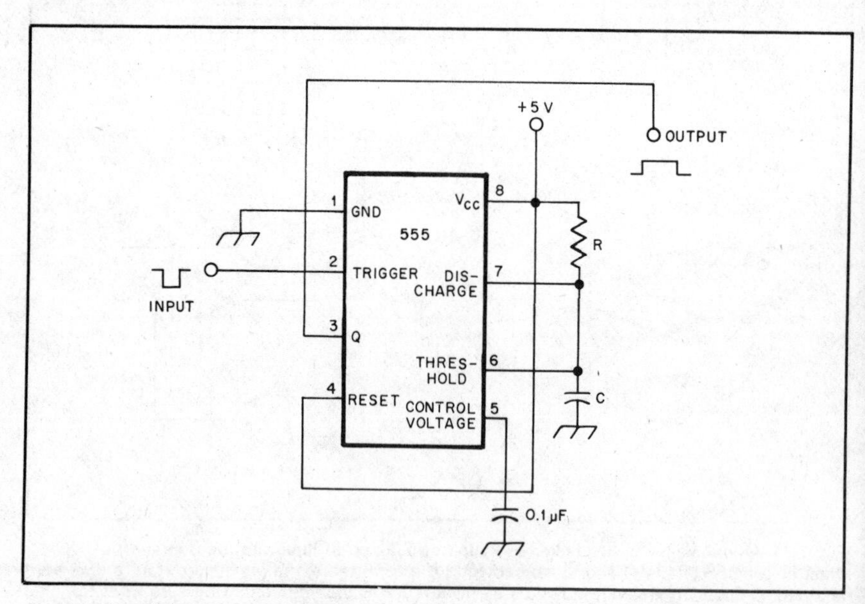

Fig. 7-2—A 555 timer IC connected as a one-shot multivibrator. See text for formula to calculate values for R and C.

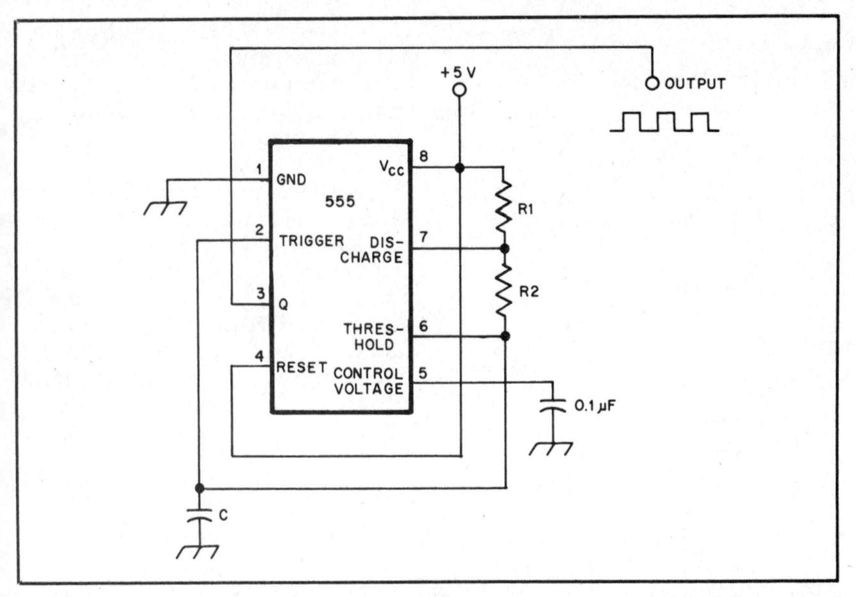

Fig. 7-3—A 555 timer IC connected as an astable multivibrator. See text for formula to calculate values of R1, R2 and C.

Astable Multivibrator

An *astable* or *free-running multivibrator* is a circuit that alternates between two unstable states. It can be synchronized by an input signal of a frequency that is slightly higher than the astable multivibrator free-running frequency.

An astable multivibrator circuit using the 555 timer is shown in Fig. 7-3. Capacitor C charges to two-thirds V_{cc} through R1 and R2, and discharges to one-third V_{cc} through R2. The ratio (R1:R2) sets the duty cycle. The frequency is determined by:

$$f = \frac{1.46}{(R1 + (2 \times R2))\,C} \qquad \text{(Eq. 7-2)}$$

where
R1 and R2 = resistances in ohms
C = capacitance in farads

[Now study FCC examination questions with numbers that begin 4BG-1. Review this section as needed. Note that there are a few questions in this section dealing with digital logic gates. That material was thoroughly covered in Chapter 6. If you need to review the schematic symbols for the logic gates, go back to the appropriate section of that chapter.]

DIGITAL FREQUENCY-DIVIDER CIRCUITS

A *counter, divider* or *divide-by-n counter* is a circuit that stores pulses and produces an output pulse when a specified number (n) of pulses is stored. In a counter consisting of flip-flops connected in series, when the first stage changes state it affects the second stage and so on.

Table 7-3
Binary-Coded Decimal (BCD) Code

Decimal Number	*Binary Weights* 8	4	2	1
0	0	0	0	0
1	0	0	0	1
2	0	0	1	0
3	0	0	1	1
4	0	1	0	0
5	0	1	0	1
6	0	1	1	0
7	0	1	1	1
8	1	0	0	0
9	1	0	0	1

Note: BCD does not use hexadecimal (base 16) numbers A (1010), B (1011), C (1100), D (1101), E (1110), or F (1111).

A ripple, ripple-carry or asynchronous counter passes the count from stage to stage; each stage is clocked by the preceding stage. In a synchronous counter, each stage is controlled by a common clock.

Weighting is the count value assigned to each counter output. In a four-stage binary ripple counter, the output of the first, second, third and fourth stages would have weights of 1, 2, 4 and 8, respectively. Decade (base 10) counters consist of divide-by-2 and divide-by-5 counters (encoded as binary-coded decimal or BCD values), as shown in Table 7-3. A base-12 counter employs a divide-by-2 and a divide-by-6 counter.

Most counters have the ability to clear the count to 0. Some counters can also be preset to a desired count. Some counters may count up (increment) and some down (decrement). Up counters count in increasing weighted values (for example: 0000, 0001, 0010, . . ., 1111). Down counters work in the reverse order. Up/down counters are also available. They are able to count in either direction, depending on the status of a control line.

Circuit Applications

Counter or divider circuits find application in various forms. Common uses for these circuits are in marker generators and frequency counters.

The regulations governing amateur operation require that your transmitted signal be maintained inside the limits of certain frequency bands. The exact frequency need not be known, as long as it is not outside the limits. Staying inside the limits is not difficult to do, and requires only a marker generator or frequency counter and some care.

Marker Generator

Many receivers and transceivers include a *crystal-controlled marker generator.* This circuit employs a high-stability crystal-controlled oscillator that generates a series of signals which, when detected in the receiver, mark the exact edges of the amateur frequency assignments. Most U.S. amateur band limits are exact multiples of 25 kHz, whether at the band extremes or at points marking the subdivisions between types of emission and license-class restrictions. A 25-kHz fundamental frequency will produce the desired marker signals, provided that the oscillator harmonics are strong enough to be heard throughout the desired range. But if the receiver calibration is not accurate enough to positively identify which harmonic you are hearing, there may still be a problem in determining how close to the band edge you are operating.

Rather than using a 25-kHz oscillator, an oscillator frequency of 100 kHz is often used. A divider circuit coupled to the oscillator provides markers at increments of other than 100 kHz. In the circuit of Fig. 7-4, two divide-by-2 stages are switch selected to produce markers at 50- and 25-kHz points. U3 is a 4013 CMOS dual D-type flip-flop. A 4001 CMOS quad NOR gate and a diode matrix provide the required switching and signal routing.

When a D flip-flop is wired with the Q NOT ($\overline{Q}$) output to the D input, it forms a T flip-flop, or complementing flip-flop. The timing diagram of Fig. 7-5 shows that the flip-flop output changes state with each positive clock pulse. So if the output

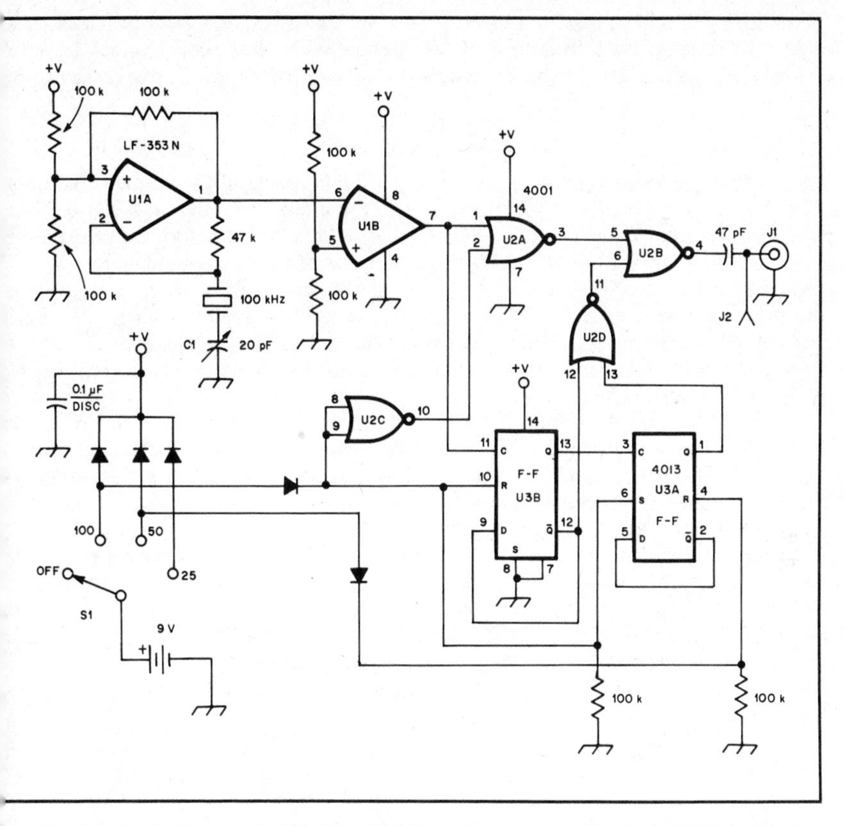

ig. 7-4—Schematic diagram of a 100, 50 and 25-kHz marker generator. Two switch-selected divide-y-two stages produce the 50- and 25-kHz markers.

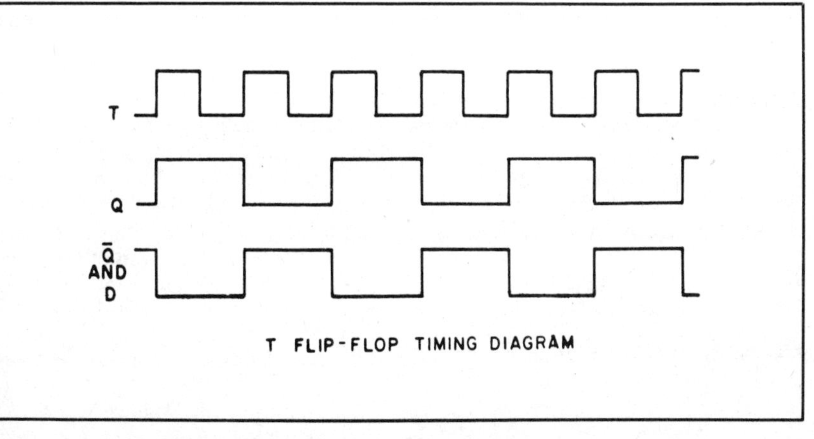

Fig. 7-5—The timing diagram for a T flip-flop is shown. As you can see, the T flip-flop serves as a divide-by-two counter.

is 0 initially, it will change to a 1 on the leading edge of the first positive clock pulse and it will change back to 0 on the leading edge of the next positive clock pulse. A T-flip-flop, then, is a divide-by-2 counter—just what we need for the marker-generator circuit.

Frequency Counters

One of the most accurate means of measuring frequency is the frequency counter. This instrument actually counts the pulses of the input signal over a specified time, and displays the frequency of the signal, usually on a digital readout. For example if an oscillator operating at 14.230 MHz is connected to the counter input, 14.230 would be displayed. Some counters are usable well up into the gigahertz range. Most counters that are used at high frequencies make use of a *prescaler* ahead of a basic low-frequency counter. The prescaler divides the high-frequency signal by increments of 10, 100, 1000 or some other amount so that the low-frequency counter can display the input frequency.

Frequency-counter accuracy depends on an internal crystal-controlled reference oscillator. The more accurate the crystal reference, the more accurate the counter readings will be. A crystal frequency of 1 MHz has become more or less standard for use in the reference oscillator. A 1-MHz AT-cut crystal will have excellent temperature stability. This means the oscillator frequency will not change appreciably with temperature changes, a very important consideration for a crystal-controlled marker generator. Another advantage of using a 1-MHz crystal is that it makes it relatively easy to compare harmonics of the crystal-oscillator signal to WWV or WWVH signals at 5, 10 or 15 MHz, for example, and to adjust it to zero beat (that is, the frequencies of the beating signals or their harmonics are equal).

[At this time you should turn to Chapter 10 and study FCC examination questions with numbers that begin 4BG-2. Review this section as needed.]

OPERATIONAL AMPLIFIERS

In Chapter 6 we discussed the basic operational-amplifier circuit configurations and you learned the voltage-gain relationships for those configurations. In this chapter we will study some of the circuits that can use these op-amp configurations. As you remember, an op amp is a high-gain, direct-coupled differential amplifier whose characteristics are chiefly determined by components external to the amplifier unit. Op amps can be assembled from discrete transistors, but better thermal stability results from fabricating the circuit on a single silicon chip. IC op amps are manufactured with bipolar, JFET and MOSFET devices, either exclusively or in combination.

The most common application for op amps is in negative feedback circuits operating from dc to perhaps a few hundred kilohertz. Provided the device has sufficient open-loop gain, the amplifier transfer function is determined almost solely by the external feedback network. (The terms open loop and closed loop refer to whether or not there is a feedback path.) The differential inputs allow for both inverting and noninverting circuits. Fig. 7-6 reviews how an op amp can be connected in practical circuits to form an inverting amplifier, a noninverting amplifier and a differential amplifier. The transfer function for an inverting amplifier was given in Chapter 6 as Eq. 6-1 and the voltage gain as Eq. 6-3. They are repeated here for convenience.

$$V_{out} = \frac{Rf}{R1} V_{in} \qquad \text{(Eq. 7-3)}$$

$$V_{gain} = \frac{Rf}{R1} \qquad \text{(Eq. 7-4)}$$

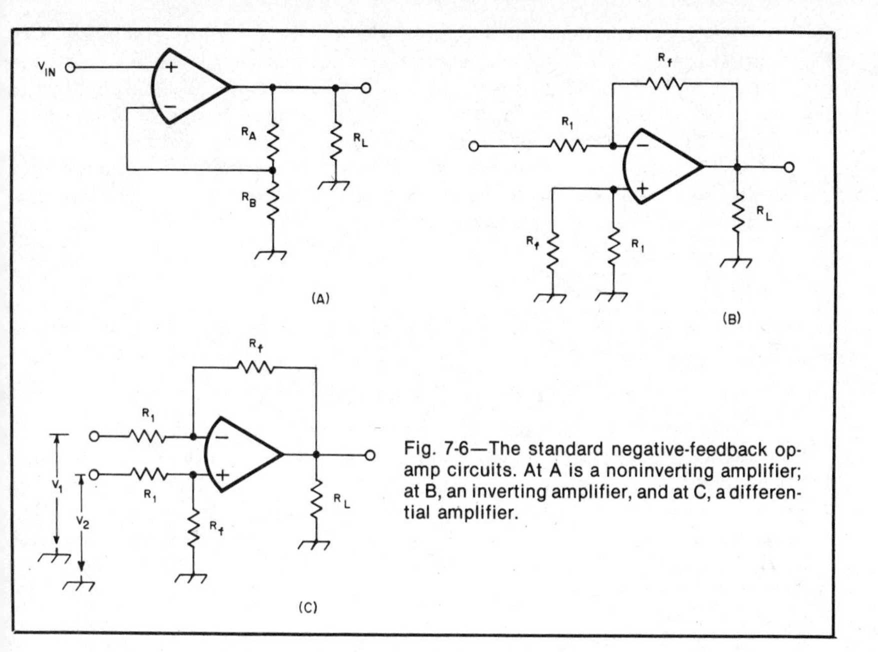

Fig. 7-6—The standard negative-feedback op-amp circuits. At A is a noninverting amplifier; at B, an inverting amplifier, and at C, a differential amplifier.

The transfer function for a noninverting amplifier was given as Eq. 6-5 and the voltage gain as Eq. 6-6.

$$V_{out} = \frac{R1 + Rf}{R1} V_{in} \qquad \text{(Eq. 7-5)}$$

$$V_{gain} = \frac{R1 + Rf}{R1} \qquad \text{(Eq. 7-6)}$$

The transfer function for a difference amplifier was given in Chapter 6 as Eq. 6-7.

$$V_{out} = \left(\frac{R1 + Rf}{R2 + R3}\right)\left(\frac{R3}{R1}\right)(V2) - \left(\frac{Rf}{R1}\right)(V1) \qquad \text{(Eq. 7-7)}$$

In most practical difference amplifiers, R2 = R1 and R3 = Rf. By making these two substitutions, we can simplify Eq. 7-7 to

$$V_{out} = \frac{Rf}{R1}(V2 - V1) \qquad \text{(Eq. 7-8)}$$

Since $V_{in} = V2 - V1$, the voltage gain for a difference amplifier is also given by Eq. 7-4 (and Eq. 6-3).

You can see that the gain of any op-amp circuit is set by the value of a few resistors. Notice that the output, or load, resistance does not appear in the equations, implying that the output impedance is zero. This condition results from the application of heavy negative feedback. Most IC op amps have built-in current limiting. This feature not only protects the IC from damage caused by short circuits, but also limits the values of load resistance for which the output impedance is zero. Most op amps work best with load resistances of at least 2 kΩ.

Since the op amp magnifies the difference between the voltages applied to its inputs, applying negative feedback has the effect of equalizing the input voltages.

In the inverting amplifier configuration, the feedback action combined with Kirchhoff's current law establishes a zero impedance, or virtual ground at the junction of Rf and R1. The circuit input impedance is just R1. Negative feedback applied to the noninverting configuration causes the input impedance to approach infinity.

Voltage gain and voltage output for an op amp can be determined easily by means of the above equations. For example, assume that we have an inverting op-amp circuit similar to the one shown in Fig. 7-6B. Calculate the circuit gain if R1 is 1 kΩ and Rf is 100 kΩ. Substitute those values into Eq. 7-4:

$$V_{gain} = \frac{Rf}{R1} = \frac{100\ k\Omega}{1\ k\Omega} = 100$$

If a 10-mV signal were applied to the input, the output voltage would be 1000 mV or 1 V.

Op Amp Specifications

Offset voltage is the potential between the amplifier input terminals in the closed-loop condition. Ideally, this voltage should be zero. Offset results from imbalance between the differential input transistors. Offset-voltage values range from millivolts in ordinary consumer-grade devices to only nanovolts in premium Milspec units. The temperature coefficient of offset voltage is called *drift*. Drift is usually considered in relation to time. Heat generated by the op-amp itself or by associated circuitry will cause the offset voltage to change over time. A few microvolts per degree Celsius (at the input) is a typical drift specification.

There are two types of noise associated with op amps. Burst or *popcorn noise* is a low-frequency pulsing, usually below 10 Hz. The amplitude of this noise is approximately an inverse function of temperature. That means that as the temperature increases, popcorn noise is less severe. The other noise, sometimes called *flicker noise,* and is a wideband signal whose amplitude varies inversely with frequency. Again, as you increase the frequency, flicker noise becomes less severe. For some analytical purposes, drift is considered as a very low frequency noise component. Op amps that have been optimized for offset, drift and noise are called instrumentation amplifiers.

The small-signal bandwidth of an op amp is the frequency range over which the

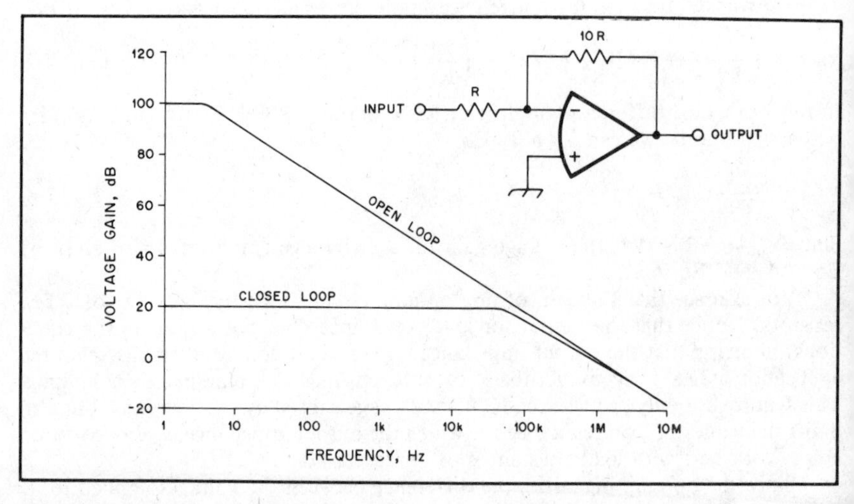

Fig. 7-7—Open-loop gain and closed-loop gain as a function of frequency. The vertical distance betwen the curves is the feedback or gain margin.

open-loop voltage gain is at least unity. This specification depends mostly on the frequency compensation scheme used. Fig. 7-7 shows how the maximum closed-loop gain varies with frequency. The power bandwidth of an op amp is a function of *slew rate,* and is always less than the small-signal value. Slew rate is a measurement of maximum output voltage swing per unit time. Values from 0.8 to 13 volts per microsecond are typical of modern devices.

Op Amps as Audio Filters

Although there are numerous applications for RC active filters, their principal use in amateur work is to establish selectivity at audio frequencies. Op amps have the distinct advantage of providing gain and variable parameters when used as audio filters. Passive filters that contain inductive and capacitive elements are generally designed for some fixed frequency and exhibit an insertion loss. Also, LC filters are usually physically larger and heavier than their op-amp counterparts.

The use of an op amp IC, such as a type 741, results in a compact filter pole that provides stable filter operation. Only five connections are made to the IC. The gain of the filter sections and the frequency characteristics are determined by the choice of components external to the IC.

Fig. 7-8 shows a single-pole band-pass filter. To select the component values for a specific filter, you must first select the desired filter Q, voltage gain, A_V, and operating frequency, f_o. Next you should choose a value for C1 and C2. They have an equal value and should be high-Q, temperature-stable components. Polystyrene capacitors are excellent for such use. Disc-ceramic capacitors are not recommended. R4 and R5 are equal in value and are used to establish the op amp reference voltage, which is $V_{cc} / 2$. Then:

$$R1 = \frac{Q}{2 \pi f_o A_V C1} \quad \text{(Eq. 7-9)}$$

$$R2 = \frac{Q}{(2Q^2 - A_V)(2 \pi f_o C_1)} \quad \text{(Eq. 7-10)}$$

$$R3 = \frac{2Q}{2\pi f_o C1} \quad \text{(Eq. 7-11)}$$

$$R4 = R5 \approx 0.02 \times R3 \quad \text{(Eq. 7-12)}$$

Single-pole filter sections can be cascaded for greater selectivity. One or two poles may be used as a band-pass or low-pass section for improving the audio-channel passband characteristics during SSB or AM reception. Up to four filter poles are frequently used to obtain selectivity for CW or RTTY reception. The greater the number of poles, up to a practical limit, the sharper the filter skirt response. Not only does a well-designed RC filter help to reduce QRM, but it also improves the signal-to-noise ratio in some receiving systems.

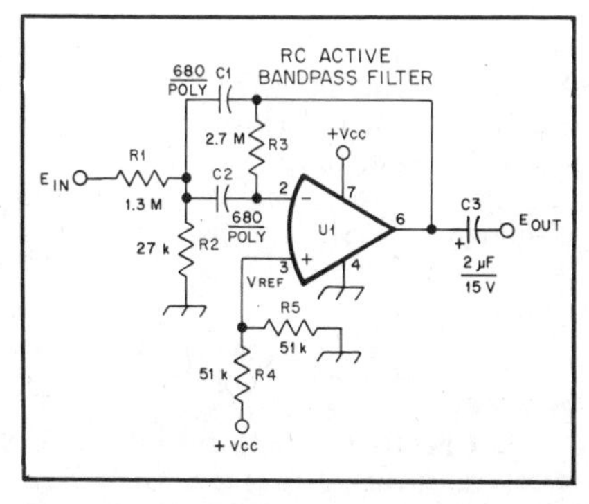

Fig. 7-8—A basic circuit for a single-pole, RC active band-pass filter.

The component values shown on Fig. 7-8 illustrate the design of a single-pole band-pass filter. An arbitrary f_o

of 900 Hz has been used in the calculation, but for CW reception the operator may prefer frequencies between 200 and 700 Hz. An A_v (gain) of 1 and a Q of 5 were chosen for this example. Both the gain and the Q can be increased for a single-section filter if desired, but for a multisection RC active filter, it is best to restrict the gain to 1 or 2 and use a maximum Q of 5; this helps prevent unwanted filter "ringing" and audio instability.

Standard-value 680-pF capacitors are chosen for C1 and C2. For certain design parameters and C1-C2 values, unwieldy resistance values may result. If this happens, select a new value for C1 and C2. Use Eqs. 7-9 through 7-12 to calculate the required resistance values.

$$R1 = \frac{Q}{2\pi\ f_o\ A_v\ C1} = \frac{5}{6.28 \times 900\ \text{Hz} \times 1 \times 680 \times 10^{-12}\ \text{F}}$$

$$R1 = \frac{5}{3.84 \times 10^{-6}} = 1.30 \times 10^{6}\ \Omega = 1.3\ \text{M}\Omega$$

$$R2 = \frac{Q}{(2Q^2 - A_v)\ (2\pi\ f_o\ C1)}$$

$$R2 = \frac{5}{(2 \times 5^2 - 1) \times 6.28 \times 900\ \text{Hz} \times 680 \times 10^{-12}\ \text{F}}$$

$$R2 = \frac{5}{(2 \times 25 - 1) \times 3.84 \times 10^{-6}} = \frac{5}{49 \times 3.84 \times 10^{-6}}$$

$$R2 = \frac{5}{1.88 \times 10^{-4}} = 2.65 \times 10^{4}\ \Omega = 26.5\ \text{k}\Omega$$

$$R3 = \frac{2Q}{2\pi\ f_o\ C1} = \frac{10}{6.28 \times 900\ \text{Hz} \times 680 \times 10^{-12}\ \text{F}}$$

$$R3 = \frac{10}{3.84 \times 10^{-6}} = 2.60 \times 10^{6}\ \Omega = 2.60\ \text{M}\Omega$$

$$R4 = R5 \approx 0.02 \times R3 = 0.02 \times 2.60 \times 10^{6}\ \Omega = 5.20 \times 10^{4}\ \Omega = 52.0\ \text{k}\Omega$$

The resistance values assigned to R1 through R5, inclusive, are the nearest standard values to those obtained from the equations. The principal effect of this is a slight alteration of f_o and A_v.

In use, the RC active filter should be inserted in the low-level audio stages. This prevents overloading the filter during strong-signal reception, such as would occur if the filter were placed at the audio output, just before the speaker or phone jack. The receiver AF gain control should be placed between the audio preamplifier and the input of the RC active filter for best results. If audio-derived AGC is used in the receiver, the RC active filter will give its best performance when it is contained within the AGC loop.

[Before proceeding to the next section, turn to Chapter 10 and study FCC examination questions with numbers that begin 4BG-3 and 4BG-7. Review this section as needed.]

FET AMPLIFIER CHARACTERISTICS

In Chapter 6, you learned the basic properties of the two main types of field-effect transistors, JFETs and MOSFETs. In this section, we will cover the dynamic characteristics of these FETs, and the conditions that affect the circuits they are used in.

Characteristic curves for JFETs and MOSFETs are shown in Figs. 7-9 and 7-10.

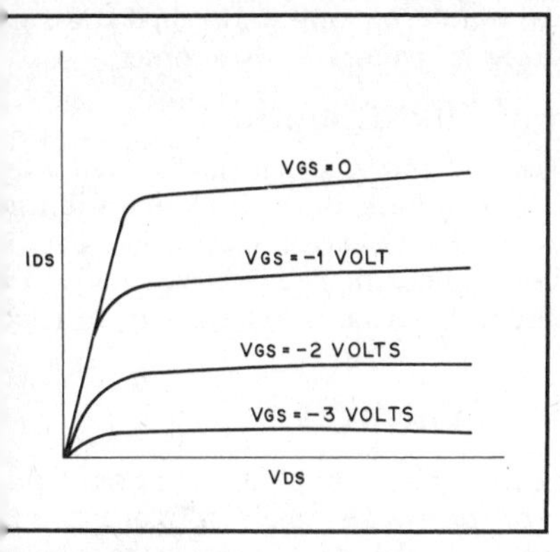

Fig. 7-9—Typical JEFT characteristic curves.

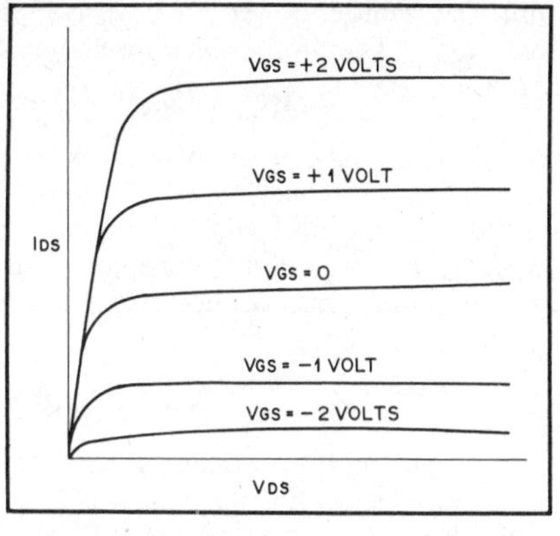

Fig. 7-10—Typical characteristic curves for a MOSFET.

Drain-source current (I_{DS}) is plotted against drain-source voltage (V_{DS}) for given gate voltages (V_{GS}).

The dynamic characteristics of an FET are most heavily influenced by dynamic mutual conductance or transconductance (G_m). This parameter is defined as the ratio of drain current change to the small gate-to-source voltage change that caused it. Mathematically, the relationship is expressed as:

$$G_m = \frac{\Delta I_D}{\Delta E_{GS}} \quad \text{(Eq. 7-13)}$$

where Δ = a small change or increment in the variable.

Typical general-purpose JFETs for small-signal RF and audio work have G_m values of about 5000 microsiemens (μS), while some units designed for CATV service feature transconductances over 13,000 μS. A transconductance of 40,000 μS is representative of the hottest dual-gate MOSFETs. Some JFETs intended for use in analog switching circuits also have transconductances of 40 μS or more to provide a low ON resistance. The newer power FETs boast transconductance figures on the order of 1 siemen.

Transconductance is of great importance in calculating the gain and impedance values of FET circuits. In common-source and common-gate amplifiers with no degeneration, the numerical voltage gain is given by:

$$A_v = G_m R_L \quad \text{(Eq. 7-14)}$$

where

A_v = the voltage gain
G_m = the transconductance in siemens
R_L = the drain load resistance in ohms

For a common-source amplifier circuit, the input impedance is controlled by the gate biasing network. The output impedance of the common-source amplifier is essentially determined by the drain resistor.

[Study FCC examination questions with numbers that begin 4BG-8. Review this section as needed.]

HIGH-PERFORMANCE-RECEIVER CONCEPTS

Receivers are an important part of any Amateur Radio station. No matter how much transmitter power you have available to you, "You can't work 'em if you can' hear em!" Therefore, some discussion of receiver parameters is in order.

It's What's Up Front That Counts

The critical portion of a receiver is the front end, that part that precedes th main selectivity-determining elements. Distortion effects in the front end will lea to blocking, the production of intermodulation products and cross modulation. Carefu design is required if these problems are to be minimized. The receiver front end i where the RF amplification takes place, before the signal is converted to an inter mediate frequency.

Noise Figure and Sensitivity

One of the basic receiver specifications is *noise figure*. Implicit in the noise-figur concept is the fact that the sensitivity or *minimum discernible signal (MDS)* of a receive depends not only on the amount of noise generated by the active devices in the receiver but also on the bandwidth of the system.

Let's consider the noise-figure measurement of an amplifier or receiver. By defini tion, the noise factor of the amplifier is the input signal-to-noise ratio divided b the output signal-to-noise ratio:

$$NF = \frac{\frac{S_{in}}{N_{in}}}{\frac{S_{out}}{N_{out}}} = \frac{S_{in}\ N_{out}}{S_{out}\ N_{in}} \qquad \text{(Eq. 7-15}$$

The terms in the equation are noise or signal powers, and the noise factor is an algebrai ratio. If the ratio is expressed in decibels (dB), as is often done with other powe ratios, the result is called noise figure.

Because the input and output noise powers depend on what is hooked to th input of the amplifier or receiver, the noise figure is a nebulous number. To mak the noise figure a standard measure of the noise in an amplifier or receiver, the inpu noise is assumed to be the noise power available from a resistor that has a temperatur of 290 kelvins. Using this value for T_o, the noise power is given as:

$$P_n = k\ T_o\ B \qquad \text{(Eq. 7-16}$$

where

T_o = 290 K
B = bandwidth in Hz
k = Boltzmann's constant (1.38×10^{-23} joules / kelvin)

It is convenient to use logarithmic units and to note that in a receiver with a band width of 1 Hz, $P_n = -174$ dBm.

$$P_n = 1.38 \times 10^{-23}\ J / K \times 290\ K \times 1\ Hz = 4.00 \times 10^{-21}\ W$$

$$dBm = 10 \log \left(\frac{P_n}{P_0}\right) = 10 \log \left(\frac{4.00 \times 10^{-21}\ W}{1.00 \times 10^{-3}\ W}\right) = 10 \log (4.00 \times 10^{-18})$$

$$dBm = 10 \times -17.4 = -174$$

This figure is considered to be the theoretical best value of MDS that a receiver ca have. In other words, if the receiver were the very best imaginable, any received signa would have to be a small amount stronger than −174 dBm.

Consider a receiver with a bandwidth of 500 Hz. The bandwidth is greater than 1 Hz by a factor of 500, or 27 dB (10 log 500). Hence, with a 500-Hz bandwidth, the power available from the standard resistor is −174 dBm + 27 dB = −147 dBm. This is the theoretical limit of the noise floor for this receiver. If you terminate the receiver input with a 50-ohm resistor, and then measure the audio output, you will have a measure of the noise generated in the receiver. If you then connect a signal generator to the receiver input, and measure the signal required to generate this same audio output power, you can calculate the noise figure. Suppose it takes a −140 dBm signal to generate the same audio power. The receiver noise figure is the difference between this value and the theoretical best MDS: −140 dB − (−147 dB) = 7 dB. The larger the noise figure, the more noise that is generated in the receiver itself, and the higher the noise floor. (A higher noise floor means the value is closer to zero, since the numbers are negative.) The MDS, or noise floor, of this receiver, is −140 dBm. What this means in performance, then, is that the sensitivity of the receiver is controlled by the receiver noise figure and the bandwidth of the receiver filters. The noise floor is a figure that describes the sensitivity limit of the receiver.

Third-Order Distortion Products

Consider now the case in which two relatively strong signals are injected simultaneously into the input of a 20-dB-gain amplifier, such as you might find in the RF or IF stages of your receiver. Assume that two input signals of −50 dBm are placed at the input of the amplifier at frequencies f_1 and f_2. Analysis of the amplifier will show that distortion in the amplifier will give rise to outputs not only at the desired input frequencies (f_1 and f_2), but also at $2f_1 - f_2$ and $2f_2 - f_1$. For example, if the input frequencies were 14,040 and 14,060 kHz, the distortion products would appear at 14,020 and 14,080 kHz. In the amplifier described, the desired outputs would be 20 dB above the −50-dBm input signal levels, or −30 dBm, and the third-order distortion products might be at −130 dBm. In this case, the distortion-product level will be 100 dB down from the desired output levels.

An interesting and significant characteristic of class-A linear amplifiers is that while the desired output levels vary linearly with changes in the input signal levels, the dominant (third-order) distortion products vary as the cube of the input-power levels. Hence, if the signal levels driving the input are increased to −40 dBm, the output power of the desired signals will be −20 dBm for each of the desired input tones. Although the level of the desired frequencies increased by 10 dB, the output power of the distortion products increased by 30 dB to −100 dBm. The distortion-product levels are now only 80 dB below the desired signal levels.

Blocking and IMD Dynamic Range

While sensitivity is a primary consideration if you want to hear DX stations, a receiver must also be able to function in the presence of strong signals. The receiver must be able to respond to a desired weak signal in the presence of a much stronger signal on a nearby frequency.

Another consideration, then, is the *dynamic range* of the receiver. This is a number that provides a measure of the input-signal levels that may be present at the receiver antenna terminals while no undesired responses are created in the output.

If we connect two signal generators to our receiver input, we can measure the blocking dynamic range. Set one generator to a low signal level (perhaps −110 dBm) and tune the receiver to that frequency, then adjust the other generator to a frequency 20 kHz away. Gradually increase the input level of the second generator until the receiver output drops 1 dB. At that point the second signal is causing the receiver AGC to limit the output on the desired signal. The signal is desensitizing the receiver, or *blocking* it. Suppose the second generator is putting a signal of −33 dBm into

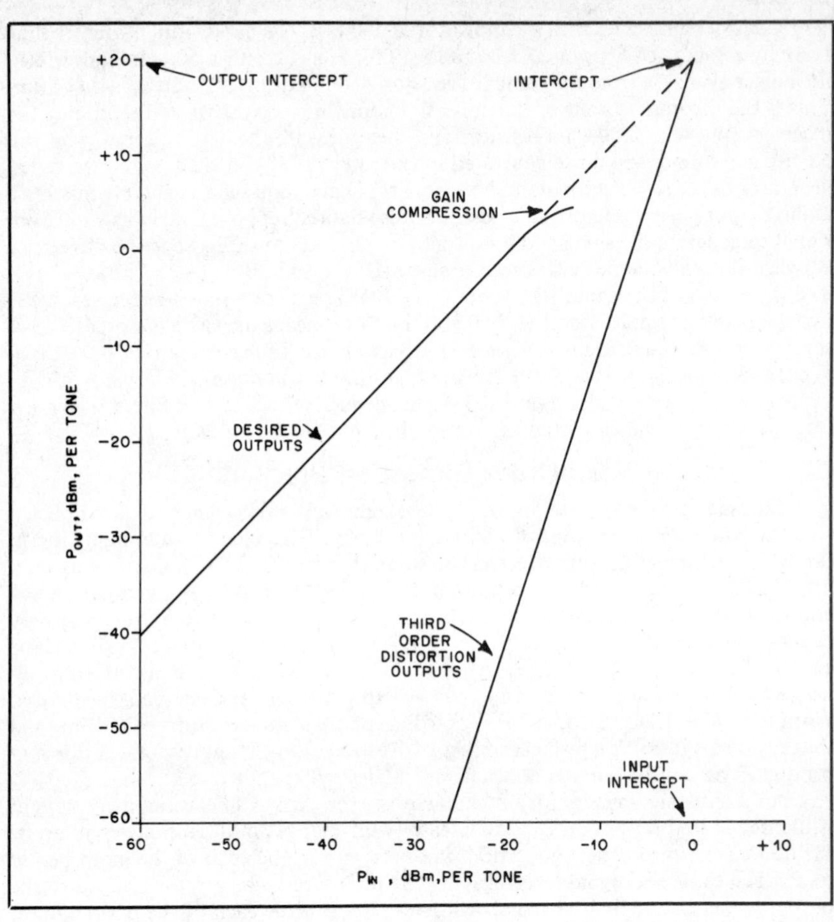

Fig. 7-11—A plot example showing signal power versus distortion products as a function of input power of two identical input signals.

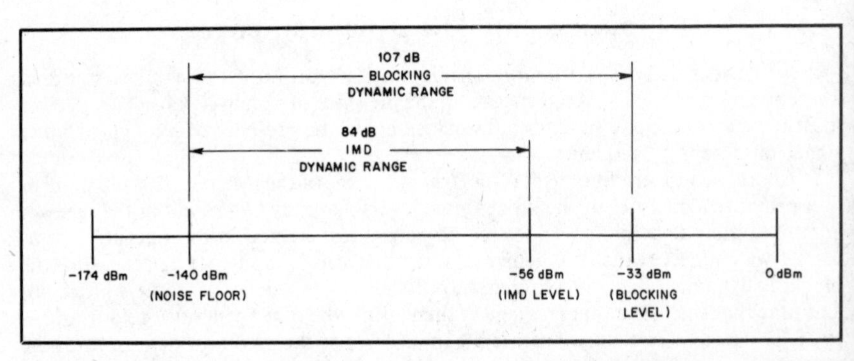

Fig. 7-12—A graph that displays the performance of a hypothetical (though typical) receiver under dynamic range testing. The noise floor is −140 dBm, blocking level −33 dBm and the IMD level is −56 dBm. This corresponds to a receiver blocking dynamic range of 107 dB and an IMD dynamic range of 84 dB.

the receiver when this occurs. This can be called the blocking level. The blocking dynamic range can be calculated by subtracting the blocking level from the noise-floor value.

Blocking dynamic range = noise floor − blocking level (Eq. 7-17)
Blocking DR = −140 dBm − (−33 dBm) = −107 dBm

This value is usually taken as an absolute value, so we would say our receiver has a blocking dynamic range of 107 dB.

We can also measure the two-tone intermodulation-distortion dynamic range of our receiver. This measurement is made by putting two equal-level signals spaced 20 kHz apart into the receiver. The two signals are strong (around −10 dBm each), and will produce quite a bit of IMD. Tune the receiver to either third-order product, and then use an attenuator to adjust the input level to produce an output at the speaker that is 3 dB above the noise. If the input signal has been attenuated 46 dB, then the input level is −56 dBm. If this level is subtracted from the noise floor, we have an IMD dynamic range measurement.

IMD DR = noise floor − IMD level (Eq. 7-18)
IMD DR = −140 dBm − (−56 dBm) = −84 dB

Refer to Fig. 7-11. This plot shows the power of the desired output and the output power of the distortion products as a function of the input-tone power levels. Eventually, the input signal levels get large enough so the desired output levels no longer increase linearly. This effect is called gain compression or blocking. Blocking may be defined as desensitization to a desired signal caused by a nearby strong interfering signal.

The linear portion of the curves of Fig. 7-11 may be extended or extrapolated to higher power levels even though the receiver is not capable of operating properly at these levels. If this is done (as shown by the dashed line), the two curves will eventually cross each other. At some usually unattainable output power level, the distortion-product level equals that of the desired output levels. This point is commonly referred to as the amplifier *intercept point*. Specifically, the output power level at which the curves intersect is called the amplifier output intercept. Similarly, the input power level corresponding to the point of intersection is called the input intercept.

Fig. 7-12 illustrates the relationship between the input signal levels, noise floor, blocking dynamic range and IMD dynamic range. With the present state of the art, receiver blocking dynamic ranges on the order of 100 dB are possible, and this is an acceptable dynamic range figure. So the receiver we have used in this example could be any typical modern receiver.

[At this time, you should turn to Chapter 10 and study FCC examination questions with numbers that begin 4BG-4 and 4BG-6. Review this section as needed.]

Selectivity

Many amateurs regard the expression *selectivity* as the ability of a receiver to separate signals. This is a fundamental truth, particularly with respect to IF selectivity that has been established by means of high-Q filters (LC, crystal, monolithic or mechanical). But in a broader sense, selectivity can be employed to reject unwanted signal energy in any part of a receiver—the front end, IF section, audio circuit or local-oscillator (LO) chain. Selectivity is a relative term, since the bandwidth can vary from a few hertz to more than one megahertz, depending on design objectives. Therefore, it is not uncommon to hear terms like "broadband filter" or "narrow-band filter" used in relation to receiver selectivity.

The degree of selectivity is determined by the filter-network bandwidth. The bandwidth is normally specified for the −3-dB points on the filter response curve; the frequencies at which the filter output power is half the peak output power elsewhere

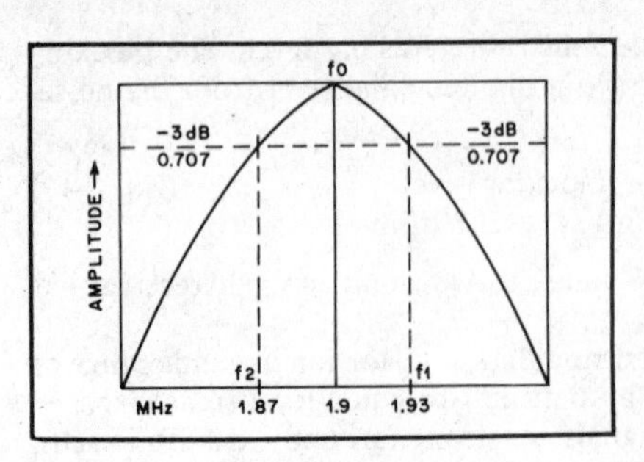

Fig. 7-13—A curve for determining the unloaded Q of a tuned circuit.

in the passband (Fig. 7-13). The difference in frequency between a −3-dB point and the filter center frequency is known as Δf. The filter bandwidth then is 2 Δf. Fig. 7-13 also illustrates how the unloaded Q of a tuned circuit or resonator relates to the bandwidth characteristic.

If a tuned circuit is used as a filter, the higher its loaded Q, the greater the selectivity. To make the skirts of the response curve steeper, several high-Q resonators can be used in cascade. This aids the overall selectivity by providing greater rejection of signals close in frequency to the desired one.

An ideal receiver would provide selectivity at significant points and might employ:

1) A selective front end for rejecting out-of-band signals to prevent overloading and spurious responses.

2) A selective IF circuit with a choice of two or more IF filters with different bandwidths: 2.4 kHz for SSB, 300 Hz for CW and RTTY, 10 kHz for double-sideband AM, and 15 kHz for FM.

3) RC active or passive audio filters to reduce wide-band noise and provide audio selectivity in the range from 300 to 2700 Hz (SSB), 2100 to 2300 Hz (or 1250 to 1450 Hz) for RTTY, and a narrow-bandwidth filter (750 to 850 Hz) for CW.

4) Selective circuits or filters in the LO chain to reject all mixer injection energy other than the desired frequency.

Front-End Selectivity

Front-end selectivity at the receiver operating frequency is provided by resonant networks placed before and after the RF amplifier stage. This part of the receiver is often called the preselector.

The band-pass characteristics of the input tuned circuits are of considerable significance if strong out-of-band signals are to be rejected—an ideal design criterion. Many commercially made receivers available to the radio amateur use tuned circuits that can be adjusted from the front panel. The greater the network Q, the sharper the frequency response and the better the adjacent-frequency rejection.

A typical preselector circuit is shown in Fig. 7-14. It's basically a parallel-tuned LC circuit—L1, C_F and C_V. C_V is a small variable capacitor that permits tuning the network over a particular frequency range. We can determine the tuning range of the preselector by using the formula:

$$f_o = \frac{1}{2\pi \sqrt{L\,C}} \qquad \text{(Eq. 7-19)}$$

where

f_o = the resonant frequency of the network, in hertz
L = the inductance of L1 in henrys
C = the total capacitance in the circuit ($C_F + C_V$) in farads

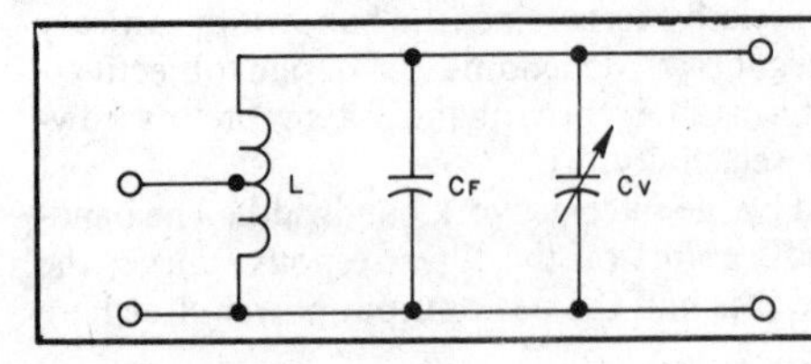

Fig. 7-14—A representative preselector circuit. C_V permits tuning the circuit over a range of frequencies.

To determine the low- and high-frequency ends of the tuning range, two calculations must be made. For the low-end frequency, the capacitance of C_F is added to the maximum capacitance of C_V. To find the high end of the tuning range, the capacitance of C_F is added to the minimum capacitance of C_V.

For example, suppose you have an inductor of 10 μH, C_F equal to 156 pF, and a tuning capacitor that has maximum and minimum values of 50 and 2 pF, respectively. Find the preselector tuning range.

First, calculate the maximum and minimum capacitance values. $C_{max} = 156 \text{ pF} + 50 \text{ pF} = 206 \text{ pF}$. $C_{min} = 156 \text{ pF} + 2 \text{ pF} = 158 \text{ pF}$. Then:

$$f_{max} = \frac{1}{2\pi \sqrt{10 \times 10^{-6} \text{ H} \times 158 \times 10^{-12} \text{ F}}}$$

$$f_{max} = \frac{1}{2\pi \sqrt{1.58 \times 10^{-15}}} = \frac{1}{2.50 \times 10^{-7}}$$

$$f_{max} = 4.00 \times 10^{6} \text{ Hz} = 4.00 \text{ MHz}$$

$$f_{min} = \frac{1}{2\pi \sqrt{10 \times 10^{-6} \text{ H} \times 206 \times 10^{-12} \text{ F}}}$$

$$f_{min} = \frac{1}{2.85 \times 10^{-8}} = 3.51 \times 10^{6} \text{ Hz} = 3.51 \text{ MHz}$$

So our example is for an 80-meter preselector circuit.

IF Selectivity

Assuming that all stages are designed properly, IF selectivity is the most significant type of selectivity in a receiver. In the past, receivers contained many stages of LC filtering to achieve high degrees of selectivity. Modern receivers use IFs above 500 or 600 kHz, and it is common practice to use crystal or ceramic filters. Below these frequencies, mechanical filters are usually employed.

Crystal Filters

Some simple crystal IF filters that may be homemade by amateurs are shown in Fig. 7-15. The single-crystal filter at A is best suited for simple receivers intended mainly for CW use. C1 is adjusted to provide the band-pass characteristic shown next to the circuit diagram. When the BFO frequency is placed on the part of the low-frequency slope (left) that gives the desired beat note respective to f_o (approximately 800 Hz), single-signal reception results. To the right of f_o, the response drops sharply to reduce output on the upper-frequency side of zero beat. If no IF filter was used, or if the BFO frequency fell at f_o, nearly equal signal amplitudes would exist on either side of zero beat, resulting in a double-signal response. Signals on the upper-frequency side of the IF passband would interfere with reception (QRM). The single-crystal filter shown is capable of at least 30 dB of rejection on the high-frequency side of zero beat.

A half-lattice filter is shown at B of Fig. 7-15. The response curve is symmetrical, and there is a slight dip at the center frequency. The dip is minimized by proper selection of R_T. In multiple-crystal filters, there may be many minor dips; this is referred to as passband ripple; good filters exhibit a minimum amount of ripple. In the filter shown, Y1 and Y2 are separated in frequency by the amount needed to obtain the desired selectivity. The bandwidth at the 3-dB points will be approximately 1.5 times the crystal-frequency spacing. For upper- or lower-sideband reception, Y1 and Y2 would be 1.5 kHz apart, yielding a 3-dB bandwidth of approximately 2.25 kHz. For CW work, a crystal spacing of 0.4 kHz results in a bandwidth of roughly 600 Hz.

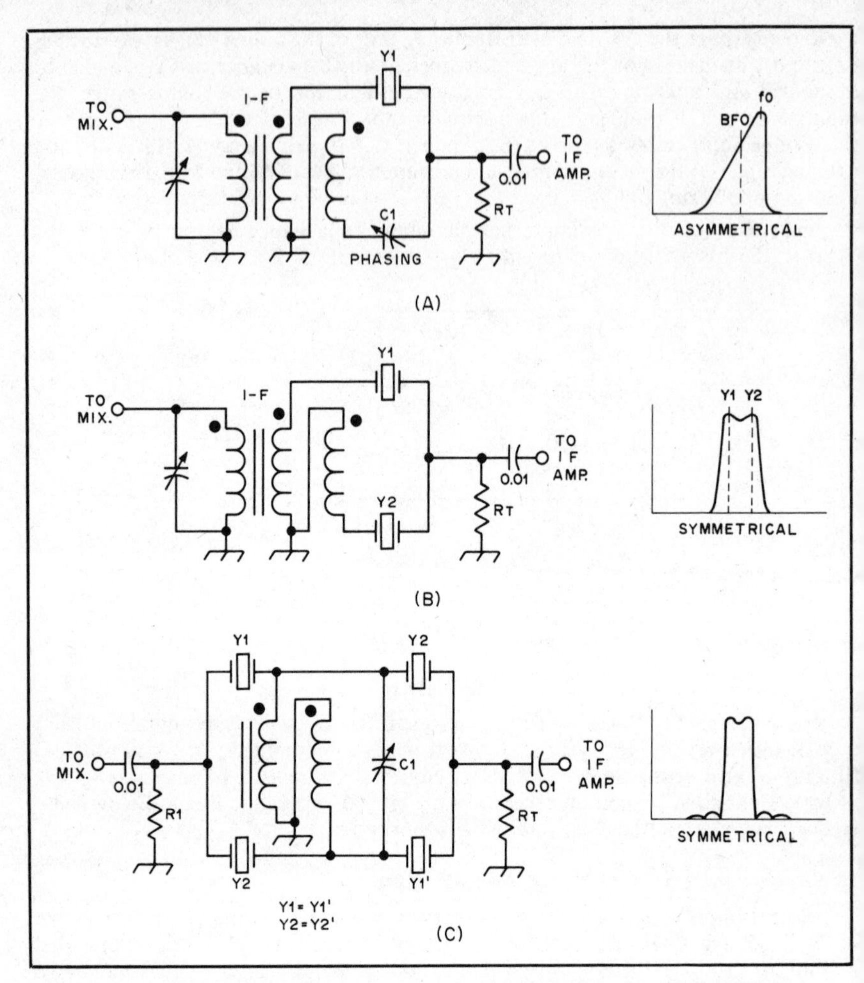

Fig. 7-15—Several simple crystal filters. Selectivity is increased as crystals are added.

The skirts of the curve are fairly wide with a single half-lattice filter that uses crystals in the HF range.

The filter skirts can be steepened by placing two half-lattice filters in cascade, as shown in Fig. 7-15C. R1 and R_T must be selected to provide minimum ripple at the center of the passband. The same rule applies for frequency spacing between the crystals. C1 is adjusted for a symmetrical response. Commercially made receivers and transceivers in use today have crystal filters with eight or more filter sections or poles; such filters have relatively steep skirts and excellent ultimate attenuation.

Mechanical Filters

At low IFs (typically 455 kHz), mechanical filters are usually employed. An illustration of how a mechanical filter operates is shown in Fig. 7-16. As the incoming IF signal passes through the input transducer, the signal is converted to mechanical

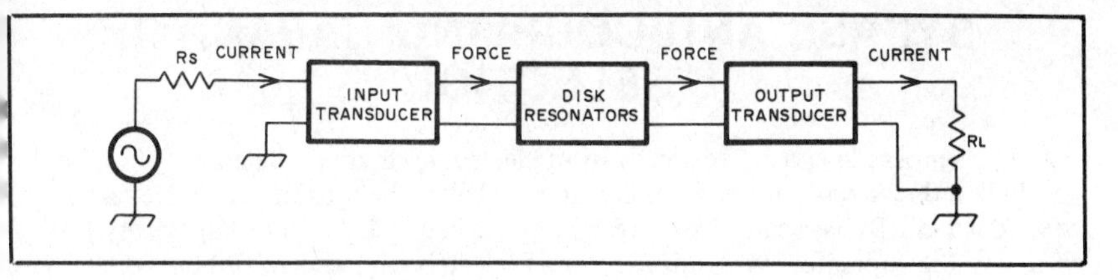

Fig. 7-16—Block diagram of a mechanical filter.

energy. This energy is passed through the disk resonators to filter out the undesired frequencies, then through the output transducer where the mechanical energy is converted back to the original electrical form. The transducers serve a second function: They reflect the source and load impedances into the mechanical portion of the circuit, thereby providing a termination for the filter.

Mechanical filters require external resonating capacitors, which are placed across the transducers. If the filters are not resonated, there will be an increase in insertion loss and passband ripple. The exact amount of capacitance will depend on the filter model used. The manufacturer's data sheet specifies the proper capacitance values.

Mechanical filters have excellent frequency-stability characteristics. This makes it possible to fabricate them for fractional bandwidths of a few hundred hertz. Bandwidths of 0.1 percent can be obtained with these filters. This means that a filter having a center frequency of 455 kHz could have a bandwidth as small as 45.5 Hz.

Modern Filter Methods

Most modern receivers and transceivers have selectable IF filters to provide suitable bandwidths for SSB, CW and RTTY. Filter bandwidths of 200 to 600 Hz for CW and 1.8 to 2.7 kHz for SSB are common. Generally, 2.4-kHz SSB and 500-Hz CW filters are used. For RTTY, the CW or SSB filter can be used depending on the signal bandwidth. Most amateur Baudot RTTY operation is done with 170-Hz shift. Therefore, a 500-Hz CW filter could be used. If a separate RTTY filter is included, it would have an even narrower response, perhaps around 300 Hz. With 300-baud ASCII transmissions or wider Baudot RTTY shifts, a larger filter bandwidth is needed; 1200 Hz would suffice for such use.

IF-filter bandwidths for FM depend on the particular application. At frequencies below 29.0 MHz, the bandwidth of FM emissions must be limited to that of an AM transmission having the same audio characteristics. Above 29.0 MHz, wider FM bandwidths are permitted. The deviation standard now employed is ±5 kHz for narrow-band FM (NBFM). An FM-bandwidth rule-of-thumb determination is

$$BW = 2M + 2DK \qquad \text{(Eq. 7-20)}$$

where

BW = bandwidth
M = maximum modulation frequency (3 kHz typical)
D = deviation frequency (5 kHz typical)
K = 1

Thus, the bandwidth of such a signal would be 2(3 + 5) or 16 kHz. Wide-band systems need a 36-kHz bandwidth.

[Study FCC examination questions with numbers that begin 4BG-5 and questions 4BG-9.1 and 4BG-9.2. Review this section as needed.]

BYPASS AND COUPLING CAPACITOR SELECTION

Capacitors play an active part in most electronic circuits. The ability of capacitors to block a dc current and pass an ac current makes them usable as *bypass* and *coupling* devices. (Bypass capacitors are sometimes called decoupling capacitors.) In addition to the capacitor type (mica, ceramic, electrolytic, and so on) and the voltage and current ratings of the capacitor, there are three other important factors that we must deal with when designing electronic circuits using capacitors. These factors are: the total value of circuit capacitance, its corresponding reactance and the operating or signal frequencies at which the capacitors must operate. The relationship of these factors is expressed by:

$$X_C = \frac{1}{2\pi f C} \qquad \text{(Eq. 7-21)}$$

where
- C = capacitance in farads
- X_C = reactance in ohms
- f = operating frequency

Capacitive reactance varies inversely with frequency—the higher the frequency, the lower the reactance for a given capacitor value. When selecting capacitance values, a rule of thumb is to choose a value of capacitive reactance that is one-tenth the value of the circuit impedance in which the capacitor will be used. This value ensures that the reactance of the capacitor is small enough to get the signal from one circuit point to another with little or no attenuation.

Actually, coupling and bypass (or decoupling) capacitors operate in exactly the same manner. The terminology that's used to describe their specific purpose tells whether we're dealing with desired signals (coupling) or undesired signals (decoupling or bypass). In the case of bypass capacitors, we're "bypassing" unwanted signals to ground to get rid of them. With coupling capacitors, we're passing desired signals between two or more circuit points.

In many circuits, there are areas that we wish to keep free of certain signals. Bypass capacitors are one means of achieving this goal. The name given to this capacitor action is self-descriptive: The capacitor allows certain signal frequencies to bypass a specific circuit area while rejecting higher-frequency energy. You can also think of bypass capacitors as filter capacitors. Although you may not have thought of them in this fashion, power-supply filter capacitors are bypass capacitors; they serve to route low-frequency ac-signals (60 and 120 Hz, for example) to ground. They keep ac hum from affecting the circuits deriving their power from the supply. Because the signal frequencies and load impedances encountered in power-supply circuits are low, the capacitance values are generally high, anywhere from 1 μF to 100,000 μF or more.

Amplifier Frequency Response

The circuit of Fig. 7-17 is a simple two-stage FET amplifier. In this circuit, R1 and R4 are gate resistors that determine the input impedance of each stage. R2 and R5 are source resistors that are used to bias the individual FETs at a specific point. R3 is the load resistor for the first stage. C1 and C3 are source-bypass capacitors and C2 is the coupling capacitor. The overall frequency response of this circuit is affected strongly by the choice of capacitance values for C1, C2 and C3.

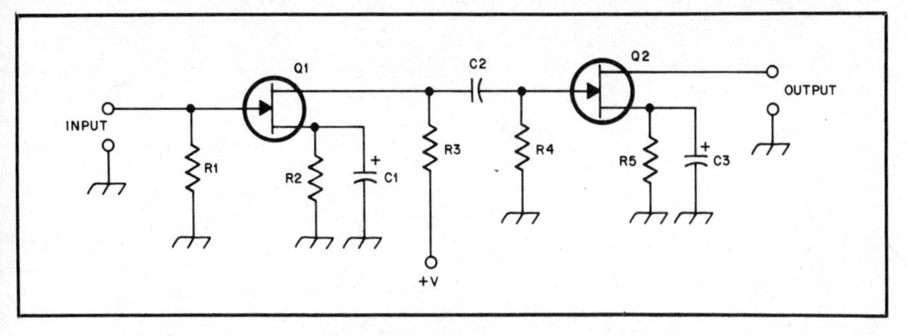

Fig. 7-17—A simple two-stage FET amplifier circuit, illustrating the use of coupling and bypass capacitors. C2 is a coupling capacitor; C1 and C3 are bypass capacitors.

In order to ensure a minimum attenuation of the desired frequencies by the presence of the source resistors, C1 and C3 are chosen to have a value large enough to provide a signal path around R2 and R5 at the lowest frequency of interest. If this amplifier stage was designed to pass frequencies from 300 to 3000 Hz and R2 and R5 were each 100 ohms, a value of 50 μF or more for C1 and C3 would be appropriate (X_C = 10.6 Ω). If C1 and C3 were small-value capacitors, say, 1 μF, the signal output of the stages at low frequencies would be reduced. This occurs because the capacitive reactance of the source-bypass capacitors is large (531 Ω), forcing some of the signal current to pass through the source resistors. The signal current creates a voltage at the source that is in phase with the voltage applied to the gate of Q1. This reduces the gate-to-source voltage, decreasing signal output.

The capacitive reactance of C2 must likewise be chosen to be small at the lowest frequency of interest. If the reactance of C2 is too large at the lower frequencies, it will act as a voltage divider with R4, reducing the signal level applied to the gate of Q2. By using a large-value coupling capacitor, we can ensure that the low-frequency response will not be hampered. Because large-value capacitors also have a larger capacitance to ground, however, there is a limit as to how large C2 can be made without affecting high-frequency response. In general, by using larger values of coupling and bypass capacitance we can improve the circuit response to low-frequency signals.

[This completes the material for this chapter. To complete your study, however, you should now turn to Chapter 10. Study FCC examination questions 4BG-9.3 through 4BG-9.8. Review this section as needed.]

Key Words

Amplitude-compandored single sideband (ACSSB)—An SSB system using a logarithmic amplifier to compress voice signals at the transmitter and an inverse logarithmic amplifier to expand the voice signals in the receiver.

Bandwidth—As related to a transmitted signal, that frequency range around a center frequency that the signal occupies. Bandwidth increases with increasing information rate.

Baud—A unit of signalling speed, used to describe data transmission rates. One bit per second for single-channel binary-coded signals.

Compandoring—In an ACSSB system, the process of compressing voice signals in a transmitter and expanding them in a receiver.

Peak negative value—On a signal waveform, the maximum displacement from the zero line in the negative direction.

Peak positive value—On a signal waveform, the maximum displacement from the zero line in the positive direction.

Peak-to-peak value—On a signal waveform, the maximum displacement between the peak positive value and the peak negative value.

Pilot tone—In an ACSSB system, a 3.1-kHz tone transmitted with the voice signal to allow a mobile receiver to lock onto the signal, and also used to control the inverse logarithmic amplifier gain.

Pulse-amplitude modulation—A pulse-modulation system in which the amplitude of a standard pulse is varied in relation to the information-signal amplitude at any instant.

Pulse-code modulation—A modulation process in which an information waveform is converted from analog to digital form by sampling the information signal, and these values are encoded in some form, such as binary-coded decimal.

Pulse modulation—Modulation of an RF carrier by a series of pulses. These pulses convey the information that has been sampled from an analog signal.

Pulse-position modulation—A pulse-modulation system in which the position (timing) of the pulses is varied from a standard value in relation to the information-signal amplitude at any instant.

Pulse-width modulation—A pulse-modulation system in which the width of a pulse is varied from a standard value in relation to the information-signal amplitude at any instant.

Chapter 8

Signals and Emissions

The FCC syllabus for Element 4B specifies a number of loosely related topics, such as pulse-modulation methods, digital-communications methods, amplitude-compandored single sideband and values relating to the peak values of signal voltages. When you have studied the information in each section, use the examination questions listed in Chapter 10 as directed, to review your understanding of the material.

Pulse Modulation

There are several methods used for transmitting information as a series of brief RF pulses. These methods are collectively known as *pulse modulation*, and they vary in how the modulating signal is conveyed by the pulses. In *pulse-width modulation (PWM)*, or pulse-duration modulation (PDM), the duration of the transmitted pulses varies with the applied modulation. In *pulse-position modulation (PPM)* the modulation is produced by varying the position of each pulse with respect to some fixed timebase. With either system, you must have a way to sample the information signal. This can be done by using a sawtooth waveform with a rapid rise time and a frequency that is several times higher than the information-signal frequency. Fig. 8-1 shows how this sawtooth waveform can be used to trigger a sampling circuit and

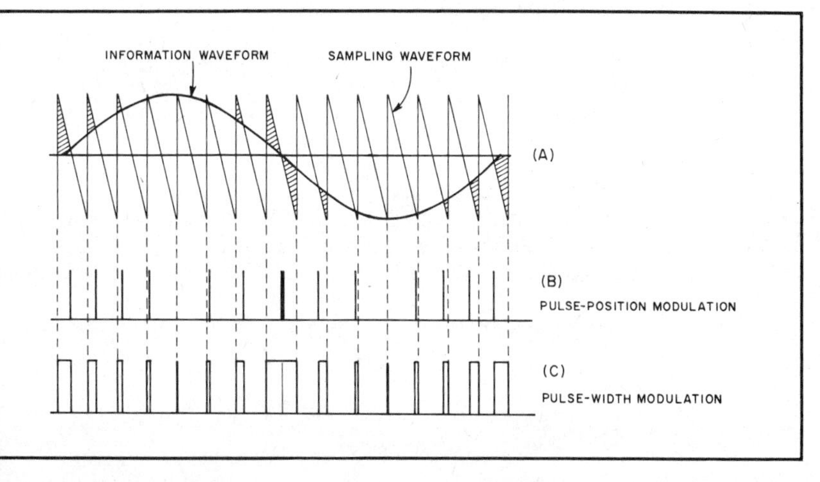

Fig. 8-1—An information signal and a sampling waveform are shown at A. B shows how the sampled information signal can produce pulse-position modulation, and C shows how it can produce pulse-width modulation.

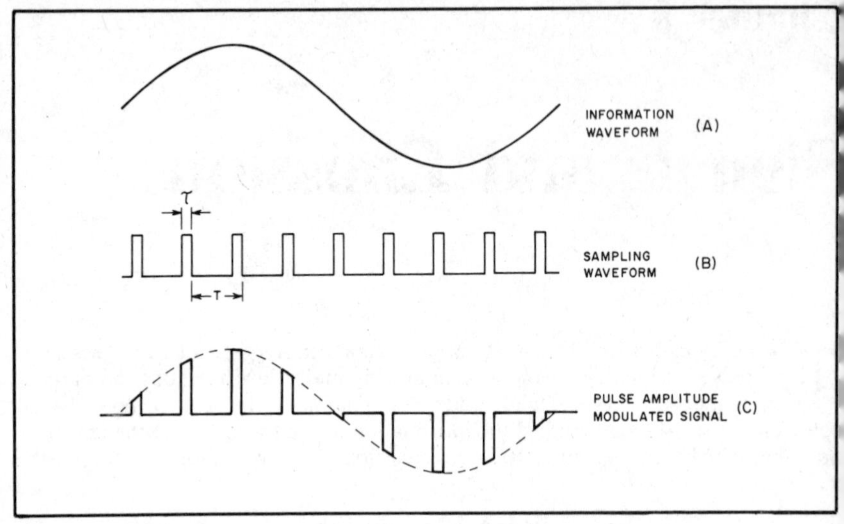

Fig. 8-2—A shows an information signal, B is a sampling waveform and C is a pulse-amplitude-modulated signal.

measure the difference between the information signal amplitude and the sampling wave.

While both PWM and PPM use pulses of uniform amplitude, the third common form of pulse modulation, *pulse-amplitude modulation (PAM)*, changes the amplitude of each pulse to reflect the modulating signal amplitude. Fig. 8-2 illustrates how PAM signals are generated. PAM is less immune to noise than is either of the other methods, and so it is seldom considered for communication systems, although it may be the easiest method to visualize and understand.

Notice that in all three cases, the duty cycle of the transmission is very low. This causes the peak power of a pulse-modulated signal to be much greater than its average power.

If the amplitude or other characteristic of a pulse-modulated signal is converted to a binary code before transmission, the system is known as *pulse-code modulation (PCM)*. This encoding can be accomplished in a number of ways, but the net result is a signal that is extremely immune to noise or other interference.

Our examples have used a sine-wave-like information signal. If the information signal that is being sampled to produce pulse modulation is a voice, then the transmitted signal will be a representation of that speech pattern. When the signal is demodulated in the receiver, the voice audio will be reproduced. A voice signal can be made to vary the duration, position, amplitude or code of a standard pulse to produce a pulse-modulated signal.

[Study FCC examination questions with numbers that begin 4BH-1. Review this section as needed.]

FCC EMISSION DESIGNATORS

The FCC uses a special system to specify the types of signals (emissions) permitted to amateurs and other users of the radio spectrum. Each emission designator has three digits; Table 8-1 shows what each character in the designator stands for. The

Table 8-1
Partial List of WARC-79 Emissions Designators

(1) First Symbol—Modulation Type	
Unmodulated carrier	N
Double sideband full carrier	A
Single sideband reduced carrier	R
Single sideband suppressed carrier	J
Vestigial sideband	C
Frequency modulation	F
Phase modulation	G
Various forms of pulse modulation	P, K, L, M, Q, V, W, X
(2) Second Symbol—Nature of Modulating Signals	
No modulating signal	Ø
One modulating channel containing quantized or digital information without the use of a modulating subcarrier	1
A single channel containing quantized or digital information with the use of a modulating subcarrier	2
A single channel containing analog information	3
Two or more channels containing quantized or digital information	7
Two or more channels containing analog information	8
(3) Third Symbol—Type of Transmitted Information	
No information transmitted	N
Telegraphy—for aural reception	A
Telegraphy—for automatic reception	B
Facsimile	C
Data transmission, telemetry, telecommand	D
Telephony	E
Television	F

designators begin with a letter that tells what type of modulation is being used. The second character is a number that describes the signal used to modulate the carrier, and the third character specifies the type of information being transmitted.

Some of the more common combinations are:

- NØN—Unmodulated carrier
- A1A—Morse code telegraphy using amplitude modulation
- A3E—Double-sideband, full-carrier, amplitude-modulated telephony
- J3E—Amplitude-modulated, single-sideband, suppressed-carrier telephony
- F3E—Frequency-modulated telephony
- G3E—Phase-modulated telephony
- F1B—Telegraphy using frequency-shift keying without a modulating audio tone (FSK RTTY). F1B is designed for automatic reception.
- F2B—Telegraphy produced by modulating an FM transmitter with audio tones (AFSK RTTY). F2B is also designed for automatic reception.
- F1D—FM data transmission, such as packet radio

If you got your amateur license before 1983 or so, you may be familiar with a different set of emission designators. Even if you got your license recently, you have probably heard other hams on the air using a set of two-letter emission designators, or seen them in an Amateur Radio book or magazine. Under this old system, for example, CW was designated A1 emission, FM telephony was F3, AM double-sideband, full-carrier telephony was A3, and amplitude-modulated television was A5. After the World Administrative Radio Conference (WARC) in 1979, The FCC began a gradual phase-in of the new emission designators, and in 1985 Part 97 was revised to express the amateur mode allocations in the new designators. These new designators allow much more specific description of the signals and the method used to produce a given type of emission.

DIGITAL SIGNALS

The duration of a data bit in an asynchronous character is, by definition, the inverse of the transmission speed in *bauds*. Baud rate (or bauds) is the number of signal elements sent per second (in a single-channel transmission). This can be expressed mathematically as:

$$T_b = \frac{1}{B} \quad \text{(Eq. 8-1)}$$

where

B = the speed in bauds

T_b = the duration of a bit in seconds.

Remember, however, that an asynchronous character begins with a special start pulse and ends with a stop pulse. (Asynchronous means that the character timing is not controlled by a separate "clock" signal.) These pulses are not necessarily the same length as the data bits. In common "60 speed" Baudot RTTY, the start pulse is the same duration as a data pulse, but the stop pulse is 1.42 times that long. Table 8-2 shows that 60-speed Baudot is really sent at 61.33 WPM, and the "standard speed" for all Baudot RTTY around 60 WPM is 45.45 bauds. For calculations, round this to 45 bauds. So the data and start bits have a duration of 1 / 45 = 0.022 second = 22 ms. Stop bits have a duration of 1.42 × 22 = 31 ms.

Unfortunately, not all Baudot RTTY uses a 1.42 bit stop pulse and a start pulse that is the same as a data bit. This may lead to confusion when making calculations and it may result in erratic operation of mechanical teleprinters. With a computer system set up for RTTY operation, however, the computer program can be changed to accept just about any combination of start, data and stop bits. So as more operators use computers for RTTY transmission and reception this problem will become less important.

[Now turn to Chapter 10 and study FCC examination questions with numbers that begin 4BH-2. Review this section as needed.]

Table 8-2
Baudot Signaling Rates and Speeds

Signaling Rate (bauds)	*Data Pulse (ms)*	*Stop Pulse (ms)*	*Speed (WPM)*	*Common Name*
45.45	22.0	22.0	65.00	Western Union, "60 speed" or 45 bauds
	22.0	31.0	61.33	
	22.0	33.0	60.61	
50.00	20.0	30.0	66.67	European or 50 bauds
56.92	17.57	25.00	76.68	"75 speed" or 57 bauds
	17.57	26.36	75.89	
74.20	13.47	19.18	100.00	"100 speed" or 74 bauds
	13.47	20.21	98.98	
100.0	10.00	15.00	133.33	100 bauds

Information Rate and Bandwidth

Bandwidth of CW Signals

The *bandwidth* of a CW signal is determined by two factors: the speed of the CW being sent and the shape of the keying envelope. The usual equation for this calculation is:

$$Bw = B \times K \qquad \text{(Eq. 8-2)}$$

where

Bw = the bandwidth of the signal
B = the speed of the transmission in bauds
K = a factor relating to the shape of the keying envelope

To solve this equation, we must find values for B and K.

Morse code speed is usually expressed in words per minute (WPM), so we need to convert WPM to bauds. Since bauds equals signal elements per second, we must determine the number of signal elements in a Morse code word. (A signal element is the time of one dot or one interelement space. Dashes count as three signal elements and the space between words is seven elements.) The standard word used for this calculation is "PARIS," which contains 50 signal elements. This results in the equation:

$$\frac{1\ \text{word}}{\text{min}} = \frac{50\ \text{elements}}{\text{min}} = \frac{50\ \text{elements}}{60\ \text{s}} = 0.83\ \text{bauds} \qquad \text{(Eq. 8-3)}$$

From that we derive:

$$\text{WPM} \times 0.83 = \text{bauds} \qquad \text{(Eq. 8-4)}$$

which can also be expressed as:

$$\text{bauds} = \frac{\text{WPM}}{1.2} \qquad \text{(Eq. 8-5)}$$

Then, for CW signals Eq. 8-2 becomes:

$$Bw = \left(\frac{\text{WPM}}{1.2}\right) \times K \qquad \text{(Eq. 8-6)}$$

The second variable, K, is a measure of keying shape. As CW rise and fall times get shorter (more abrupt, harder keying), K gets larger. This is because the short rise and fall times contain more harmonics than longer, softer envelopes. The more harmonics in the keying envelope, the greater the bandwidth of the resulting CW signal. On paths with good strong signals, soft keying can be used, and K will be around 3. On fading paths, harder keying is necessary, resulting in a K of 5.

So, to find the bandwidth of a signal with soft keying:

$$Bw = \text{WPM} \times (3 / 1.2) = \text{WPM} \times 2.5 \qquad \text{(Eq. 8-7)}$$

and for hard keying:

$$Bw = \text{WPM} \times (5 / 1.2) = \text{WPM} \times 4.2 \qquad \text{(Eq. 8-8)}$$

Since most Amateur Radio contacts are made over paths with at least some fading, amateur transmitters are usually adjusted to have keying rise and fall times on the order of 5 milliseconds. For this keying envelope, the K factor is about 4.8, and the resulting equation is:

$$Bw = \text{WPM} \times 4 \qquad \text{(Eq. 8-9)}$$

This equation is a good rule of thumb for calculating the bandwidth of an amateur CW transmission.

As an example, suppose you are sending Morse code at a speed of 30 WPM. What is the bandwidth of your transmitted signal?

$$Bw = WPM \times 4 = 30 \times 4 = 120 \text{ Hz.}$$

Bandwidth of Binary Data Signals

Most amateur Baudot, AMTOR and ASCII transmissions employ frequency shift keying (FSK). In FSK systems, the transmitter uses one frequency to represent a 0 and another frequency to represent a 1. By shifting between these two frequencies (called the mark and space frequencies), the transmitter sends binary data. The difference between the mark frequency and the space frequency is called the shift. FSK signals can be generated either by shifting a transmitter oscillator or by injecting two audio tones, separated by the correct shift, into a single-sideband transmitter. The emission designators for these two methods of FSK generation are F1B and J2B, respectively. Both methods generate the same signal, and the bandwidth of that signal is determined by the frequency shift used and the speed at which data is transmitted. The bandwidth is not affected by the code that is being transmitted. The equation relating bandwidth to shift and data rate is:

$$Bw = (K \times \text{Shift}) + B \qquad \text{(Eq. 8-10)}$$

where

Bw = the necessary bandwidth in hertz

K = a constant that depends on the allowable signal distortion and transmission path. For most practical Amateur Radio communications, K = 1.2.

Shift = the frequency shift in hertz

B = the data rate in bauds

For example, the bandwidth of a 170-Hz shift, 45-baud signal is:

$$Bw = (1.2 \times 170 \text{ Hz}) + 45 = 249 \text{ Hz} \qquad \text{(Eq. 8-11)}$$

What if the data rate is given as "speed" in words per minute, rather than bauds? Table 8-2 shows the relationship between some common RTTY speeds and their signaling rates in bauds. For example, if we want to know the bandwidth required by a 75-WPM, 170-Hz-shift emission F1B transmission, we look at the table and see that "75-speed" is 57 bauds. We then use Eq. 8-10 to calculate the required bandwidth:

$$Bw = (1.2 \times 170 \text{ Hz}) + 57 = 261 \text{ Hz}$$

[Before proceeding to the next section, turn to Chapter 10. Study FCC examination questions with numbers that begin 4BH-4. Review this section as needed.]

AMPLITUDE-COMPANDORED SINGLE SIDEBAND

Human speech has a dynamic range greater than 40 dB. That is, the booming voice of a lecturer may be 10,000 times louder than the voice of one whispering in the audience. Even the peaks and nulls of normal conversation exhibit a dynamic range of 20 dB. Modulation systems must be able to accurately encode and later decode these amplitude variations to recover all of the original voice information. Failure to do so results in amplitude distortion, and loss of intelligibility.

Communications systems must also contend with noise. Noise masks desired signals, reducing intelligibility. In any communications system, the S/N (signal-to-noise) ratio plays an important part in overall performance—the higher the S/N ratio, the better a system can accurately convey desired information.

In amplitude-modulation systems, S/N ratio can be improved by increasing the desired signal level relative to the noise. One obvious way to accomplish this is by increasing transmitter power. Another, more economical approach results from

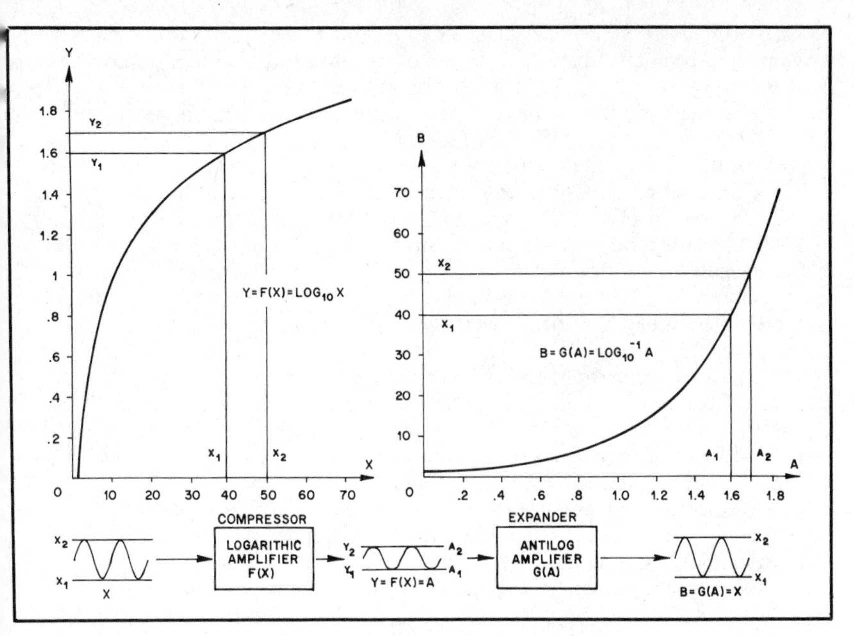

Fig. 8-3—Mathematical relationships in an amplitude-compandoring amplifier. See text for discussion.

analysis of speech-waveform amplitude patterns. Since the average energy content of human speech is much less than its peak energy, a modulation method that decreases the peak-to-average ratio of transmitted speech would improve the S/N ratio.

Logarithmic amplifiers are often used to perform this task. Fig. 8-3 is a graphical representation of a logarithmic amplifier. Assume the input signal, X, and output, Y, are mathematically related by the function Y = f(X). Further, assume this function is logarithmic. As X changes from amplitude level x1 to x2, Y changes from y1 to y2. Clearly, the output amplitude, Y, is smaller than the input amplitude, X; Y is said to be compressed. After transmission, the compressed signal is expanded back to its original shape. In the receiver, this is done with an expanding amplifier (g(A)). It performs an inverse log function to get X back. The combined compression and expansion process is known as *compandoring*. When incorporated into a single-sideband transmission system, *amplitude-compandored single sideband (ACSSB)* offers significant S/N ratio improvement.

ACSSB was originally developed to provide narrowband channels in the land-mobile VHF bands. It can be transmitted in a much narrower bandwidth than FM—in fact, a single FM channel can accommodate five ACSSB signals.

Baseband (audio) analog signal processing is at the heart of

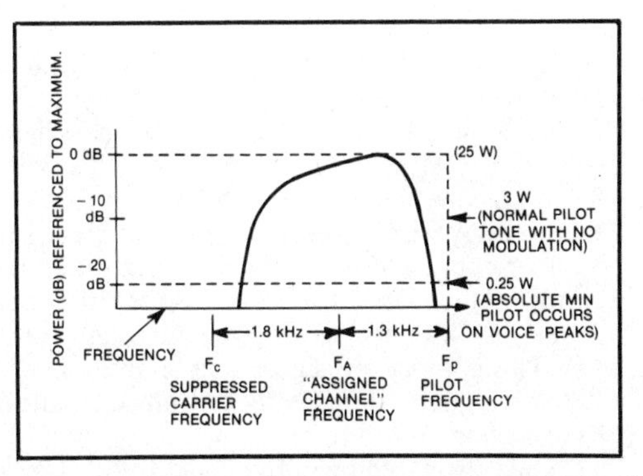

Fig. 8-4—Frequency spectrum of an (upper-sideband) ACSSB signal.

ACSSB. Fig. 8-4 represents a typical ACSSB signal. The passband is not flat. Preemphasis has been introduced (12 dB per octave) in the transmitter audio stages to accentuate higher frequencies. Conversely, the receiver uses a built-in deemphasis network to effectively flatten the passband. This process is similar to the preemphasis and deemphasis found in typical FM systems, and has the same aim—to improve the S/N ratio by reducing high-frequency hiss.

ACSSB systems use a *pilot tone* located 3.1 kHz away from the suppressed carrier. It is generated as an audio tone, and mixed with compressed voice information to form the baseband ACSSB signal. Steep-skirted low-pass filters keep higher-frequency audio components from interfering with the pilot; a cutoff frequency of approximately 2.8 kHz is used. On receive, audio and pilot information is processed separately. The strength of the recovered pilot tone is used to adjust the gain of audio expansion amplifiers.

Whenever the microphone PTT button is pressed, a full-power pilot tone is transmitted for ¼ second (250 ms). During this time, audio from the microphone is disabled. This allows an ACSSB receiver to lock quickly to the pilot tone, automatically tuning the receiver to the proper frequency. This is especially helpful in mobile systems. After ¼ second, the pilot tone amplitude drops approximately 9 dB, and voice transmission begins. The amplitude of the pilot varies with the information being sent—as the instantaneous voice signal becomes greater, the pilot is reduced in amplitude. Similarly, low-level voice energy causes the pilot to increase amplitude. The receiver uses this pilot tone to expand the signal, restoring the original dynamic range. For instance, a weak incoming pilot will drive the expansion amplifier in an ACSSB receiver audio stage to maximum gain, while a stronger pilot reduces gain, and thus intensity, of the audio output.

The effects of compandoring are most pronounced on weak or fading signals. Rapid fading causes received signal strength to drop too sharply for normal AGC circuits to respond. When this happens in an ACSSB receiver, the recovered audio signal and the pilot tone fade equally. Expansion amplifiers in the ACSSB receiver interpret this diminished pilot tone as a command to increase stage gain. Thus, momentarily weak signals are boosted, virtually eliminating flutter.

When extremely weak ACSSB signals are being received, however, both the recovered voice information and pilot are heavily shrouded in noise. Since the pilot requires less bandwidth than the audio signal, narrow pilot filters are used. These filters pass less noise than the wider audio filters, so the pilot tone S/N ratio will be higher. This, in turn, means the decoded pilot retains more information. Even though the audio channel may be very noisy, clean pilot tone information properly adjusts the gain of the logarithmic audio expansion amplifiers to accurately reconstruct the dynamic range of the original voice information. The result is gruff-sounding, because the missing voice information (tonal texture) has been replaced by noise. Still, this weak ACSSB signal is intelligible, while an equivalent FM signal would be lost in the noise.

Fig. 8-5 illustrates how audio compandoring can improve amateur satellite communications. At present, optimum satellite S/N ratios are on the order of 22 dB. This means the "loudest" signal appearing on the satellite transponder will be only 5 S units above the noise. At a typical amateur station, the recovered S/N ratio may be considerably worse. An 8-dB S/N ratio, as shown in A, leaves most of the SSB downlink signal buried beneath the noise. If ACSSB is used with 4:1 logarithmic compression, as in B, the 30-dB-dynamic-range audio signal will compress to less than 8 dB. This compressed signal rides entirely above the system (satellite transponder and receiver) noise floor. After downlink expansion, all the original information can be recovered.

Aside from compandoring, ACSSB boasts another useful feature—AFC (automatic frequency control). AFC circuits enable ACSSB receivers to scan for a pilot tone and lock to it, eliminating drift, Doppler shift and poor tuning.

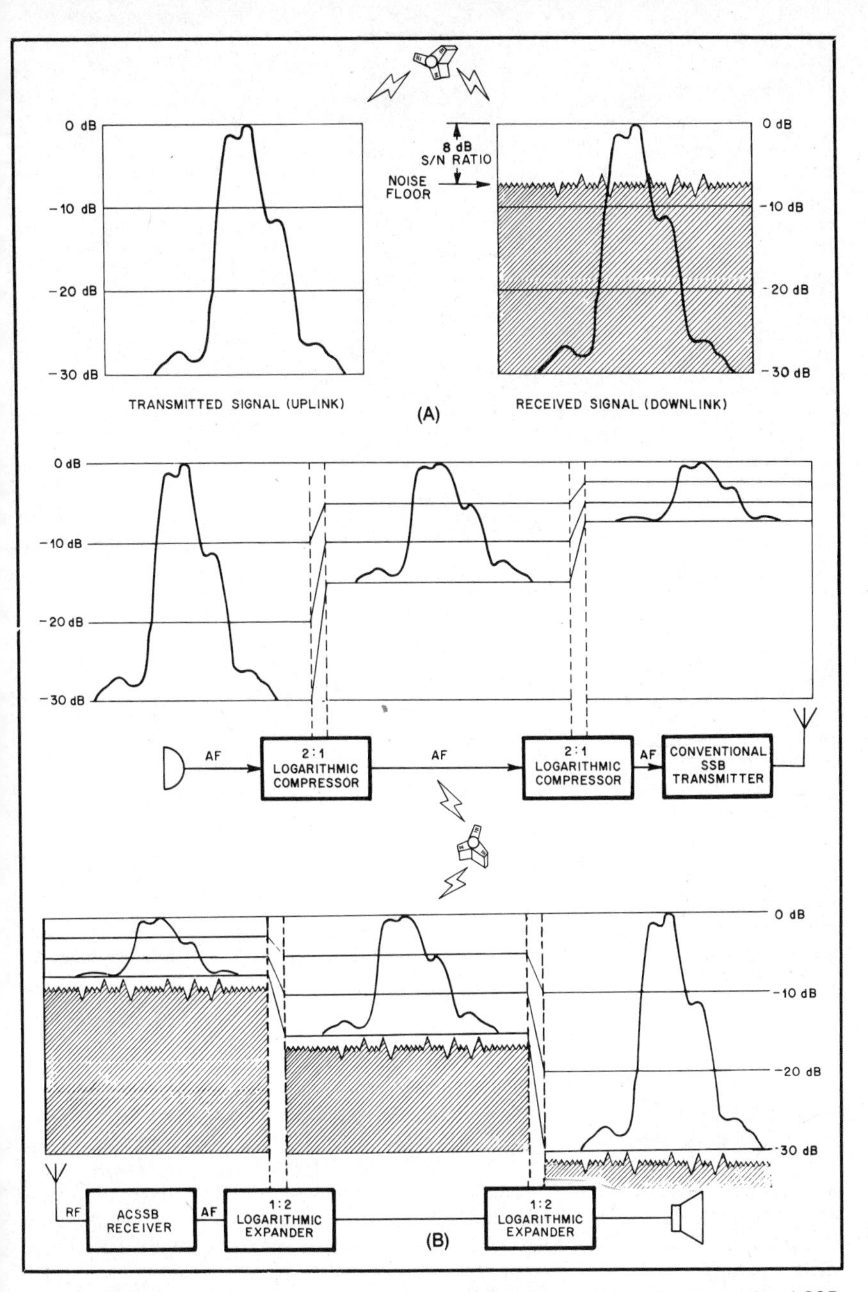

Fig. 8-5—SSB transmission through an amateur satellite. At A, most of the conventional SSB signal is below the noise. With ACSSB (B), it is possible to recover the audio information completely.

To perform this track-and-lock function, the AFC circuitry requires a suitable pilot tone. See Fig. 8-6. "Wide" and "narrow" pilot filters are used. First, the scanning receiver passes noise and signal information from the wide pilot filter to a phase detec-

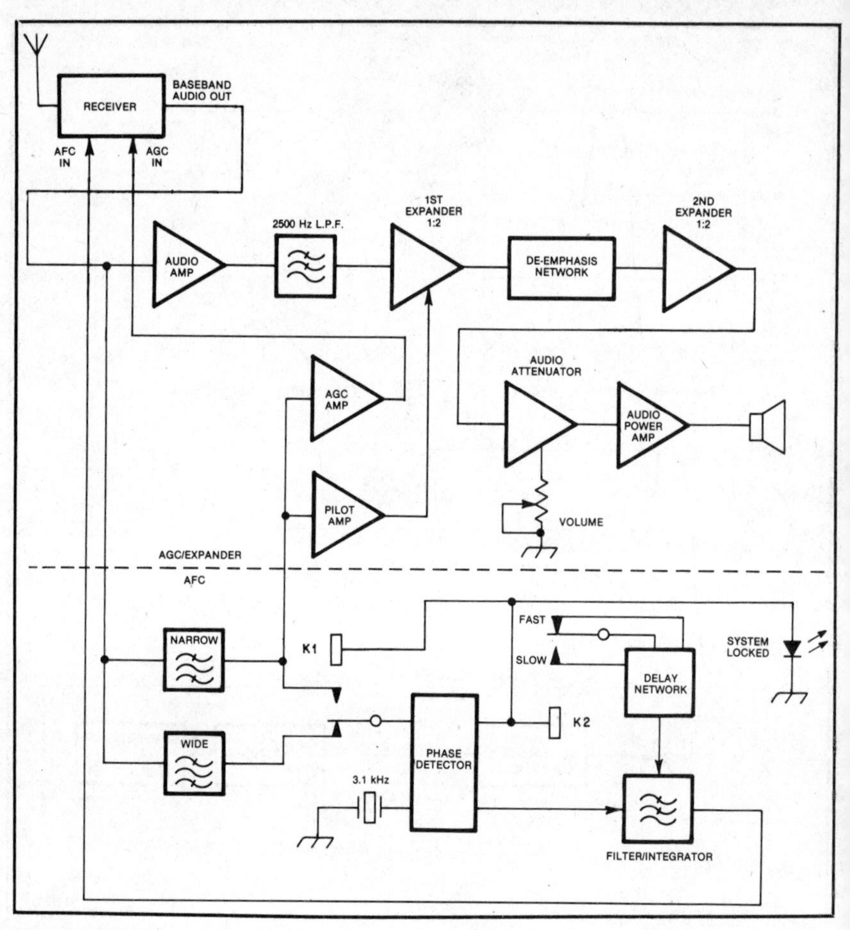

Fig. 8-6—ACSSB systems use baseband (audio) signal processing. Here, the recovered pilot tone makes automatic gain and frequency control possible. See text for details.

tor. Output from the phase detector passes through an integrating low-pass filter. This produces a constantly changing voltage that permits the receiver to scan a given frequency channel. When the phase detector senses an incoming signal, it produces pulses that make the receiver quickly converge to the 3.1-kHz pilot frequency. Once converged, the receiver becomes "locked" to the pilot—a narrow pilot filter automatically replaces the wide filter, and the loop low-pass filter time response is electronically slowed down. This keeps noise bursts and other extraneous signals from disturbing the phase-locked AFC loop. Any frequency changes caused by transmitter drift or Doppler shift will be tracked automatically.

[Turn to Chapter 10 and study FCC examination questions with numbers that begin 4BH-3. Review this section as needed.]

PEAK AMPLITUDE AND PEAK-TO-PEAK VALUES

When an oscilloscope is used to view varying voltages, the easiest dimension to

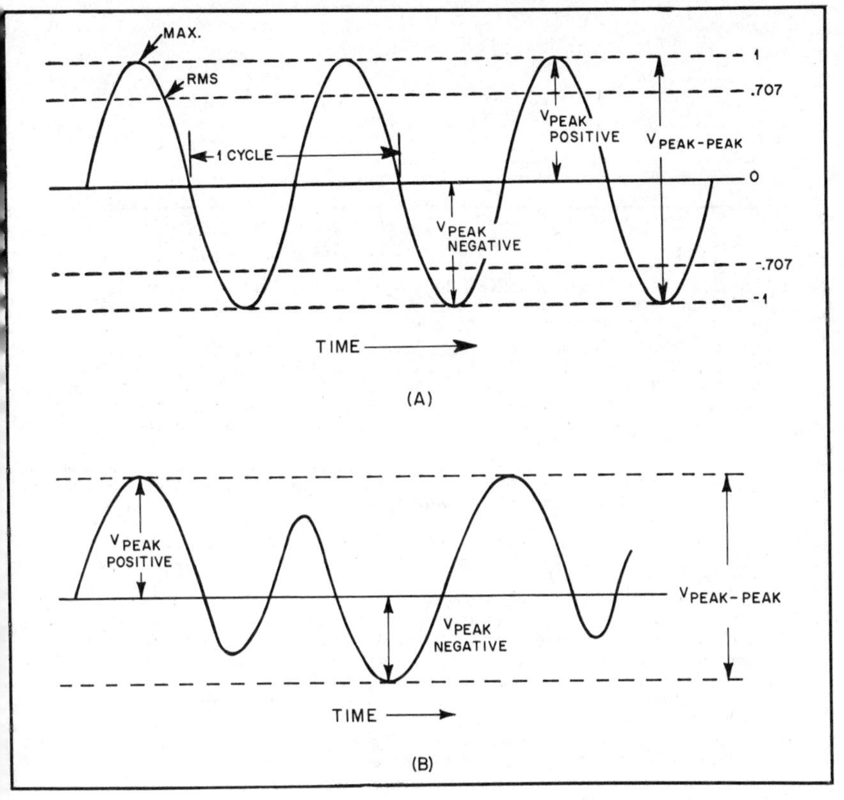

Fig. 8-7—It is easy to measure the peak negative, peak positive and peak-to-peak values of a waveform on an oscilloscope display. A shows a sine waveform and B shows a complex waveform.

measure is the total vertical displacement, or *peak-to-peak (P-P) voltage*. The maximum positive or negative displacement from the center line is called peak voltage. Fig. 8-7 shows that the *peak positive* and *peak negative voltages* need not have the same value. Of course, we can apply the same terminology to current or power values; we are not just talking about voltage values.

$$\text{peak-peak} = \text{peak positive} - \text{peak negative} \qquad \text{(Eq. 8-12)}$$

Waveforms for which peak positive plus peak negative equals zero are called symmetrical waveforms. For such waveforms:

$$\text{peak-to-peak} = 2 \times \text{peak positive} = -2 \times \text{peak negative} \qquad \text{(Eq. 8-13)}$$

Peak-to-peak voltage measurements are particularly useful for the characterization of linear devices. Generally, "linear" circuits are linear only over some finite range of input voltages. Inputs outside of this range will produce unwanted, spurious outputs. Because it is important to avoid such nonlinear operation, the maximum peak-to-peak input amplitude is an important criterion for evaluating linear amplifiers.

[You have now studied all of the material about signals and emissions as required for the Element 4B, Extra Class exam. Before you go on to Chapter 9, however, you should turn to Chapter 10 and study FCC examination questions with numbers that begin 4BH-5 and 4BH-6. Review this section as needed.]

Key Words

Beamwidth—As related to directive antennas, the width (measured in degrees) of the major lobe between the two directions at which the relative power is one half (−3 dB) its value at the peak of the lobe.

Circular polarization—Describes an electromagnetic wave in which the electric and magnetic fields are rotating. If the electric field vector is rotating in a clockwise sense, then it is called right-hand circular polarization and if the electric field is rotating in a counterclockwise sense, it is called left-hand circular polarization. Note that the polarization sense depends on whether you are standing behind the antenna for a signal being transmitted, or in front of it for a signal being received.

Delta match—A method for impedance matching between an open-wire transmission line and a half-wave radiator that is not split at the center. The feed-line wires are fanned out to attach to the antenna wire symmetrically around the center point. The resulting connection looks somewhat like a capital Greek delta.

Dipole antenna—An antenna with two elements in a straight line that are fed in the center; literally, two poles. For amateur work, dipoles are usually operated at half-wave resonance.

Effective isotropic radiated power (EIRP)—A measure of the power radiated from an antenna system. EIRP takes into account transmitter output power, feed-line losses and other system losses, and antenna gain as compared to an isotropic radiator.

Gamma match—A method for matching the impedance of a half-wave radiator that is split in the center (as a dipole) and a feed line. It consists of an adjustable arm that is mounted close to the driven element and in parallel with it near the feed point. The connection looks somewhat like a capital Greek gamma.

Horizontal polarization—Describes an electromagnetic wave in which the electric field is horizontal, or parallel to the earth's surface.

Isotropic radiator—An imaginary antenna in free space that radiates equally in all directions (a spherical radiation pattern). It is used as a reference to compare the gain of various real antennas.

Matching stub—A section of transmission line used to tune an antenna element to resonance or to aid in obtaining an impedance match between the feed point and the feed line.

Nonresonant rhombic antenna—A diamond-shaped antenna consisting of sides that are each at least one wavelength long. The feed line is connected to one end of the diamond, and there is a terminating resistance of approximately 800 ohms at the opposite end. The antenna has a unidirectional radiation pattern.

Parabolic (dish) antenna—An antenna reflector that is a portion of a parabolic curve. Used mainly at UHF and higher frequencies to obtain high gain and narrow beamwidth when excited by one of a variety of driven elements placed in the plane of and perpendicular to the axis of the parabola.

Resonant rhombic antenna—A diamond-shaped antenna consisting of sides that are each at least one wavelength long. The feed line is connected to one end of the diamond, and the opposite end is left open. The antenna has a bidirectional radiation pattern.

Vertical polarization—Describes an electromagnetic wave in which the electric field is vertical, or perpendicular to the earth's surface.

Chapter 9

Antennas and Feed Lines

There are several sections to this chapter, covering material you need to know in order to pass your Extra Class exam. But as you learn this material you will be able to do far more than simply pass an exam. The more you learn about antennas and feed lines, the better able you will be to experiment with new antenna and impedance-matching ideas. Experimentation with various types of antennas will enhance your enjoyment of our hobby. There are many more types of antennas than can be described in this book. Of the antenna types mentioned here, little detail is given about some. There is enough information to enable you to pass the Element 4B exam, but not so much as to be confusing. When you need to learn more about a particular antenna type, you will want to refer to the *ARRL Antenna Book* or one of the other good antenna reference books that are available.

After you have studied the information in each section of this chapter, use the examination questions in Chapter 10 to check your understanding of the material. If you are unable to answer a question correctly, go back and review the appropriate part of this chapter.

THE ISOTROPIC RADIATOR

The radiation from a practical antenna never has the same intensity in all directions. The intensity may even be zero in some directions from the antenna; in others it will probably be greater than you would expect from an antenna that did radiate equally well in all directions. Even though no actual antenna radiates with equal intensity in all directions, it is, nevertheless, useful to assume that one exists. Such a hypothetical antenna is called an *isotropic radiator*.

As you might expect, the solid pattern of an isotropic radiator is a sphere, since the field strength is the same in all directions. In any plane containing the isotropic antenna (which may be considered as a point in space) the pattern is a circle with the antenna at its center. From a theoretical point of view, the isotropic radiator is useful as a comparison standard.

By contrast, the solid pattern of a *dipole antenna* resembles a doughnut

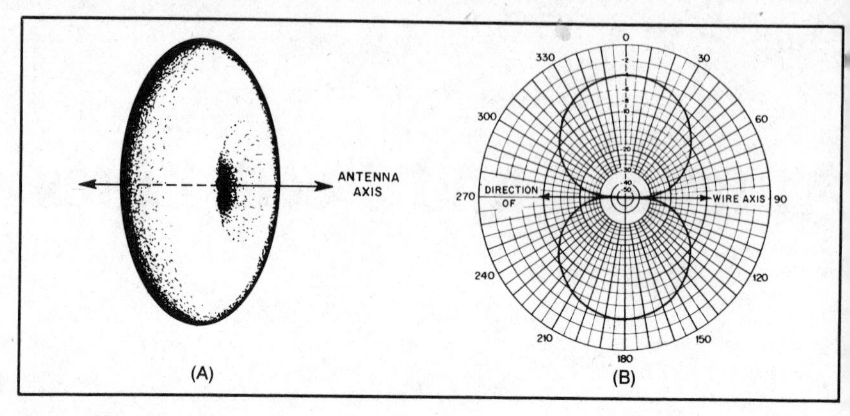

Fig. 9-1—At A, the solid (three dimensional) directive pattern of a dipole. At B, the plane directive diagram of a dipole. The solid line shows the direction of the wire. All values on the pattern shown at B have been reduced by 4 dB, for comparison with the pattern shown in Fig. 9-2.

(Fig. 9-1). Radiation in the main lobe of a dipole is 2.14 dB greater than would be expected from an isotropic radiator (assuming the same power to the antennas). While the isotropic radiator is a handy mathematical tool, the dipole is a simple antenna that can be constructed and tested on an antenna range. For that reason, the dipole is also used as a comparison standard.

So we have a situation where there are two standard references: one for the theoretical case, and one for the physical universe. To convert from one reference to the other is quite simple. Just remember that a dipole has 2.14-dB gain over an isotropic radiator.

$$\text{dBd} = \text{dBi} - 2.14 \quad \text{(Eq. 9-1)}$$

$$\text{dBi} = \text{dBd} + 2.14 \quad \text{(Eq. 9-2)}$$

where

dBd is antenna gain over a reference dipole
dBi is antenna gain over an isotropic radiator

For example, a three-element Yagi might have 7 dB more gain than a dipole. In other words it has a gain of 7 dBd, or 9.14 dBi.

[Now study FCC examination questions with numbers that begin 4BI-2. Review this section as needed.]

ANTENNAS FOR SPACE RADIO COMMUNICATIONS

Chapter 2 covers the use of Amateur Radio satellites. Here we consider those special requirements for antennas used in space radio communications. But first, let's quickly review some antenna fundamentals.

Why Antennas Have Gain

Practical antennas radiate power unequally in the various directions. A dipole in free space has a radiation pattern that resembles a doughnut, as shown in Fig. 9-1A. In Fig. 9-1B you will find the directive pattern for a half-wave dipole as it is commonly drawn.

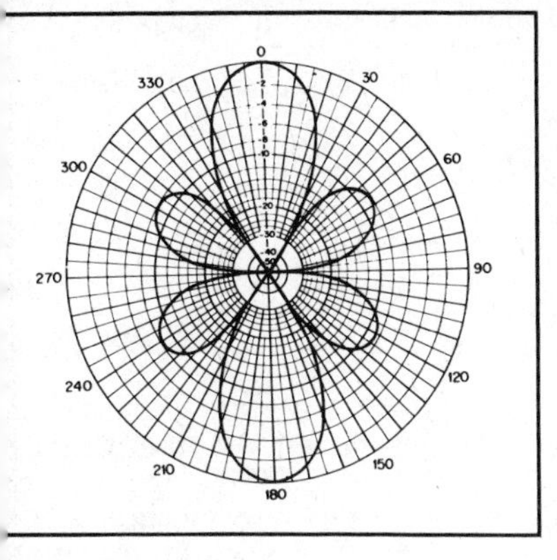

ig. 9-2—Directive pattern for an extended ouble Zepp.

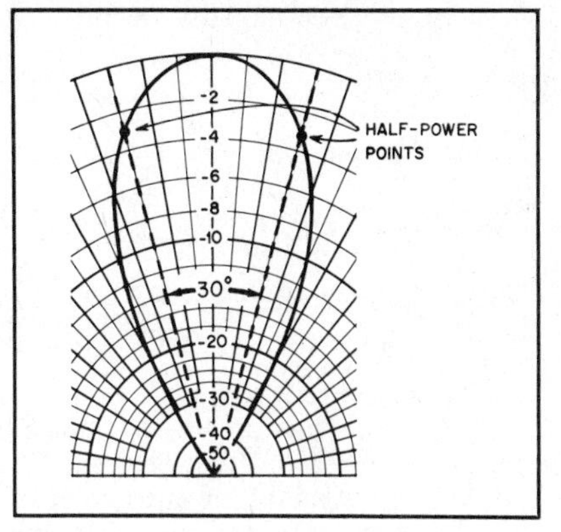

Fig. 9-3—The width of a beam is the angular distance between the directions at which the received or transmitted power is one half the maximum power (−3 dB).

If each leg of the dipole was lengthened from 0.25 to 0.64 wavelength, the antenna would have the directive pattern shown in Fig. 9-2. That antenna is called an extended double Zepp. The Zepp concentrates more power in the favored directions, giving it a gain of approximately 3 dB as compared to a dipole. That advantage holds for ransmitting and receiving.

Compare the directive patterns for the dipole and the extended double Zepp. Can you see that the radiation from the Zepp is more tightly focused? That is how gain is realized in an antenna. The more tightly the radiation is focused, the greater he antenna gain.

To determine the *beamwidth* of an antenna, first locate the point of maximum signal on the pattern. Then look to either side of that maximum point, and find the wo spots where the field intensity is 3 dB less than the maximum. These points represent a signal that has half the power of the maximum, so they are often called "half-power points." Fig. 9-3 is an example of a lobe having a beamwidth of 30 degrees.

Gain and Antenna Size

"The larger the antenna (in wavelengths) the greater the gain" is a good guideline. For a properly designed Yagi, that means the longer the boom, the greater the gain.

The gain of a *parabolic*, or "dish," *antenna* is directly proportional to the square of the dish diameter, and directly proportional to the square of the frequency. That means a 6-dB increase in gain will result if either the reflector diameter or the operating frequency is doubled.

How Much Gain is Required?

You might think more gain is better. That may be true when you are transmitting in a fixed direction. If you are trying to communicate through a rapidly moving satellite, however, it is not true. The sharper pattern of a higher-gain antenna can cause aiming problems, so a smaller gain with a wider beamwidth may be more to your advantage. You will want to find the "right" compromise between gain and beamwidth.

Effective isotropic radiated power (EIRP) is the standard measure of power radiation for space communications. It takes into account the antenna gain and the amount of transmitter power available at the antenna. EIRP is measured in dBW (decibels referenced to 1 watt), and an isotropic radiator is used for comparison. For example, a station has a 13.2-dBi-gain antenna, 2.2-dB feed-line loss and 10-W transmitter output power. You should remember that a decibel is the logarithm of a power ratio.

$$dB = 10 \log \left(\frac{P2}{P1}\right) \quad \text{(Eq. 9-3)}$$

For our example, the reference power is 1 W, so P1 = 1 W.

$$dB = 10 \log \left(\frac{10\ W}{1\ W}\right) = 10 \times 1 = 10\ dBW$$

The EIRP of our station is the sum of all of these factors.

$$EIRP = +13.2\ dBi - 2.2\ dB + 10\ dBW = +21\ dBW$$

The same EIRP could be obtained at a station with a 10.3-dBi-gain antenna, 2.3-dB feed-line loss and 20-W transmitter output. Using Eq. 9-3 to express the transmitter power in dBW, we get +13 dBW. The advantage of the station in the second example is that the antenna beamwidth is not as sharp, so aiming is easier. There are many combinations of antenna gain and transmitter power that will work for space communications, as long as you meet the minimum EIRP requirements.

Several factors determine the minimum EIRP you'll need for your earth station. To start with, there is the satellite. The spacecraft altitude and the types of antennas on board are important factors. The frequency in use determines the path loss; the higher the frequency the greater the loss. For receiving, your receiver sensitivity must be considered; transmitter output power should be considered when transmitting to a satellite. The same factors apply for telecommand (remote control) except that higher signal-to-noise ratios are usually desirable. Once these factors are known, minimum antenna gain can be calculated.

How to Calculate Gain

If the radiation beam is well defined (not much power in minor lobes), then there is an approximate formula relating the antenna gain to the measured half-power beamwidth of the horizontal- and vertical-plane radiation patterns. This formula uses a lossless isotropic radiator as the gain reference. The formula is

$$\text{Gain Ratio} = \frac{41{,}253}{\theta_E \times \theta_H} \quad \text{(Eq. 9-4)}$$

where θ_E and θ_H are the half-power beamwidths, in degrees, of the E- and H-plane patterns, respectively. Gain refers to a gain ratio, and is not expressed in decibels. The gain reference can be changed to that of an actual dipole, rather than for a lossless isotropic radiator, by changing the constant 41,253 to 25,000. Thus, the formula becomes

$$\text{Gain Ratio} = \frac{25{,}000}{\theta_E \times \theta_H} \quad \text{(Eq. 9-5)}$$

It is not hard to solve this equation for beamwidth, as long as we make some simplyifying assumptions. If the pattern is symmetrical around the E and H planes, we can set the two angles equal. This gives a θ^2 term in the denominator, so it is basic algebra to solve for beamwidth:

$$\text{Gain Ratio} = \frac{41{,}253}{\theta^2} \quad \text{(Eq. 9-6)}$$

$$\text{Beamwidth} = \sqrt{\frac{41{,}253}{\text{Gain Ratio}}} = \frac{203}{\sqrt{\text{Gain Ratio}}} \quad \text{(Eq. 9-7)}$$

where the half-power beamwidth is in degrees and the gain is a ratio (not expressed in decibels). Again, this equation is an approximation for an antenna with symmetrical horizontal- and vertical-plane patterns and no significant minor lobes. From this equation, it is easy to see that as antenna gain increases, the beamwidth becomes narrower.

If we are given gain in dB, we must first convert to a ratio. Equation 9-3 tells us:

$$\text{dB} = 10 \log \left(\frac{\text{P2}}{\text{P1}}\right)$$

So if we divide both sides by 10 we have

$$\text{dB}/10 = \log (\text{ratio})$$

To get rid of the logarithm we take the antilog of both sides. The antilog is the "inverse logarithm." The log of 20 is 1.3, so the antilog of 1.3 is 20. We can also think of the antilog of a number as 10 raised to the number ($10^{1.3} = 20$). So if we take the antilog of both sides of our equation, we have

$$10^{(\text{dB}/10)} = \text{Gain Ratio}$$

This means that if we know the gain of an antenna is 20 dB, we can solve for the ratio:

$$\text{Gain ratio} = 10^{(20/10)} = 10^2 = 100$$

And we use Equation 9-7 for the beamwidth:

$$\text{Beamwidth} = \frac{203}{\sqrt{\text{Gain Ratio}}}$$

$$\text{Beamwidth} = \frac{203}{\sqrt{100}}$$

$$\text{Beamwidth} = \frac{203}{10}$$

$$\text{Beamwidth} = 20.3 \text{ degrees}$$

What About Polarization?

Best results in space radio communication are obtained not by using *horizontal* or *vertical polarization*, but by using a combination of the two called *circular polarization*. In horizontal and vertical polarization, the electric field builds to a maximum in one direction, subsides to zero, builds to a maximum in the opposite direction, subsides to zero and continues as shown in Fig. 9-4A and 9-4B. Vertical polarization means the electric field is oriented perpendicular to the earth and horizontal polarization means the electric field is oriented parallel to the earth. When two equal signals, one horizontally polarized and one vertically polarized, are combined with a phase difference of 90°, the result is a circularly polarized wave. See Fig. 9-4C. A circularly polarized antenna can be constructed from two dipoles or Yagis mounted at 90° with respect to each other and fed 90° out of phase.

[Before proceeding to the next section, turn to Chapter 10 and study FCC examination questions with numbers that begin 4BI-1. Review this section as needed.]

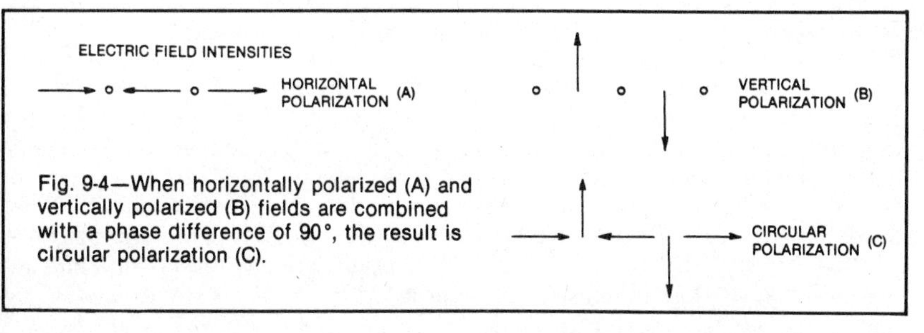

Fig. 9-4—When horizontally polarized (A) and vertically polarized (B) fields are combined with a phase difference of 90°, the result is circular polarization (C).

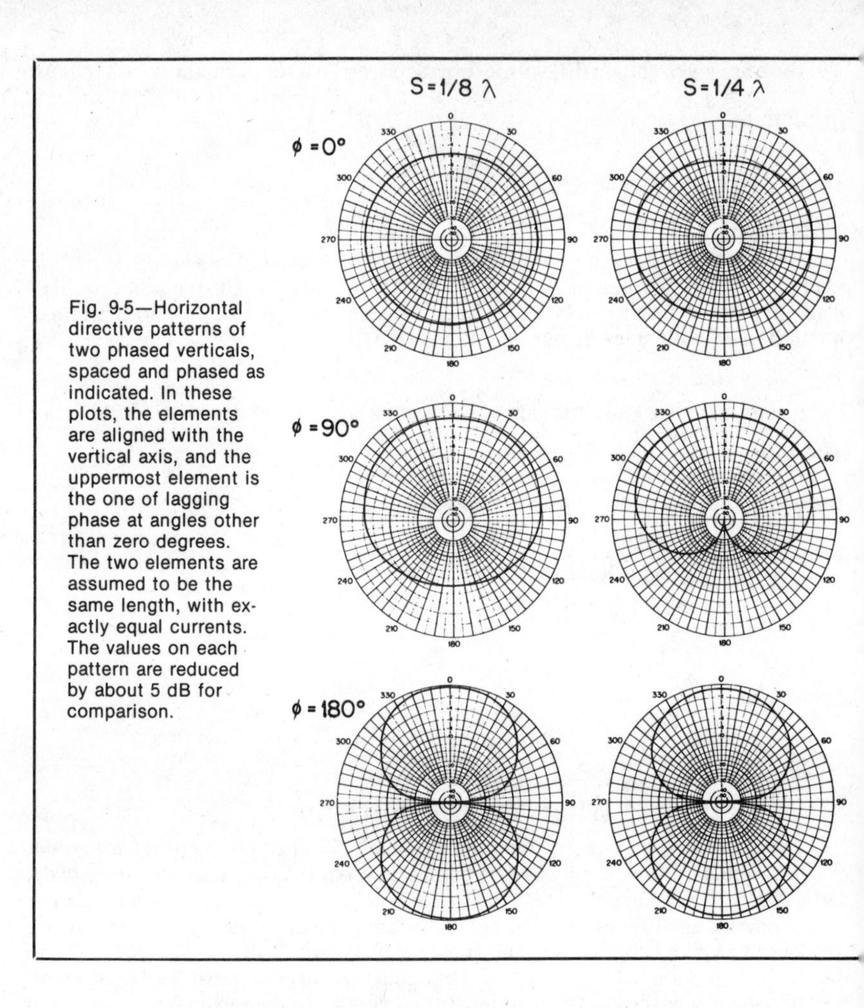

Fig. 9-5—Horizontal directive patterns of two phased verticals, spaced and phased as indicated. In these plots, the elements are aligned with the vertical axis, and the uppermost element is the one of lagging phase at angles other than zero degrees. The two elements are assumed to be the same length, with exactly equal currents. The values on each pattern are reduced by about 5 dB for comparison.

PHASED VERTICAL ANTENNAS

For your Extra Class exam, you will need to know the various pattern shapes that can be obtained using two vertical antennas. Fig. 9-5 shows patterns for a number of common spacings. The patterns assume that the antennas are identical and are fed with equal current. You should be able to predict patterns at other spacings after studying the general trends illustrated in Fig. 9-5. The two antennas are arranged along the vertical axis for each of those figures. The antenna toward the top of the pattern has the lagging excitation.

By studying the patterns shown in Fig. 9-5 you can see that when the two antennas are fed in phase, a pattern that is broadside to the elements always results. At spacings of less than 5/8 wavelength, with the elements fed 180° out of phase, the maximum radiation lobe is in line with the antennas. With intermediate amounts of phase difference, the results cannot be so simply stated. Patterns evolve that are not symmetrical in all four quadrants. A more complete table of patterns is given in *The ARRL Antenna Book*.

[Now turn to Chapter 10 and study FCC examination questions with numbers that begin 4BI-3. Review this section as needed.]

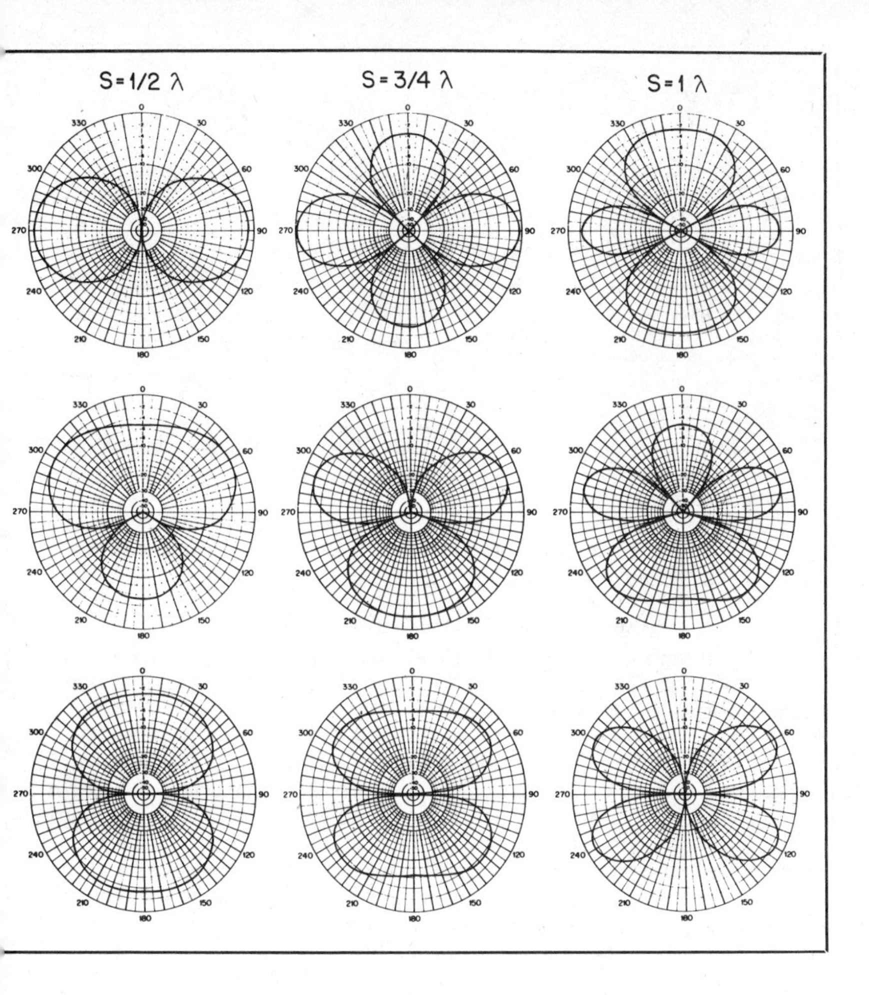

RHOMBIC ANTENNAS

There are two major types of *rhombic antennas: resonant* and *nonresonant*. You will need to know the differences between the two and the advantages and disadvantages of each.

The diamond-shaped rhombic antenna can be looked upon as two acute-angle Vs placed end-to-end; there are four equal-length legs and the opposite angles are equal. Each leg is at least one wavelength long. Rhombics are usually characterized as high-gain antennas. They may be used over a wide frequency range, and the directional pattern is essentially the same over the entire range. This is because a change in frequency causes the major lobe from one leg to shift in one direction, while the lobe from the opposite leg shifts the other way. Like other long-wire antennas the rhombic occupies a large area.

Resonant Rhombic Antennas

The direction of maximum radiation with a resonant rhombic antenna is given by the arrows in Fig. 9-6. As you can see, the antenna is bidirectional, although the

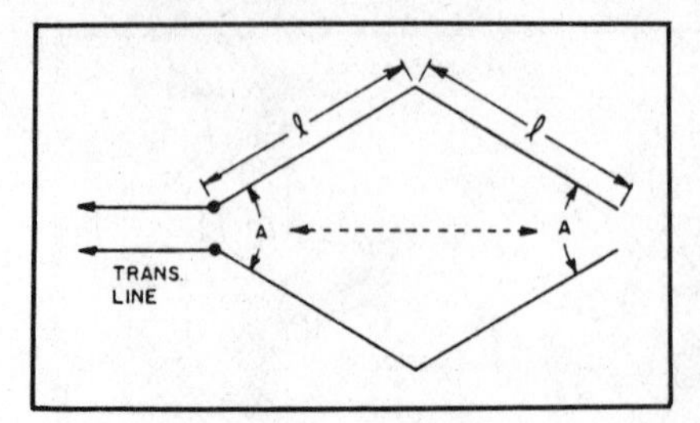

Fig. 9-6—The resonant rhombic is a diamond-shaped antenna. All legs are the same length, and opposite angles of the diamond are equal.

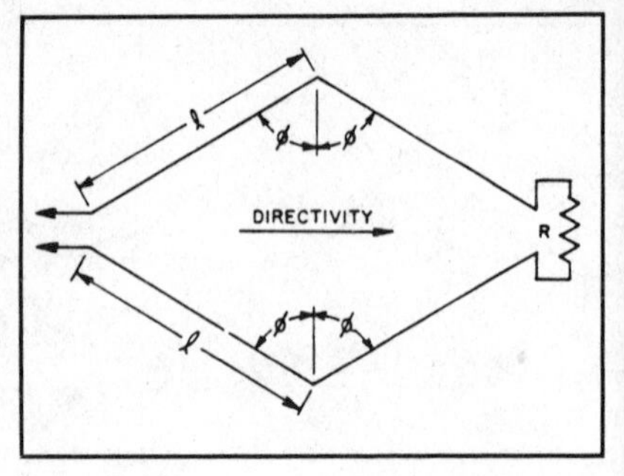

Fig. 9-7—The nonresonant rhombic antenna.

pattern will not be symmetrical. There are minor lobes in other directions; their number and intensity depend on the length of the legs. Notice that the wires at the end opposite the feed point are open.

This antenna has the advantage of simplicity as compared to other rhombics. Input impedance varies considerably with input frequency, as it would with any long-wire antenna. Resonant rhombic antennas are not widely used.

Nonresonant Rhombic Antennas

The most sophisticated type of long-wire antenna is the nonresonant rhombic. As you can see in Fig. 9-7, a terminating resistor has been added to the resonant rhombic to form the nonresonant version.

Although there is no marked difference in the gain obtainable with resonant and nonresonant rhombics of comparable design, the nonresonant antenna has the advantage that it presents an essentially resistive and constant load over a wide frequency range. In addition, nonresonant operation makes the antenna substantially unidirectional, while the unterminated or resonant rhombic is always bidirectional. In a sense, you can consider the power to be dissipated in the terminating resistor as simply power that would have been radiated in the other direction had the resistor not been there, so the fact that some of the power (about one-third) is used up in heating the resistor does not mean an actual loss in the desired direction.

The characteristic impedance of an ordinary nonresonant rhombic antenna, looking into the input end, is on the order of 700 to 800 ohms when properly terminated in a resistance at the far end. The required terminating resistance is usually slightly higher than the input impedance. The correct value normally will be about 800 ohms; for most work this value will ensure that the operation will not be far from optimum.

A nonreactive resistor should be used to terminate a rhombic antenna. Wire-wound resistors are not suitable because they have far too much inductance and distributed capacitance. To allow a safety factor, the total rated power dissipation of the resistor or resistors should be equal to half the power output of the transmitter.

[Study FCC examination questions with numbers that begin 4BI-4. Review this section as needed.]

MATCHING ANTENNAS TO FEED LINES

There are many techniques for matching transmission lines to antennas. The Extra Class syllabus includes delta, gamma and stub matching, so you will have to be familiar with those techniques.

The Delta Match

An amateur faces a problem when trying to feed a half-wave dipole with an open-wire feed line. The center impedance of the dipole is too low to be matched directly by any practical type of air-insulated parallel-conductor line. However it is possible to find, between two points, a value of impedance that can be matched to such a line when a "fanned" section or *delta match* is used to couple the line and antenna. This principle is illustrated in Fig. 9-8.

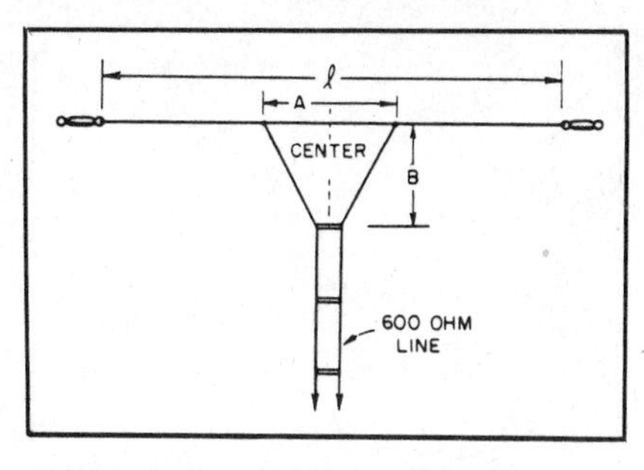

Fig. 9-8—The delta matching system.

When the proper dimensions are unknown, the delta match is awkward to adjust because both the length and width of the delta must be varied. An additional disadvantage is that there is always some radiation from the delta. This is because the conductors are not close enough together to meet the requirement (for negligible radiation) that the spacing should be very small in comparison with the wavelength.

The Gamma Match

A commonly used method for matching a coaxial feed line to the driven element of a parasitic array is the *gamma match*. Shown in Fig. 9-9, the gamma match has considerable flexibility in impedance matching ratio. Because this match is inherently unbalanced, no balun is needed.

Electrically speaking, the gamma conductor and the associated antenna conductor can be looked upon as a section of transmission line shorted at the end. Since it is shorter than ¼ wavelength, it has inductive reactance; this means that if the antenna itself is exactly resonant at the operating frequency, the input impedance of the gamma will show inductive reactance as well as resistance. Any reactance must be tuned out if a good match to the transmission line is to be realized. This can be done in two ways. The antenna can be shortened to obtain a value of capacitive reactance that will reflect through the matching system to cancel the inductive reactance at the input terminals, or a capacitance of the proper value can be inserted in series at the input terminals as shown in Fig. 9-9.

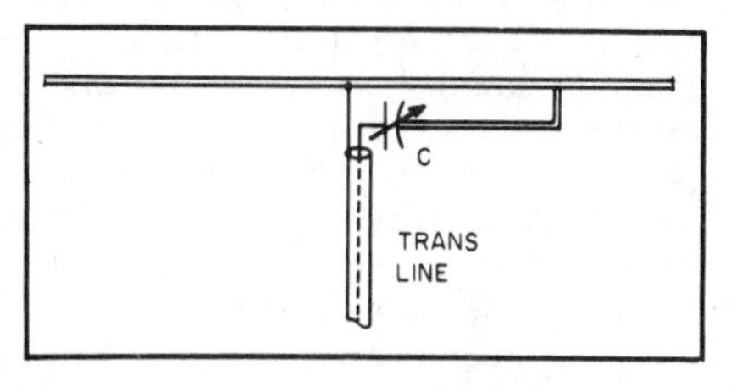

Fig. 9-9—The gamma matching system. As a rule, the gamma-matching capacitor value should be approximately 7 pF per meter of wavelength at the operating frequency (about 70 pF for a 10-meter antenna).

Gamma matches have been widely used for matching coaxial cable to all-metal parasitic beams for a number of years. Because this technique is well suited to "plumber's delight" construction, in which all the metal parts are electrically and mechanically connected, it has become quite popular for amateur arrays.

Because of the many variable factors—driven-element length, gamma rod length, rod diameter, spacing between rod and driven element, and value of series capacitance—a number of combinations will provide the desired match. The maximum capacitance value should be approximately 7 pF per meter of wavelength. For example, 140 pF would be the right value for 20-meter operation. A more detailed discussion of the gamma match can be found in *The ARRL Antenna Book*.

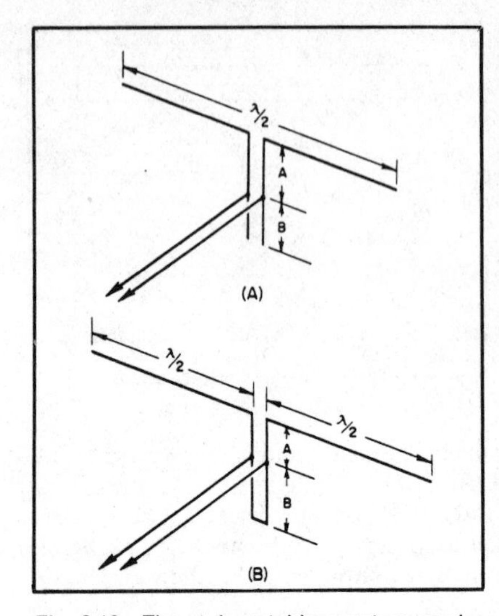

Fig. 9-10—The stub matching system used with common antenna types.

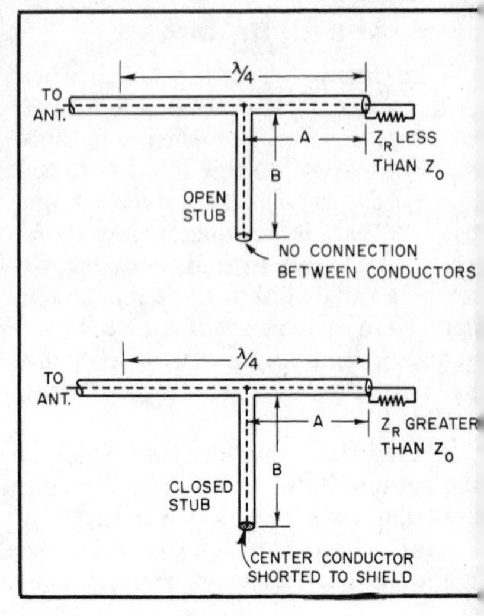

Fig. 9-11—Open and closed stubs used for matching to coaxial lines.

The Stub Match

In some cases, it is possible to match a transmission line and antenna by con necting an appropriate reactance in parallel with them. Reactances formed from sec tions of transmission line are called *matching stubs*. Those stubs are designated eithe as open or closed, depending on whether the free end is open or short circuited. Im pedance matching can be obtained by connecting the feed line at an appropriate poin along the matching stub, as shown in Fig. 9-10. The system illustrated here is sometime referred to as the universal stub system.

Matching stubs have the advantage that they can be used even when the loa is considerably reactive. That is a particularly useful characteristic when the antenn is not a multiple of a quarter-wavelength long, as in the case of a 5/8-wavelengt radiator.

A stub match can be used with coaxial cable, as illustrated in Fig. 9-11. In practical installation, the junction of the transmission line and stub would be T connector.

[Turn to Chapter 10 at this time, and study FCC examination questions wit numbers that begin 4BI-5. Review this section as needed.]

PROPERTIES OF OPEN AND SHORTED FEED-LINE SECTIONS

For this section you will need to study Figs. 9-12 and 9-13. In those diagram you will see the relationships between impedance, voltage and current illustrated fo open and shorted sections of various lengths of feed line. The impedance "seen" looking into various lengths of line are indicated directly above the chart. Curve above the axis marked with R, X_L and X_C indicate the relative value of the im pedance presented at the input. Circuit symbols indicate the equivalent circuits fo the lines at that particular length. Standing waves of voltage (E) and current (I) ar

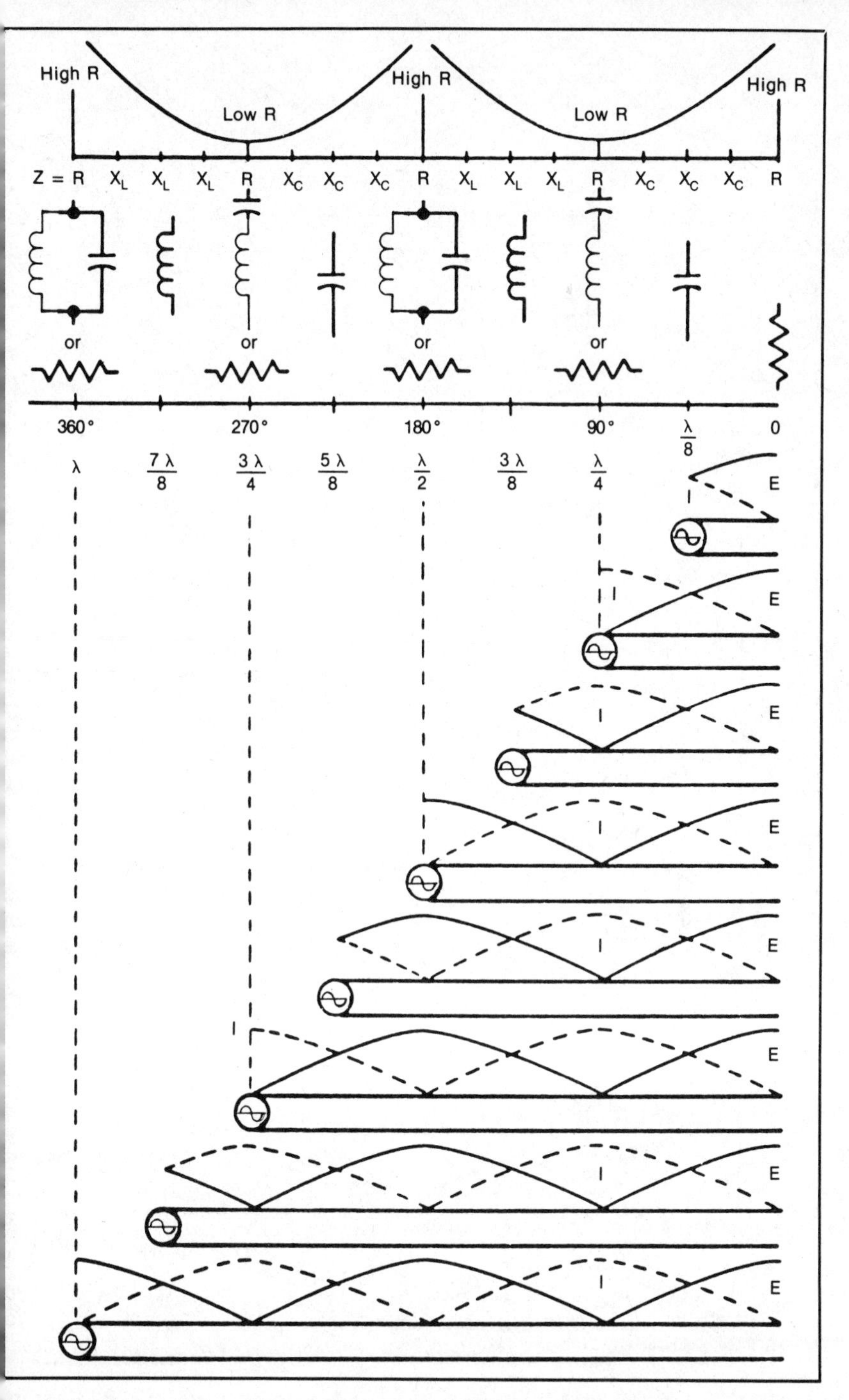

Fig. 9-12—Characteristics of open-ended transmission lines. Voltage standing waves are shown as solid lines above each length of cable, and current standing waves are shown as dashed lines.

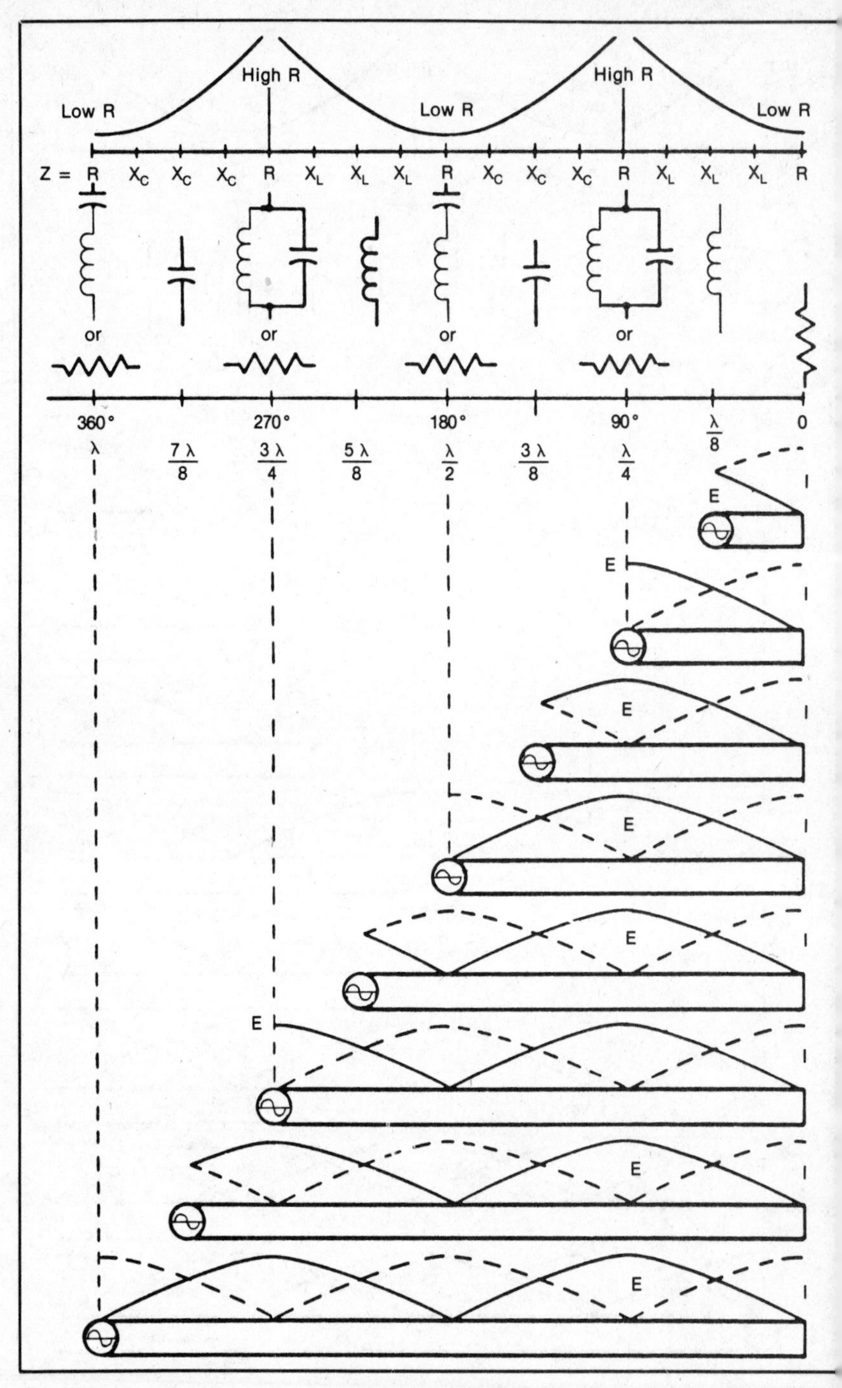

Fig. 9-13—Characteristics of short-circuited transmission lines. Voltage standing waves are shown as solid lines above each length of cable, and current standing waves are shown as dashed lines.

hown above each line. Remember that Z = E / I, by Ohm's Law, so you can use he curves above each piece of line to estimate the input impedance of a given line ngth.

Current is maximum and impedance is minimum at all odd quarter-wavelength oints (¼, ¾ and so on) as measured from the input end along an open line. At ll even quarter-wavelength points (1/2, 1, 3/2, and so on), the voltage is maximum nd the impedance is maximum. At points in between quarter-wavelength marks, he impedance is either capacitive or inductive, as shown near the top of Fig. 9-12 nd 9-13. For example, between 0 and ¼ wavelength the impedance is capacitive. t exactly the 1/8-wavelength point, the impedance is capacitive and is equal to the haracteristic impedance of the line. Between ¼ and ½ wavelength the impedance s inductive. At exactly the 3/8-wavelength point, the impedance is inductive and equal o the characteristic impedance of the line. As you can see on the graph, the impedance lternates between capacitive and inductive values.

Shorted lines can be considered in a similar manner. See Fig. 9-13. You can find nore information on transmission lines in *The ARRL Antenna Book*.

[Now complete your study of this chapter by turning to Chapter 10 and studyng FCC examination questions with numbers that begin 4BI-6. Review this section s needed. That completes your study of the technical material for your Extra Class xamination. Be sure you have also studied the FCC Rules and Regulations and can nswer all of the questions in section 4BA of Chapter 10.]

Chapter 10

Element 4B Question Pool

Don't Start Here!

Before you read the questions and answers printed in this chapter, be sure to read the appropriate text in the previous chapters. Use these questions as review exercises, when suggested in the text. You should not attempt to memorize all of the questions and answers. The material presented in this book has been carefully written and scientifically prepared to guide you step by step through the learning process. By understanding the electronics principles and Amateur Radio concepts as they are presented, your insight into our hobby and your appreciation for the privileges granted by each license class will be greatly enhanced.

This chapter contains the complete Question Pool for the Amateur Extra exam, Element 4B. The FCC specifies the minimum number of questions for an Element 4B exam, and also specifies that a certain percentage of the questions from each subelement must appear on the exam. Most VECs are now giving 40-question Element 4B exams, and based on that number, there must be eight questions from the Rules and Regulations section, subelement 4AB, four questions from the Operating Procedures section, subelement 4BB, and so on. The number of questions to be selected from each section is printed at the beginning of each subelement, and is summarized in Table 10-1.

The FCC now allows Volunteer-Examiner Teams to select the questions that will be used on amateur exams. If your test is coordinated by the ARRL/VEC, however, your test will be prepared by the VEC. The multiple-choice answers and distractors printed here, along with the answer key, have been carefully prepared and evaluated by the ARRL staff and volunteers in the Field Organization of the League. Most Volunteer-Examiner Coordinators, including the ARRL/VEC, agreed to use the question pool printed here until November 1988. If your exam is coordinated by the ARRL/VEC or one of the other VECs using this pool, the questions and multiple-choice answers on your test will appear just as they do in this book. Some VECs use the questions printed here with different answers; check with the VEC coordinating your test session.

Table 10-1
Extra Class Exam Content

Subelement	4BA	4BB	4BC	4BD	4BE	4BF	4BG	4BH	4BI
Number of Questions	8	4	2	4	6	4	4	4	4

SUBELEMENT 4BA—Commission's Rules (8 questions)

4BA-1.1 What exclusive frequency privileges in the 80 meter band are authorized to Amateur Extra control operators?

A. 3525-3775 kHz
B. 3500-3525 kHz
C. 3700-3750 kHz
D. 3500-3550 kHz

4BA-1.2 What exclusive frequency privileges in the 75 meter band are authorized to Amateur Extra control operators?

A. 3750-3775 kHz
B. 3800-3850 kHz
C. 3775-3800 kHz
D. 3800-3825 kHz

4BA-1.3 What exclusive frequency privileges in the 40 meter band are authorized to Amateur Extra control operators?

A. 7000-7025 kHz
B. 7000-7050 kHz
C. 7025-7050 kHz
D. 7100-7150 kHz

4BA-1.4 What exclusive frequency privileges in the 20 meter band are authorized to Amateur Extra control operators?

A. 14.100-14.175 MHz and 14.150-14.175 MHz
B. 14.000-14.125 MHz and 14.250-14.300 MHz
C. 14.025-14.050 MHz and 14.100-14.150 MHz
D. 14.000-14.025 MHz and 14.150-14.175 MHz

4BA-1.5 What exclusive frequency privileges in the 15 meter band are authorized to Amateur Extra control operators?

A. 21.000-21.200 MHz and 21.250-21.270 MHz
B. 21.050-21.100 MHz and 21.150-21.175 MHz
C. 21.000-21.025 MHz and 21.200-21.225 MHz
D. 21.000-21.025 MHz and 21.250-21.275 MHz

4BA-1.6 What frequencies are available to amateur radio operators in th 75 meter band in I.T.U. Region 3?

A. 3.6 - 3.65 MHz
B. 3.5 - 3.9 MHz
C. 3.85 - 3.9 MHz
D. 3.75 - 3.9 MHz

4BA-1.7 What frequencies are available to amateur radio operators in th 40 meter band in I.T.U. Region 3?

A. 7.0 - 7.1 MHz
B. 7.025 - 7.1 MHz
C. 7.05 - 7.15 MHz
D. 7.1 - 7.2 MHz

4BA-1.8 What are the purposes of the Amateur-Satellite Service?

A. It is a radionavigation service using stations on earth satellites for the same purposes as those of the Amateur Radio Service
B. It is a radiocommunication service using stations on earth satellites for weather information
C. It is a radiocommunication service using stations on earth satellites for the same purpose as those of the Amateur Radio Service
D. It is a radiolocation service using stations on earth satellites for amateur radio operators engaged in satellite radar experimentation

4BA-2.1 Which amateur stations are eligible for space operation?

A. Any licensee except Novice
B. General, Advanced and Extra class licensees only
C. Advanced and Extra class licensees only
D. Amateur Extra class licensees only

4BA-2.2 Which amateur stations are eligible for earth operation?

A. Any amateur radio station
B. Amateur Extra class licensees only
C. Any licensee except Novice
D. A special license issued by the FCC is required before any amateur radio station is placed in earth operation

4BA-2.3 Which amateur stations are eligible for telecommand operation?

A. Any amateur radio licensee except Novice
B. Amateur Extra class licensees only
C. Telecommand operation is not permitted in the amateur satellite service
D. Any amateur radio station designated by the space station licensee

4BA-2.4 What are the requirements for space operation?

A. The station must be licensed to a holder of any class FCC amateur license except Novice; must restrict emission to frequencies specified by FCC for space operation; must have proper notifications on file with FCC; and must be able to effect cessation of emissions at any time
B. The station must be licensed to an Amateur Extra Class license-holder; must restrict emission to frequencies specified by FCC for space operation; must have proper notification on file with FCC; and must be able to effect cessation of emissions at any time
C. The station must be licensed to a holder of an Amateur Extra Class license, must operate only on amateur bands with the type of emission appropriate to those bands, and must be able to effect cessation of emissions at any time
D. The station must be licensed to a holder of any class amateur license except Novice or Technician, must restrict emissions to frequencies specified by FCC for space operation, and must be able to effect cessation of emissions at any time

4BA-2.5 What frequencies are available for space operation?
A. 7.0-7.1, 14.00-14.25, 21.00-21.45, 24.890-24.990, 28.00-29.70, 144-146, 435-438 and 24,000-24,050 MHz
B. 7.0-7.3, 14.00-14.35, 21.00-21.45, 28.00-29.70, 144-146, 432-438 and 24,000-24,050 MHz
C. All frequencies available to the Amateur Radio Service, providing license-class, power and emission-type restrictions are observed
D. Only frequencies available to Amateur Extra Class licensees

4BA-2.6 When must the licensee of a station scheduled for space operation give the FCC written pre-space notification?
A. 3 months to 72 hours prior to initiating space operation
B. 6 months to 3 months prior to initiating space operation
C. 12 months to 3 months prior to initiating space operation
D. 27 months to 3 months prior to initiating space operation

4BA-2.7 In what time period is the licensee of a station in space operation required to give the FCC written in-space notification?
A. No later than 24 hours following initiation of space operation
B. No later than 72 hours following initiation of space operation
C. No later than 7 days following initiation of space operation
D. No later than 30 days following initiation of space operation

4BA-2.8 In what time period is the licensee of a station in space operation required to give the FCC written post-space notification?
A. No later than 48 hours after termination is complete, under normal circumstances
B. No later than 72 hours after termination is complete, under normal circumstances
C. No later than 7 days after termination is complete, under normal circumstances
D. No later than 3 months after termination is complete, under normal circumstances

4BA-3.1 How much must the mean power of any spurious emission or radiation from an amateur transmitter be attenuated when the carrier frequency is below 30-MHz?
A. For mean power equal to or greater than 5 watts, at least 30 dB below the mean power of the fundamental, and less than 25 mW
B. For mean power equal to or greater than 5 watts, at least 40 dB below the mean power of the fundamental, and less than 50 mW
C. For mean power equal to or greater than 5 watts, at least 30 dB below the mean power of the fundamental, and less than 50 mW
D. For mean power equal to or greater than 5 watts, at least 40 dB below the mean power of the fundamental, and less than 25 mW

4BA-3.2 How much must the mean power of any spurious emission or radiation from an amateur transmitter be attenuated when the carrier frequency is above 30-MHz?

A. At least 30 dB below mean power of the fundamental, if mean power output is greater than 25 watts and frequency is below 225 MHz
B. At least 40 dB below mean power of the fundamental, if mean power is greater than 25 watts and frequency is below 225 MHz
C. At least 50 dB below mean power of the fundamental, if mean power is greater than 25 watts and frequency is below 225 MHz
D. At least 60 dB below mean power of the fundamental, if mean power is greater than 25 watts and frequency is below 225 MHz

4BA-3.3 What is a spurious emission or radiation?

A. As defined by Section 97.73, any emission or radiation falling outside the amateur band being used
B. As defined by Section 97.73, any emission or radiation other than the fundamental that exceeds 25 microwatts, regardless of frequency
C. As defined by Section 97.73, any emission or radiation other than the fundamental that exceeds 10 microwatts, regardless of frequency
D. As defined by Section 97.73, any emission or radiation falling outside the amateur band that exceeds 25 microwatts

4BA-3.4 What can the FCC require the licensee to do if any spurious radiation from an amateur station causes harmful interference to the reception of another radio station?

A. Reduce the spurious emissions to 0 dB below the fundamental
B. Observe quiet hours and pay a fine
C. Forfeit the station license and pay a fine
D. Eliminate or reduce the interference

4BA-4.1 What are the special conditions an amateur station operated aboard a ship or aircraft must comply with?

A. Approval by master of the ship or captain of the aircraft; legal and safe installation of amateur radio equipment, which must be separate from all other radio equipment aboard and not interfere with the operation of that equipment
B. Approval by master of the ship or captain of the aircraft; legal and safe installation of amateur radio equipment which must be integrated with other radio equipment aboard and must not interfere with the operation of other radio equipment aboard
C. Approval by master of the ship or captain of the aircraft; FCC marine endorsement on Amateur license; legal and safe installation of amateur radio equipment, which must be separate from all other radio equipment aboard and must not interfere with that equipment
D. Approval by master of the ship or captain of the aircraft; FCC marine endorsement on amateur license; legal and safe installation of amateur radio equipment so that it does not interfere with other radio equipment aboard

4BA-4.2 What types of licenses or permits are required for an amateur station to transmit from a vessel registered in the United States?

A. No amateur license is required outside of international waters
B. Any amateur radio license or reciprocal permit issued by the FCC
C. Only amateur licensees General Class or above may transmit on a vessel registered in the US
D. Only an Extra Class licensee may operate aboard a vessel registered in the US

4BA-5.1 What is RACES?

A. An amateur radio network for providing emergency communications during long-distance athletic contests
B. The Radio Amateur Civil Emergency Service
C. The Radio Amateur Corps for Engineering Services
D. An amateur radio network providing emergency communications for transoceanic boat races

4BA-5.2 What is the purpose of RACES?

A. To provide civil-defense communications during emergencies
B. To provide emergency communications for transoceanic boat or aircraft races
C. To provide routine and emergency communications for long-distance athletic events
D. To provide routine and emergency communications for large-scale international events, such as the Olympic games

4BA-5.3 Which amateur stations may be operated in RACES?

A. Only Extra Class amateur radio stations
B. Any licensed amateur radio station except a station licensed to a Novice
C. Any licensed amateur radio station certified by the responsible civil defense organization
D. Any licensed amateur radio station other than a station licensed to a Novice, providing the station is certified by the responsible Civil Defense organization

4BA-5.4 What persons may be control operators of RACES stations?

A. Anyone who holds a valid FCC amateur operator's license other than Novice
B. Only Amateur Extra class licensees
C. Anyone who holds an FCC amateur radio license other than Novice and is certified by a civil defense organization
D. Anyone who holds an FCC amateur radio license and is certified by a Civil Defense organization

4BA-5.5 What are the points of communications for amateur stations operated in RACES and certified by the responsible civil defense organization as registered with that organization?

A. Any RACES, Civil Defense, or Disaster Communications Service station
B. Any RACES stations and any FCC licensed amateur stations except stations licensed to Novices
C. Any FCC licensed amateur station or a station in the Disaster Communications Service
D. Any FCC licensed amateur station except stations licensed to Novices

4BA-5.6 What are permissible communications in RACES?

A. Any communications concerning local traffic nets
B. Any communications concerning the Amateur Radio Emergency Service
C. Any communications concerning national defense and security or immediate safety of people and property that are authorized by the area Civil Defense organization
D. Any communications concerning national defense or security or immediate safety of people or property but only when a state of emergency has been declared by the President, the governor, or other authorized official, and then only so long as the state of emergency endures

4BA-6.1 What are the points of communication for an amateur station?

A. Other amateur stations only
B. Other amateur stations and other station authorized by the FCC to communicate with amateurs
C. Other amateur stations and stations in the Personal Radio Service
D. Other amateur stations and stations in the Aviation or Private Land Mobile Radio Services

4BA-6.2 With which stations may an amateur station communicate?

A. Amateur, RACES, FCC Monitoring stations
B. Amateur stations and any other station authorized by the FCC to communicate with amateur stations
C. Amateur stations only
D. Amateur stations and US Government stations

4BA-6.3 Under what circumstances, if any, may an amateur station communicate with a non-amateur station?

A. Only during emergencies and when the Commission has authorized that station to communicate with amateur stations
B. Under no circumstances
C. Only when the state governor has authorized that station to communicate with amateurs
D. Only during Public Service events in connection with REACT groups

4BA-7.1 How can a person who is not a citizen of the United States obtain an FCC amateur license?

A. The person cannot obtain an amateur license unless he/she is a citizen of the United States
B. The person can take an amateur license examination if he/she is not an agent of a foreign government
C. The person can take an amateur license examination if he/she has the proper visas
D. The person can take an amateur license examination only if he/she is a permanent resident of the United States

4BA-7.2 What is a Reciprocal Operating Permit?

A. An FCC authorization to a holder of an amateur license issued by certain foreign governments to operate an amateur radio station in the United States and its possessions
B. An FCC permit to allow a United States licensed amateur to operate his station in a foreign nation, except Canada
C. An FCC permit allowing a foreign licensed amateur to handle traffic between the United States and his own nation, subject to FCC rules on traffic-handling and third-party messages
D. An FCC permit to a commercial telecommunications company allowing that company to pay amateurs to handle traffic during emergencies

4BA-7-3 Who is eligible for a Reciprocal Operating Permit?

A. Anyone holding a valid amateur radio license issued by a foreign government
B. Anyone holding a valid amateur radio license issued by a foreign government with which the United States has a reciprocal operating agreement, providing that person is not a United States citizen
C. Anyone who holds a valid amateur radio license issued by a foreign government with which the United States has a reciprocal operating agreement
D. Anyone other than a United States citizen who holds a valid amateur radio or shortwave listeners license issued by a foreign government

4BA-7.4 Under what circumstances, if any, is a citizen of the United States eligible to obtain a Reciprocal Operating Permit?

A. A US Citizen is not eligible to obtain a Reciprocal Operating Permit for use in the United States
B. Only if the applicant brings his or her equipment with them to the foreign country
C. Only if that person is unable to qualify for a United States amateur license
D. If the applicant is unlicensed at the time of application, only if he or she held a United States amateur license other than Novice class less than 10 years before the time of application

4BA-7.5 Under what circumstances, if any, is a person who is not a citizen of the country which issued the foreign amateur radio license eligible for a Reciprocal Operating Permit?

A. Only if the applicant has been living for at least one continuous year in the country whose government issued the license, and is not a United States citizen
B. Only if the applicant's spouse is a citizen of the nation whose government issued the license
C. Only if the applicant's spouse is a United States citizen
D. Only if the applicant holds a valid amateur operator and station license issued by the foreign government and is not a United States citizen

4BA-7.6 What are the operator frequency privileges authorized by a Reciprocal Operating Permit?

A. Those authorized to a holder of the equivalent United States amateur license, unless the FCC specifies otherwise by endorsement on the permit
B. Those that the holder of the Reciprocal Operating Permit would have if he were in his own country
C. Only those frequencies permitted to United States amateurs that the holder of the Reciprocal Operating Permit would have in his own country, unless the FCC specifies otherwise
D. Only those frequencies approved by the International Amateur Radio Union, unless the FCC specifies otherwise

4BA-7.7 How does the alien amateur identify his/her station when operating under a Reciprocal Operating Permit?

A. By using only his own call
B. By using his own call, followed by the city and state in the United States or possessions closest to his present location
C. By using his own call, followed by the letter(s) and number indicating the United States call-letter district of his location at the time of the contact, with the city and state nearest his location specified once during each contact
D. By using his own call sign, followed by the serial number of his Reciprocal Operating Permit and the call-letter district number of his present location

4BA-8.1 What is a Volunteer-Examiner Coordinator?

A. A person authorized to administer FCC amateur radio license examinations to candidates for the Novice license
B. A person, club, or organization authorized to administer FCC amateur radio examinations for any class of license other than Novice
C. A club or organization that has entered into an agreement with the FCC to coordinate the efforts of volunteer examiners in preparing and administering exams for amateur radio operator licenses
D. A club or organization that has entered into an agreement with the FCC to coordinate efforts of volunteer examiners in preparing and administering examinations for amateur radio operator licenses other than Novice

4BA-8.2 What are the requirements to be a VEC?

A. Be engaged in the manufacture and/or sale of amateur equipment or in the coordination of amateur activities throughout at least one call letter district, and agree to abide by FCC Rules concerning administration of amateur radio examinations
B. Be organized at least partially for the purpose of furthering amateur radio, be at least regional in scope, and agree to abide by FCC Rules concerning administration of amateur radio examinations
C. Be organized at least partially for the purpose of furthering amateur radio, be at the most, county-wide in scope, and agree to abide by FCC Rules concerning administration of amateur radio examinations
D. Be engaged in a business related to amateur radio and agree to administer amateur radio examinations in accordance with FCC Rules throughout at least one call letter district

4BA-8.3 What are the functions of a VEC?

A. Accredit volunteer examiners; assemble, print, and distribute written examinations; collect candidates' application forms, answer sheets and test results and forward the applications to the FCC; maintain pools of questions for Amateur Radio examinations; and perform other clerical tasks in accordance with FCC Rules

B. Assemble, print and sell FCC-approved examination forms, accredit volunteer examiners, collect candidates' application forms, answer sheets and test results and forward the applications to the FCC, screen applications for completeness and authenticity, and perform other clerical tasks in accordance with FCC Rules

C. Accredit volunteer examiners, certify that examiners' equipment is type-accepted by the FCC, assemble, print and distribute FCC-approved examination forms, collect candidates' application forms, answer sheets and test results and forward the applications to the FCC, and perform other clerical tasks in accordance with FCC Rules

D. Maintain pools of questions for Amateur Radio examinations; administer code and theory examinations, score and forward them to the FCC along with the candidates' application forms so that the appropriate license may be issued to each successful candidate

4BA-9.1 What is Examination Element 1(A)?

A. The Novice theory examination
B. The Extra 20 WPM code test
C. The Novice 5 WPM code test
D. The Novice examination on Rules and Regulations

4BA-9.2 What is Examination Element 1(B)?

A. The Novice theory and regulations examination
B. The Technician/General theory and regulations examination
C. The Novice 5 WPM code test
D. The General l3 WPM code test

4BA-9.3 What is Examination Element 1(C)?

A. A test of the applicant's ability to comprehend Morse code sent at a rate of 20 WPM
B. A test of the applicant's knowledge of elementary principles of electronics and FCC regulations pertaining to amateur radio
C. A test of the applicant's ability to comprehend Morse code sent at a rate of 13 WPM
D. A test of the applicant's ability to comprehend Morse code sent at a rate of 5 WPM

4BA-9.4 What is Examination Element 2?

A. A test of an applicant's ability to comprehend Morse code sent at a rate of 13 WPM
B. A test of an applicant's knowledge of general amateur practice and provisions of treaties, statutes and rules affecting amateur stations and operators.
C. A test of an applicant's knowledge of basic radio theory and FCC rules essential to beginner's operation.
D. A test of an applicant's ability to comprehend Morse code sent at a rate of 5 WPM

4BA-9.5 What are Examination Elements 3A and 3B?

A. Tests of an applicant's knowledge of propagation characteristics, advanced principles of electronics, and FCC Regulations pertaining to Amateur Radio
B. Tests of an applicant's knowledge of general amateur practice and provisions of treaties, statutes and rules affecting amateur stations and operators
C. Tests of an applicant's knowledge of basic amateur practice, advanced electronic theory and ITU Regulations pertaining to Amateur Radio
D. Tests of an applicant's knowledge of general amateur practice and DOC Regulations pertaining to Amateur Radio

4BA-9.6 What is Examination Element 4(A)?

A. A test of an applicant's knowledge of advanced radio theory and operation as it relates to modern amateur techniques.
B. A test on international regulations given to US applicants who plan to operate in certain foreign countries for three months or longer.
C. A test of an applicant's knowledge of intermediate level radio theory and operation as it relates to modern amateur techniques.
D. A test of an applicant's ability to comprehend Morse code sent at a rate of 13 WPM

4BA-9.7 What is Examination Element 4(B)?

A. A test of an applicant's ability to comprehend Morse code sent at a rate of 20 WPM
B. A test of an applicant's knowledge of advanced radio theory and operation as it relates to modern amateur techniques
C. A test of an applicant's ability to comprehend Morse code sent at a rate of 13 WPM
D. A test of an applicant's knowledge of general amateur practice and provisions of treaties, statutes, and rules affecting amateur stations and operators

4BA-10.1 What examination elements are required for an Amateur Extra operator license?

A. 1C and 4B
B. 3B, 4A and 4B
C. 1B, 2, 3A, 3B, 4A and 4B
D. 1C, 2, 3A, 3B, 4A and 4B

4BA-10.2 What examination elements are required for an Advanced operator license?

A. 1A, 2, 3A, 3B and 4A
B. 1B, 3A and 3B
C. 1B and 4A
D. 1B, 2, 3A, 3B and 4A

4BA-10.3 What examination elements are required for a General operator license?

A. 1B, 2, 3A and 3B
B. 1A, 2, 3A and 3B
C. 1A, 3A and 3B
D. 1B, 3A and 3B

4BA-10.4 What examination elements are required for a Technician operator license?
A. 1A and 2B
B. 1A and 3A
C. 1A, 2 and 3A
D. 2 and 3A

4BA-11.1 What examination credit must be given to an applicant who holds a valid Novice operator license?
A. Credit for successful completion of elements 1A and 2
B. Credit for successful completion of elements 1B and 3A
C. Credit for successful completion of elements 1B and 2
D. Credit for successful completion of elements 1A and 3A

4BA-11.2 What examination credit must be given to an applicant who holds a valid Technician operator license?
A. Credit for successful completion of elements 1A and 2
B. Credit for successful completion of elements 1A, 2 and 3A
C. Credit for successful completion of elements 1B, 2 and 3A
D. Credit for successful completion of elements 1B and 3A

4BA-11.3 What examination credit must be given to an applicant who holds a valid General operator license?
A. Credit for successful completion of elements 1B, 2, 3A, 3B and 4A
B. Credit for successful completion of elements 1A, 3A, 3B and 4A
C. Credit for successful completion of elements 1A, 2, 3A, 3B and 4B
D. Credit for successful completion of elements 1B, 2, 3A and 3B

4BA-11.4 What examination credit must be given to an applicant who holds a valid Advanced operator license?
A. Credit for successful completion of element 4A
B. Credit for successful completion of elements 1B and 4A
C. Credit for successful completion of elements 1B, 2, 3A, 3B and 4A
D. Credit for successful completion of elements 1C, 3A, 3B, 4A and 4B

4BA-11.5 What examination credit, if any, may be given to an applicant who holds a valid amateur operator license issued by another country?
A. Credit for successful completion of any elements that may be identical to those required for US licensees
B. No credit
C. Credit for successful completion of elements 1A, 1B and 1C
D. Credit for successful completion of elements 2, 3A, 3B, 4A and 4B

4BA-11.6 What examination credit, if any, may be given to an applicant who holds a valid amateur operator license issued by any other United States government agency than the FCC?
A. No credit
B. Credit for successful completion of elements 1A, 1B and 3C
C. Credit for successful completion of elements 4A and 4B
D. Credit for successful completion of element 1C

4BA-11.7 What examination credit must be given to an applicant who holds a valid FCC commercial radiotelegraph license?

A. No credit
B. Credit for successful completion of element 1B only
C. Credit for successful completion of elements 1A, 1B and 1C
D. Credit for successful completion of element 1A only

4BA-11.8 What examination credit must be given to the holder of a valid Certificate of Successful Completion of Examination?

A. Credit for previously completed written examination elements only
B. Credit for the code speed associated with the previously completed telegraphy examination elements only
C. Credit for previously completed written and telegraphy examination elements only
D. Credit for previously completed commercial examination elements only

4BA-12.1 Who determines where and when examinations for amateur operator licenses are to be administered?

A. The FCC
B. The Section Manager
C. The applicants
D. The Volunteer Examiner Team

4BA-12.2 Where must the examiner(s) be and what must the examiner(s) be doing during an examination?

A. The examiner(s) must be present and observing the candidate throughout the entire examination
B. The examiner(s) must be absent to allow the candidate to complete the entire examination in accordance with the traditional honor system
C. The examiner(s) must be present to observe the candidate throughout the administration of telegraphy examination elements only
D. The examiner(s) must be present to observe the candidate throughout the administration of written examination elements only

4BA-12.3 Who is responsible for the proper conduct and necessary supervision during an examination?

A. The VEC
B. The FCC
C. The accredited volunteer examiners
D. The candidates and the accredited volunteer examiners

4BA-12.4 What should an examiner do when a candidate fails to comply with the examiner's instructions?

A. Warn the candidate that his/her continued failure to comply with the examiner's instructions will result in termination of the examination
B. Immediately terminate the examination
C. Allow the candidate to complete the examination, but refuse to issue a certificate of successful completion of examination for any elements passed by fraudulent means
D. Immediately terminate the examination and report the violation to federal law enforcement officials

4BA-12.5 What must the candidate do at the completion of the examination?

A. Complete a brief written evaluation of the examination session
B. Return all test papers to the examiners
C. Return all test papers to the examiners and wait for them to be graded before leaving the examination site
D. Pay the $4.00 registration fee

4BA-12.6 How long must an applicant wait to retake an examination element he/she has failed?

A. 10 days
B. 15 days
C. 30 days
D. There is no FCC requirement that an applicant wait to retake an examination element that he/she has failed

4BA-13.1 Who must prepare Examination Element 1(B)?

A. Extra class licensees serving as volunteer examiners, or Volunteer Examiner Coordinators
B. Advanced class licensees serving as volunteer examiners, or Volunteer Examiner Coordinators
C. The FCC
D. The Field Operations Bureau

4BA-13.2 Who must prepare Examination Element 1(C)?

A. The FCC
B. The Field Operations Bureau
C. Advanced class licensees serving as Volunteer Examiners, or Volunteer Examiner Coordinators
D. Extra class licensees serving as Volunteer Examiners, or Volunteer Examiner Coordinators

4BA-13.3 Who must prepare Examination Elements 3A and 3B?

A. Advanced or Extra class licensees serving as Volunteer Examiners, or Volunteer Examiner Coordinators
B. The FCC or Volunteer Examiner Coordinators
C. The Field Operations Bureau
D. Advanced or General class licensees serving as Volunteer Examiners, or Volunteer Examiner Coordinators

4BA-13.4 Who must prepare Examination Element 4(A)?

A. Advanced or Extra class licensees serving as Volunteer Examiners, or Volunteer Examiner Coordinators
B. The FCC or Volunteer Examiner Coordinators
C. The Field Operations Bureau
D. Extra class licensees serving as Volunteer Examiners, or Volunteer Examiner Coordinators

4BA-13.5 Who must prepare Examination Element 4(B)?

A. Advanced or Extra class licensees serving as Volunteer Examiners, or Volunteer Examiner Coordinators
B. The FCC or Volunteer Examiner Coordinators
C. The Field Operations Bureau
D. Extra class licensees serving as Volunteer Examiners, or Volunteer Examiner Coordinators

BA-13.6 Where are the questions listed that must be used in written examinations?
A. In the appropriate VEC question pool
B. In PR Bulletin 1035C
C. In PL 97-259
D. In the appropriate FCC Report and Order

BA-14.1 When must the test papers be graded?
A. Within 5 days of completion of an Examination Element
B. Within 30 days of completion of an Examination Element
C. Immediately upon completion of an Examination Element
D. Within ten days of completion of an Examination Element

BA-14.2 Who must grade the test papers?
A. The ARRL
B. The Volunteer Examiners
C. The Volunteer Examiner Coordinator
D. The FCC

BA-14.3 How does the examiner(s) inform a candidate who does not score a passing grade?
A. Give the percentage of the questions answered correctly and return the application to the candidate
B. Give the percentage of the questions answered incorrectly and return the application to the candidate
C. Tell the candidate that he/she failed
D. Show how the incorrect answers should have been answered and give a copy of the corrected answer sheet to the candidate

4BA-14.4 What must the examiner(s) do when the candidate scores a passing grade?
A. Give the percentage of the questions answered correctly and return the application to the candidate
B. Tell the candidate that he/she passed
C. Issue the candidate an operator license
D. Issue the candidate a Certificate of Successful Completion of Examination for the exam elements

4BA-14.5 How long after the administration of a successful examination for a Technician, General, Advanced or Amateur Extra operator license must the examiners submit the candidate's application and test papers?
A. Within 10 days
B. Within 15 days
C. Within 30 days
D. Within 90 days

4BA-14.6 To whom do the examiners submit successful candidates' applications and test papers?
A. To the candidate
B. To the VEC
C. To the local radio club
D. To the regional Section Manager

4BA-15.1 What are the requirements for a volunteer examiner administering an examination for a Technician operator license?

A. The Volunteer Examiner must be a Novice class licensee accredited by a Volunteer Examiner Coordinator
B. The Volunteer Examiner must be an Advanced or Extra class licensee accredited by a Volunteer Examiner Coordinator
C. The Volunteer Examiner must be an Extra Class licensee accredited by a Volunteer Examiner Coordinator
D. The Volunteer Examiner must be a General class licensee accredited by a Volunteer Examiner Coordinator

4BA-15.2 What are the requirements for a volunteer examiner administering an examination for a General operator license?

A. The examiner must hold an Advanced class license and be accredited by a VEC
B. The examiner must hold an Extra Class license and be accredited by a VEC
C. The examiner must hold a General class license and be accredited by a VEC
D. The examiner must hold an Extra class license to administer the written test element, but an Advanced class examiner may administer the CW test element

4BA-15.3 What are the requirements for a volunteer examiner administering an examination for an Advanced operator license?

A. The examiner must hold an Advanced class license and be accredited by a VEC
B. The examiner must hold an Extra Class license and be accredited by a VEC
C. The examiner must hold a General class license and be accredited by a VEC
D. The examiner must hold an Extra class license to administer the written test element, but an Advanced class examiner may administer the CW test element

4BA-15.4 What are the requirements for a volunteer examiner administering an examination for an Amateur Extra operator license?

A. The examiner must hold an Advanced class license and be accredited by a VEC
B. The examiner must hold an Extra Class license and be accredited by a VEC
C. The examiner must hold a General class license and be accredited by a VEC
D. The examiner must hold an Extra class license to administer the written test element, but an Advanced class examiner may administer the CW test element

4BA-16.1 When an applicant passes an examination to upgrade his/her operator license, under what authority may he/she be the control operator of an amateur station with the privileges of the higher operator class?

A. The Certificate of Successful Completion of Examination
B. That of the ARRL
C. Already licensed applicants in the Amateur Radio Service may not use their newly earned privileges until they receive their permanent amateur station and operator license
D. Applicants may only use their newly earned privileges during emergencies pending issuance of their permanent amateur station and operator license

4BA-16.2 What is a Certificate of Successful Completion of Examination?

A. A document printed by the FCC
B. A document required for already licensed applicants operating with privileges of an amateur operator class higher than that of their permanent amateur operator license
C. A document a candidate may use for an indefinite period of time to receive credit for successful completion of any written element
D. A permanent amateur radio station and operator license certificate issued to a newly-upgraded licensee by the FCC within 90 days of the completion of the examination

4BA-16.3 How long may a successful applicant operate his/her station under Section 97.35 with the rights and privileges of the higher operator class for which the applicant has passed the appropriate examinations?

A. 30 days or until issuance of a permanent operator and station license, whichever comes first
B. 3 months or until issuance of the permanent operator and station license, whichever comes first
C. 6 months or until issuance of the permanent operator and station license, whichever comes first
D. 1 year or until issuance of the permanent operator and station license, whichever comes first

4BA-16.4 How must the station call sign be amended when operating under the temporary authority authorized by Section 97.35?

A. The applicant must use an identifier code as a prefix to his/her present call sign, e.g., when using voice; "interim AE KA1MJP"
B. The applicant must use an identifier code as a suffix to his/her present call sign, e.g., when using voice; "KA1MJP temporary AE"
C. By adding after the call sign, when using voice, the phrase "operating temporary Technician, General, Advanced or Extra"
D. By adding to the call sign, when using CW, the slant bar followed by the letters T, G, A or E

4BA-17.1 Who may reimburse VEs and VECs for out-of-pocket expenses incurred in preparing, processing, or administering examinations?

A. Examinees
B. FCC
C. ARRL
D. FCC and Examiners

4BA-17.2 What is the maximum amount of reimbursement fee from any examinee for any one examination at a particular session?

A. $5.00 as of 1987
B. $4.37 as of 1987
C. $3.00 as of 1987
D. $0.00 as of 1987

4BA-17.3 What action must a VEC take against a VE who accepts reimbursement and fails to provide the annual expense certification?

A. Suspend the VE's accreditation for 1 year
B. Discredit the VE
C. Suspend the VE's accreditation and report the information to the FCC
D. Suspend the VE's accreditation for 6 months

4BA-17.4 What type of expense records must be maintained by a VE who accepts reimbursement?

A. All out-of-pocket expenses and reimbursements from the examinees
B. All out-of-pocket expenses only
C. Reimbursements from examiners only
D. FCC reimbursements only

4BA-17.5 For what period of time must a VE maintain records of out-of-pocket expenses and reimbursements for each examination session for which he/she accepts reimbursement?

A. 1 year
B. 2 years
C. 3 years
D. 4 years

4BA-17.6 By what date each year must a VE forward to the VEC a certification concerning expenses for which reimbursement was accepted?

A. December 15 of each year
B. January 15 of each year
C. April 15 of each year
D. October 15 of each year

4BA-17.7 For what type of services may a VE be reimbursed for out-of-pocket expenses?

A. Preparing, processing or administering examinations above the Novice class
B. Preparing, processing or administering examinations including the Novice class
C. A VE cannot be reimbursed for out-of-pocket expenses
D. Only for preparation of examination elements

4BA-18.1 What is VE accreditation?

A. The process by which all Advanced and Extra Class licensees are automatically given permission to conduct Amateur Radio examinations
B. The process by which the FCC tests volunteers who wish to coordinate Amateur Radio license examinations
C. The process by which the prospective VE requests his/her requirements for accreditation
D. The process by which each VEC makes sure its VEs meet FCC requirements to serve as Volunteer Examiners

4BA-18.2 What are the requirements for VE accreditation?

A. Advanced class license or higher, 18 years old, not have any conflict of interest and never had his/her license suspended or revoked
B. Advanced class license or higher, 16 years old, and not have any conflict of interest
C. Extra Class license or higher, 18 years old and be a member of ARRL
D. There are no requirements for accreditation, other than a General class or higher license

4BA-18.3 The services of which persons seeking to be VEs will not be accepted by the FCC?

A. Persons with Advanced class licenses
B. Persons being between 18 and 21 years of age
C. Persons who have had their licenses suspended or revoked
D. Persons who are employees of the Federal Government

4BA-18.4 When is VE accreditation necessary?

A. Always in order to administer a Technician or higher class license examination
B. Always in order to administer a Novice or higher class license examination
C. Sometimes in order to administer an Advanced or higher class license examination
D. VE accreditation is not necessary in order to administer a General or higher class license examination

4BA-19.1 Under what circumstances, if any, may a person be compensated for services as a VE?

A. When the VE spends more than 4 hours at the test session
B. When the VE loses a day's pay to give the exam
C. When the VE spends many hours preparing for the test session
D. Under no circumstances

4BA-19.2 How much money, if any, may a person accept for services as a VE?

A. None
B. Up to a half day's pay if the VE spends more than 4 hours at the test session
C. Up to a full day's pay if the VE spends more than 4 hours preparing for the test session
D. Up to $50 if the VE spends more than 4 hours at the test session

4BA-20.1 Under what circumstance, if any, may an organization engaged in the manufacture of equipment used in connection with amateur radio transmissions be a VEC?

A. Under no circumstances
B. If the organization's amateur-related sales are very small
C. If the organization is manufacturing very specialized amateur equipment
D. Only upon FCC approval that preventive measures have been taken to preclude any possible conflict of interest

4BA-20.2 Under what circumstances, if any, may a person who is an employee of a company which is engaged in the distribution of equipment used in connection with amateur radio transmissions be a VE?

A. Under no circumstances
B. Only if the employee's work is not directly related to that part of the company involved in the manufacture or distribution of amateur equipment
C. Only if the employee has no financial interest in the company
D. Only if the employee is an Extra Class licensee

4BA-20.3 Under what circumstances, if any, may a person who owns a significant interest in a company which is engaged in the preparation of publications used in preparation for obtaining an amateur operator license be a VE?

A. Under no circumstances
B. Only if the organization's amateur related sales are very small
C. Only if the organization is publishing very specialized material
D. Only if the person is an Extra Class licensee

4BA-20.4 Under what circumstance, if any, may an organization engaged in the distribution of publications used in preparation for obtaining an amateur operator license be a VEC?

A. Under no circumstances
B. Only if the organization's amateur publishing business is very small
C. Only if the organization is selling the publication at cost to examinees
D. Only upon FCC approval that preventive measures have been taken to preclude any possible conflict of interest

4BA-21.1 Under what circumstances, if any, may a VEC discriminate in accrediting a person as a VE on the basis of race?

A. Under no circumstances
B. Only when the prospective VE is less than 21 years of age
C. Only if the VEC also files the proper statement with the Human Rights Commission
D. Only if the prospective VE also files the proper statement with the FCC

4BA-21.2 Under what circumstances, if any, may a VEC discriminate in accrediting a person as a VE on the basis of sex?

A. Under no circumstances
B. Only when the VEC conducts exams in only one state
C. Only when the exams are being held on private property
D. Only when the prospective VE less than 21 years old

4BA-21.3 Under what circumstances, if any, may a VEC discriminate in accrediting a person as a VE on the basis of national origin?

A. Under no circumstances
B. Only when the prospective VE is not a citizen of the US
C. Only when the prospective VE is from a country with which the US does not have a reciprocal agreement
D. Only when the VEC conducts exams in only one state

BA-21.4 Under what circumstances, if any, may a VEC discriminate in accrediting a person as a VE on the basis of religion?

A. Under no circumstances
B. Only when the VEC conducts exams in only one state
C. Only if the prospective VE is age 21 or younger
D. Only if the VEC files the proper statement with the Human Rights Commission

BA-21.5 Under what circumstances, if any, may a VEC refuse to accredit a person as a VE on the basis of membership in an amateur radio organization?

A. Under no circumstances
B. Only when the prospective VE is an ARRL member
C. Only when the prospective VE is not a member of the local Amateur Radio Club
D. Only when the club is at least regional in scope

BA-21.6 Under what circumstances, if any, may a VEC refuse to accredit a person as a VE on the basis of lack of membership in an amateur radio organization?

A. Under no circumstances
B. Only when the prospective VE is not an ARRL member
C. Only when the club is at least regional in scope
D. Only when the prospective VE is a not a member of the local Amateur Radio club giving the examinations

BA-21.7 Under what circumstances, if any, may a VEC discriminate in accrediting persons on the basis of their declining to accept reimbursement?

A. Under no circumstances
B. Only when the VEC conducts exams in only one state
C. Only if the VE is not a member of the local Amateur Radio Club
D. Only if the VE is not an ARRL member

BA-22.1 Who is responsible for preparing an Element 1(B) examination?

A. Extra Class licensees serving as Volunteer Examiners or Volunteer Examiner Coordinators
B. Advanced Class licensees serving as Volunteer Examiners or Volunteer Examiner Coordinators
C. The FCC
D. The Field Operations Bureau

BA-22.2 What is an Element 1(B) examination intended to prove?

A. The applicant's knowledge of Novice theory and regulations
B. The applicant's knowledge of Technician/General theory and regulations
C. The applicant's ability to send and receive Morse code at 5 WPM
D. The applicant's ability to send and receive Morse code at 13 WPM

4BA-23.1 Who is responsible for preparing an Element 1(C) examination?
A. The FCC
B. The Field Operations Bureau
C. Advanced Class licensees serving as Volunteer Examiners or Volunteer Examiner Coordinators
D. Extra Class licensees serving as Volunteer Examiners or Volunteer Examiner Coordinators

4BA-23.2 What is an Element 1(C) examination intended to prove?
A. The applicant's ability to send and receive Morse code sent at the rate of 20 WPM
B. The applicant's knowledge of elementary principles of electronics and FCC regulations pertaining to Amateur Radio
C. The applicant's ability to send and receive Morse code sent at the rate of 13 WPM
D. The applicant's ability to send and receive Morse code sent at a rate of 5 WPM

4BA-24.1 Who is responsible for the preparation of Element 3A and Element 3B examinations?
A. Advanced or Extra Class licensees serving as Volunteer Examiners, or Volunteer-Examiner Coordinators
B. The FCC or Volunteer-Examiner Coordinators
C. The Field Operations Bureau
D. General Class licensees serving as Volunteer Examiners

4BA-24.2 How are Element 3A and Element 3B examinations prepared?
A. By Advanced or Extra Class Volunteer Examiners or Volunteer-Examiner Coordinators selecting questions from the appropriate VEC question pool
B. By Volunteer-Examiner Coordinators selecting questions from the appropriate FCC bulletin
C. By Extra Class Volunteer Examiners selecting questions from the appropriate FCC bulletin
D. By the FCC selecting questions from the appropriate VEC question pool

4BA-25.1 Who is responsible for the preparation of an Element 4(A) examination?
A. Advanced Class licensees serving as Volunteer Examiners, or Volunteer-Examiner Coordinators
B. The FCC
C. The Field Operations Bureau, or Volunteer Examiner Coordinators
D. Extra Class licensees serving as Volunteer Examiners, or Volunteer-Examiner Coordinators

4BA-25.2 How is an Element 4(A) examination prepared?
A. By Extra Class Volunteer Examiners or Volunteer-Examiner Coordinators selecting questions from the appropriate VEC question pool
B. By Volunteer-Examiner Coordinators selecting questions from the appropriate FCC bulletin
C. By Extra Class Volunteer Examiners selecting questions from the appropriate FCC bulletin
D. By the FCC selecting questions from the appropriate VEC question pool

4BA-26.1 Who is responsible for the preparation of an Element 4(B) examination?
A. Advanced Class licensees serving as Volunteer Examiners, or Volunteer-Examiner Coordinators
B. The FCC
C. The Field Operations Bureau, or Volunteer-Examiner Coordinators
D. Extra Class licensees serving as Volunteer Examiners, or Volunteer-Examiner Coordinators

4BA-26.2 How is an Element 4(B) examination prepared?
A. By Extra Class Volunteer Examiners or Volunteer-Examiner Coordinators selecting questions from the appropriate VEC question pool
B. By Volunteer-Examiner Coordinators selecting questions from the appropriate FCC bulletin
C. By Extra Class Volunteer Examiners selecting questions from the appropriate FCC bulletin
D. By the FCC selecting questions from the appropriate VEC question pool

4BA-27.1 What qualifications must an organization have in order to be a VEC?
A. Be organized for the purpose of furthering the purpose of amateur radio, be at least national in scope, and agree to return any compensation beyond out-of-pocket expenses to the FCC
B. Be organized for the purpose of furthering amateur radio, be at least regional in scope, and agree not to accept any compensation except reimbursement for out-of-pocket expenses
C. Be organized for the purpose of furthering amateur radio, be at least national in scope, and agree not to accept any compensation except reimbursement for out-of-pocket expenses
D. Be organized for the purpose of furthering the purpose of amateur radio, be local in scope, and agree not to accept any compensation except reimbursement for out-of-pocket expenses

4BA-28.1 What is a volunteer examiner?
A. A General class radio amateur who is accredited by a VEC to prepare or administer examinations to applicants for amateur radio licenses
B. An amateur radio operator who is accredited by a VEC to prepare or administer examinations to applicants for amateur radio licenses
C. An amateur radio operator who prepares or administers examinations to applicants for amateur radio licenses for a fee
D. An FCC staff member who tests volunteers who want to give Amateur Radio examinations

4BA-28.2 What is a VE?

A. A General class radio amateur who is accredited by a VEC to prepare or administer examinations to applicants for amateur radio licenses
B. An amateur radio operator who is accredited by a VEC to prepare or administer examinations to applicants for amateur radio licenses
C. An amateur radio operator who prepares or administers examinations to applicants for amateur radio licenses for a fee
D. An FCC staff member who tests volunteers who want to give Amateur Radio examinations

4BA-29.1 What is a volunteer-examiner coordinator?

A. A person authorized to administer FCC license examinations to candidates for Novice licenses
B. A person, club, or organization authorized to administer FCC amateur radio examinations for any class of license other than Novice
C. A club or organization that has entered into a written agreement with the ARRL to coordinate efforts of Volunteer Examiners in preparing and administering examinations for amateur radio operator licenses
D. A club or organization that has entered into a written agreement with the FCC to coordinate efforts of Volunteer Examiners in preparing and administering examinations for amateur radio operator licenses

4BA-29.2 What is a VEC?

A. A person authorized to administer FCC license examinations to candidates for Novice licenses
B. A person, club, or organization authorized to administer FCC amateur radio examinations for any class of license other than Novice
C. A club or organization that has entered into a written agreement with the ARRL to coordinate efforts of Volunteer Examiners in preparing and administering examinations for amateur radio operator licenses
D. A club or organization that has entered into a written agreement with the FCC to coordinate the efforts of Volunteer Examiners in preparing and administering examinations for amateur radio operator licenses

UBELEMENT 4BB—Operating Procedures (4 Questions)

BB-1.1 Why does the downlink frequency appear to vary by several kHz during a low earth orbit amateur satellite pass?
- A. The distance between the satellite and ground station is changing, causing the Kepler effect
- B. The distance between the satellite and ground station is changing, causing the Bernoulli effect
- C. The distance between the satellite and ground station is changing, causing the Boyles law effect
- D. The distance between the satellite and ground station is changing, causing the Doppler effect

BB-1.2 What are the two basic types of linear transponders used in amateur satellites?
- A. The non-inverting type and the inverting type
- B. Geostationary and elliptical
- C. Phase 2 and Phase 3
- D. Amplitude modulated and frequency modulated

BB-1.3 What is a linear transponder?
- A. A repeater that passes only linear or CW signals
- B. A device that receives and retransmits signals of any mode in a certain passband
- C. An amplifier for SSB transmissions
- D. A device used to change FM to SSB

BB-1.4 What is Mode A in an amateur satellite?
- A. Operation through a 10 meter receiver on a satellite that retransmits on 2 meters
- B. The lowest frequency used in Phase 3 transponders
- C. The highest frequency used in Phase 3 translators
- D. Operation through a 2 meter receiver on a satellite that retransmits on 10 meters

BB-1.5 What is Mode B in an amateur satellite?
- A. Operation through a 28 MHz receiver on a satellite that retransmits on 2 meters
- B. Operation through a 70 centimeter receiver on a satellite that retransmits on 2 meters
- C. The beacon output
- D. A codestore device used to record messages

BB-1.6 What is Mode J in an amateur satellite?
- A. Ground transmission on 70 cm, ground reception on 2 meters from the transponder
- B. Ground transmission on 2 meters, ground reception on 70 cm from the transponder
- C. Ground transmission on 2 meters, ground reception on 10 meters from the transponder
- D. Ground transmission on 70 cm, ground reception on 10 meters from the transponder

4BB-1.7 What is Mode L in an amateur satellite?

A. Ground transmission on 70 cm, ground reception on 10 mete from the transponder
B. Ground transmission on 23 cm, ground reception on 70 cm from the transponder
C. Ground transmission on 70 cm, ground reception on 23 cm from the transponder
D. Ground transmission on 10 meters, ground reception on 70 c from the transponder

4BB-1.8 What is an ascending pass for an amateur satellite?

A. A pass from west to east during which the satellite is in rang
B. A pass from east to west during which the satellite is in rang
C. A pass from south to north during which the satellite is in range
D. A pass from north to south during which the satellite is in range

4BB-1.9 What is a descending pass for an amateur satellite?

A. A pass from north to south during which the satellite is in range
B. A pass from west to east during which the satellite is in range
C. A pass from east to west during which the satellite is in range
D. A pass from south to north during which the satellite is in range

4BB-1.10 What is the period of an amateur satellite?

A. An orbital arc that extends from 60 degrees west longitude to 145 degrees west longitude
B. The point on an orbit where satellite height is minimum
C. The amount of time it takes for a satellite to complete one orbit
D. The time it takes a satellite to travel from perigee to apogee

4BB-2.1 How often is a new frame transmitted in a fast-scan television system?

A. 30 times per second
B. 60 times per second
C. 90 times per second
D. 120 times per second

4BB-2.2 How many horizontal lines make up a fast-scan television frame?

A. 30
B. 60
C. 525
D. 1050

4BB-2.3 How is the interlace scanning pattern generated in a fast-scan television system?

A. By scanning the field from top to bottom
B. By scanning the field from bottom to top
C. By scanning even numbered lines in one field and odd numbered ones in the next
D. By scanning from left to right in one field and right to left in the next

BB-2.4 What is blanking in a video signal?

A. Synchronization of the horizontal and vertical sync-pulses
B. Turning off the scanning beam while it is traveling from right to left and from bottom to top
C. Turning off the scanning beam at the conclusion of a transmission
D. Transmitting a black and white test pattern

BB-2.5 What is the standard video voltage level between the sync tip and the whitest white at TV camera outputs and modulator inputs?

A. 1 volt peak-to-peak
B. 120 IEEE units
C. 12 volts DC
D. 5 volts RMS

BB-2.6 What is one way audio is transmitted in an amateur fast-scan television system?

A. It must be sent on the 144 MHz band
B. It has to be sent on an FM subcarrier 5.0 MHz higher than the picture carrier
C. FCC Rules require audio signals to be sent on SSB on an amateur HF band
D. It may be sent on an FM simplex channel on another amateur VHF band

BB-2.7 What is the bandwidth of a fast-scan television transmission?

A. 3 kHz
B. 10 kHz
C. 25 kHz
D. 6 MHz

BB-2.8 What is the standard video level, in percent PEV, for black?

A. 0%
B. 12.5%
C. 70%
D. 100%

BB-2.9 What is the standard video level, in percent PEV, for white?

A. 0%
B. 12.5%
C. 70%
D. 100%

BB-2.10 What is the standard video level, in percent PEV, for blanking?

A. 0%
B. 12.5%
C. 75%
D. 100%

SUBELEMENT 4BC—Radio Wave Propagation (2 Questions)

4BC-1.1 How far apart may stations communicating by moonbounce be separated?
- A. 500 miles maximum, if the moon is at perigee
- B. 2,000 miles maximum, if the moon is at apogee
- C. 5,000 miles maximum, if the moon is at perigee
- D. Any distance as long as the stations have a mutual lunar window

4BC-1.2 What characterizes libration fading of an E-M-E signal?
- A. A slow change in the pitch of the CW signal
- B. A fluttery, rapid irregular fading
- C. A gradual loss of signal as the sun rises
- D. The returning echo is several hertz lower in frequency than the transmitted signal

4BC-1.3 What are the best days to schedule E-M-E contacts?
- A. When the moon is at perigee
- B. When the moon is full
- C. When the moon is at apogee
- D. When the weather at both stations is clear

4BC-1.4 What type of receiving system is required for E-M-E communications?
- A. Equipment capable of reception on 14 MHz
- B. Equipment with very low dynamic range
- C. Equipment with very low gain
- D. Equipment with very low noise figures

4BC-1.5 What type of transmitting system is required for E-M-E communications?
- A. A transmitting system capable of operation on the 21 MHz band
- B. A transmitting system capable of producing maximum legal power
- C. A transmitting system using an unmodulated carrier
- D. A transmitting system with a high second harmonic output

4BC-2.1 When the earth's atmosphere is struck by a meteor, a cylindrical region of free electrons is formed at what layer of the ionosphere
- A. The F1 layer
- B. The E layer
- C. The F2 layer
- D. The D layer

4BC-2.2 Which range of frequencies is best suited for meteor-burst transmissions?
- A. 1.8 - 1.9 MHz
- B. 10 - 14 MHz
- C. 28 - 148 MHz
- D. 220 - 450 MHz

4BC-3.1 What is transequatorial propagation?

A. Propagation between two points at approximately the same distance north and south of the magnetic equator
B. Propagation between two points on the magnetic equator
C. Propagation between two continents by way of ducts along the magnetic equator
D. Propagation between any two stations at the same latitude

4BC-3.2 What is the maximum range for signals using transequatorial propagation?

A. About 1,000 miles
B. About 2,500 miles
C. About 5,000 miles
D. About 7,500 miles

4BC-3.3 What is the best time of day for transequatorial propagation?

A. Morning
B. Noon
C. Afternoon or early evening
D. Trans-equatorial propagation only works at night

SUBELEMENT 4BD—Amateur Radio Practice (4 Questions)

4BD-1.1 What is a spectrum analyzer?
A. A piece of test equipment used to display electrical signals in the frequency domain
B. A test instrument consisting of two RF detectors, one connected to the input of the amplifier and one to the output
C. A piece of test equipment used to display electrical signals in the time domain
D. A piece of test equipment used for determining the maximum usable frequency

4BD-1.2 What type of instrument may be used to observe electrical signals in the frequency domain?
A. An oscilloscope
B. A spectrum analyzer
C. A frequency counter
D. A linearity tracer

4BD-1.3 How does a spectrum analyzer differ from a conventional time-domain oscilloscope?
A. The oscilloscope is used to display electrical signals while the spectrum analyzer is used to measure ionospheric reflection
B. The oscilloscope is used to display electrical signals in the frequency domain while the spectrum analyzer is used to display electrical signals in the time domain
C. The oscilloscope is used to display electrical signals in the time domain while the spectrum analyzer is used to display electrical signals in the frequency domain
D. The oscilloscope is used for displaying audio frequencies and the spectrum analyzer is used for displaying radio frequencies

4BD-1.4 What does the horizontal axis of a spectrum analyzer display?
A. Amplitude
B. Voltage
C. Resonance
D. Frequency

4BD-1.5 What does the vertical axis of a spectrum analyzer display?
A. Amplitude
B. Duration
C. Frequency
D. Time

4BD-2.1 What is a logic probe?
A. A piece of test equipment used to trace AF signals
B. A piece of test equipment used to measure voltages that indicate logic states
C. An instrument used to determine whether a circuit is analog or digital
D. A software tool used to indicate bit error rate

4BD-2.2 How is a logic probe used?
A. To trace AF signals
B. To check high voltage output of a power supply
C. To indicate high or low logic states
D. As a software tool for debugging computer programs

4BD-2.3 What does a logic probe indicate?

A. Electromagnetic interference
B. Emitter-coupled logic
C. Bits per second
D. A one or zero state

4BD-2.4 What advantage does a logic probe have over a voltmeter in monitoring logic states in a circuit?

A. A logic probe has fewer leads to connect to a circuit than a voltmeter
B. A logic probe can be used to test analog and digital circuits
C. A logic probe can be powered by commercial AC lines
D. A logic probe is smaller and shows a simplified readout

4BD-3.1 What is one of the most significant deterrents to effective signal reception at a mobile station?

A. Ignition noise
B. Doppler shift
C. Radar interference
D. Mechanical vibrations

4BD-3.2 What is the proper procedure for suppressing electrical noise in a mobile station?

A. Apply shielding and filtering where necessary
B. Insulate all plane sheet metal surfaces from each other
C. Apply antistatic spray liberally to all non-metallic surfaces
D. Install filter capacitors in series with all DC wiring

4BD-3.3 How can ferrite beads be used to suppress ignition noise?

A. Install them in the resistive high voltage cable every 2 years
B. Install them between the starter solenoid and the starter motor
C. Install them in the primary and secondary ignition leads
D. Install them in the antenna lead to the radio

4BD-3.4 How can metal bonding of a vehicle reduce the level of spark plug noise?

A. It reduces the spark gap distance, causing a lower frequency spark
B. It helps radiate the spark plug noise away from the vehicle
C. It encourages lower frequency electrical resonances in the vehicle
D. It shields the engine compartment

4BD-3.5 What are the main areas of a vehicle to bond in order to reduce electrical noise?

A. All instrument wiring
B. All large metal parts
C. All non-conducting parts
D. All body parts except the engine

4BD-3.6 How can alternator whine be minimized?

A. By connecting the radio's power leads to the battery by the longest possible path
B. By connecting the radio's power leads to the battery by the shortest possible path
C. By installing a high pass filter in series with the radio's DC power lead to the vehicle's electrical system
D. By installing filter capacitors in series with the DC power lead

4BD-3.7 How can conducted noise in a vehicle be minimized?

A. By installing ferrite beads in the output of the catalytic converter
B. By installing filter capacitors in series with the battery cables
C. By shunting all instrument wiring to ground with RF chokes
D. By filtering the alternator leads

4BD-3.8 How can the electrical noise generated by automobile instruments be suppressed?

A. By installing an electrolytic capacitor in series with the signal wire from the sender unit
B. By shunting a low value RF choke from the sender output to ground
C. By installing a coaxial capacitor at the sender element
D. By using resistive wire from the sending unit to the instrument panel

4BD-3.9 How can conducted and radiated noise caused by an automobile alternator be suppressed?

A. By installing filter capacitors in series with the DC power lead and by installing a blocking capacitor in the field lead
B. By connecting the radio's power leads to the battery by the longest possible path and by installing a blocking capacitor in series with the positive lead
C. By installing a high pass filter in series with the radio's power lead to the vehicle's electrical system and by installing a low-pass filter in parallel with the field lead
D. By connecting the radio power leads directly to the battery and by installing coaxial capacitors in the alternator leads

4BD-3.10 What can cause corona-discharge noise?

A. An oversized capacitive loading hat on the tip of the antenna
B. A broad rounded tip on the antenna
C. A sharp point at the tip of the antenna
D. Inductive coupling between the mobile antenna and nearby high voltage power lines

4BD-4.1 What is the main drawback of a wire-loop antenna for direction finding?

A. It has a bidirectional pattern broadside to the loop
B. It is non-rotatable
C. It receives equally well in all directions
D. It is practical for use only on VHF bands

4BD-4.2 What antenna characteristics are desirable for direction finding?
A. A non-cardioid pattern
B. Good front-to-back and front-to-side ratios
C. Good top-to-bottom and front-to-side ratios
D. Shallow nulls

4BD-4.3 What is the triangulation method of direction finding?
A. Using the geometric angle of ground waves and sky waves emanating from the same source to locate the signal source
B. A fixed receiving station uses three beam headings to plot the signal source on a map
C. Beam headings from several receiving stations are used to plot the signal source on a map
D. The use of three vertical antennas to indicate the location of the signal source

4BD-4.4 Why is an RF attenuator desirable in a receiver used for direction finding?
A. It narrows the bandwidth of the received signal
B. It eliminates the effects of isotropic radiation
C. It reduces loss of received signals caused by antenna pattern nulls
D. It prevents receiver overload from extremely strong signals

4BD-4.5 What is a sense antenna?
A. A vertical antenna added to a loop antenna to produce a cardioid reception pattern
B. A horizontal antenna added to a loop antenna to produce a cardioid reception pattern
C. A vertical antenna added to an Adcock antenna to produce an omnidirectional reception pattern
D. A horizontal antenna added to an Adcock antenna to produce a cardioid reception pattern

4BD-4.6 What is an Adcock antenna?
A. An antenna used for ground wave reception in radio direction finding
B. An antenna used for sky wave reception in radio direction finding
C. A bidirectional antenna used for skywave reception in radio direction finding
D. An omnidirectional antenna used in radio direction finding that is immune to skywaves

4BD-4.7 What is a loop antenna?
A. A circularly polarized antenna
B. A wire loop used in tracking sky wave signals
C. A wire loop used in radio direction finding
D. An antenna coupled to the feed line through an inductive loop of wire

4BD-4.8 How can the output voltage of a loop antenna be increased?

A. By reducing the permeability of the loop shield
B. By increasing the number of wire turns in the loop while reducing the area of the loop structure
C. By reducing either the number of wire turns in the loop, or the area of the loop structure
D. By increasing either the number of wire turns in the loop, or the area of the loop structure

4BD-4.9 Why is an antenna system with a cardioid pattern desirable for a direction-finding system?

A. The broad side responses of the cardioid pattern can be aimed at the desired station
B. The deep null of the cardioid pattern can pinpoint the direction of the desired station
C. The sharp peak response of the cardioid pattern can pinpoint the direction of the desired station
D. The high radiation angle of the cardioid pattern is useful for short-distance direction finding

4BD-4.10 What type of terrain can cause errors in direction finding?

A. Homogeneous terrain
B. Smooth grassy terrain
C. Varied terrain
D. Terrain with no buildings or mountains

SUBELEMENT 4BE—Electrical Principles (6 Questions)

4BE-1.1 What is the photoconductive effect?

A. The conversion of photon energy to electromotive energy
B. The increased conductivity of an illuminated semiconductor junction
C. The conversion of electromotive energy to photon energy
D. The decreased conductivity of an illuminated semiconductor junction

4BE-1.2 What happens to photoconductive material when light shines upon it?

A. The conductivity of the material increases
B. The conductivity of the material decreases
C. The conductivity of the material stays the same
D. The conductivity of the material becomes temperature dependent

4BE-1.3 What happens to the resistance of a photoconductive material when light shines upon it?

A. It increases
B. It becomes temperature dependent
C. It stays the same
D. It decreases

4BE-1.4 What happens to the conductivity of a semiconductor junction when it is illuminated?

A. It stays the same
B. It becomes temperature dependent
C. It increases
D. It decreases

4BE-1.5 What is an optocoupler?

A. A resistor and a capacitor
B. A frequency modulated helium-neon laser
C. An amplitude modulated helium-neon laser
D. An LED and a phototransistor

4BE-1.6 What is an optoisolator?

A. An LED and a phototransistor
B. A P-N junction that develops an excess positive charge when exposed to light
C. An LED and a capacitor
D. An LED and a solar cell

4BE-1.7 What is an optical shaft-encoder?

A. An array of optocouplers chopped by a stationary wheel
B. An array of optocouplers whose light transmission path is controlled by a rotating wheel
C. An array of optocouplers whose propagation velocity is controlled by a stationary wheel
D. An array of optocouplers whose propagation velocity is controlled by a rotating wheel

4BE-1.8 What is a solid-state relay?

A. An optoisolator and a resistor
B. Two transistors in a flip-flop configuration
C. An optoisolator and a triac
D. A relay with silicon or germanium contact points

4BE-1.9 What is one of the safest ways to mate solid-state circuits operating at widely differing voltages?

A. With an optoisolator
B. With a silicon controlled rectifier
C. With a resistor
D. With an inductor

4BE-1.10 What does the photoconductive effect in crystalline solids produce a noticeable change in?

A. The capacitance of the solid
B. The inductance of the solid
C. The specific gravity of the solid
D. The resistance of the solid

4BE-2.1 What is the meaning of the term time constant of an RC circuit?

A. The time required to charge the capacitor in the circuit to 36.8% of the supply voltage
B. The time required to charge the capacitor in the circuit to 36.8% of the supply current
C. The time required to charge the capacitor in the circuit to 63.2% of the supply current
D. The time required to charge the capacitor in the circuit to 63.2% of the supply voltage

4BE-2.2 What is the meaning of the term time constant of an RL circuit?

A. The time required for the current in the circuit to build up to 36.8% of the maximum value
B. The time required for the voltage in the circuit to build up to 63.2% of the maximum value
C. The time required for the current in the circuit to build up to 63.2% of the maximum value
D. The time required for the voltage in the circuit to build up to 36.8% of the maximum value

4BE-2.3 What is the term for the time required for the capacitor in an RC circuit to be charged to 63.2% of the supply voltage?

A. An exponential rate of one
B. One time constant
C. One exponential period
D. A time factor of one

4BE-2.4 What is the term for the time required for the current in a RL circuit to build to 63.2% of the maximum value?

A. One time constant
B. An exponential period of one
C. A time factor of one
D. One exponential rate

4BE-2.5 What is the term for the time it takes for a charged capacitor in an RC circuit to discharge to 36.8% of its initial value of stored charge?

A. One discharge period
B. An exponential discharge rate of one
C. A discharge factor of one
D. One time constant

4BE-2.6 What is meant by back EMF?

A. An current equal to the applied EMF
B. An opposing EMF equal to R times C (RC) percent of the applied EMF
C. A current that opposes the applied EMF
D. A voltage that opposes the applied EMF

4BE-2.7 What units are normally used to express the time constant for an RC circuit?

A. 1/seconds
B. seconds
C. ohms
D. henrys

4BE-2.8 What units are normally used to express the time constant for an RL circuit?

A. seconds
B. 1/seconds
C. ohms
D. henrys

4BE-2.9 To what percentage of the supply voltage will a capacitor in an RC circuit be charged after two time constants?

A. 36.8%
B. 63.2%
C. 86.5%
D. 95%

4BE-2.10 To what percentage of the starting voltage will a capacitor in an RC circuit be discharged after two time constants?

A. 86.5%
B. 63.2%
C. 36.8%
D. 13.5%

4BE-3.1 What is the time constant of a circuit having a 100-microfarad capacitor in series with a 470 kilohm resistor?

A. 4700 seconds
B. 470 seconds
C. 47 seconds
D. 0.47 seconds

4BE-3.2 What is the time constant of a circuit having a 220-microfarad capacitor in parallel with a 1 megohm resistor?

A. 220 seconds
B. 22 seconds
C. 2.2 seconds
D. 0.22 seconds

4BE-3.3 What is the time constant of a circuit having two 100-microfarad capacitors and two 470 kilohm resistors all in series?

A. 470 seconds
B. 47 seconds
C. 4.7 seconds
D. 0.47 seconds

4BE-3.4 What is the time constant of a circuit having two 100-microfarad capacitors and two 470 kilohm resistors all in parallel?

A. 470 seconds
B. 47 seconds
C. 4.7 seconds
D. 0.47 seconds

4BE-3.5 What is the time constant of a circuit having two 220-microfarad capacitors and two 1 megohm resistors all in series?

A. 55 seconds
B. 110 seconds
C. 220 seconds
D. 440 seconds

4BE-3.6 What is the time constant of a circuit having two 220-microfarad capacitors and two 1 megohm resistors all in parallel?

A. 22 seconds
B. 44 seconds
C. 220 seconds
D. 440 seconds

4BE-3.7 What is the time constant of a circuit having one 100-microfarad, one 220-microfarad capacitor and one 470 kilohm and one 1 megohm resistor all in series?

A. 68.8 seconds
B. 101.1 seconds
C. 220.0 seconds
D. 470.0 seconds

4BE-3.8 What is the time constant of a circuit having a 470-microfarad capacitor and a 1 megohm resistor in parallel?

A. 0.47 seconds
B. 47 seconds
C. 220 seconds
D. 470 seconds

4BE-3.9 What is the time constant of a circuit having a 470-microfarad capacitor in series with a 470 kilohm resistor?

A. 221 seconds
B. 221000 seconds
C. 470 seconds
D. 470000 seconds

4BE-3.10 What is the time constant of a circuit having a 220-microfarad capacitor in series with a 470 kilohm resistor?

A. 103 seconds
B. 220 seconds
C. 470 seconds
D. 470000 seconds

4BE-3.11 How long does it take for an initial charge of 20-vdc to decrease to 7.36-vdc in a 0.01-microfarad capacitor when a 2 megohm resistor is across it?

A. 12.64 seconds
B. 0.02 seconds
C. 1 second
D. 7.98 seconds

4BE-3.12 How long does it take for an initial charge of 20-vdc to decrease to 2.71-vdc in a 0.01-microfarad capacitor when a 2 megohm resistor is across it?

A. 0.04 seconds
B. 0.02 seconds
C. 7.36 seconds
D. 12.64 seconds

4BE-3.13 How long does it take for an initial charge of 20-vdc to decrease to 1-vdc in a 0.01-microfarad capacitor when a 2 megohm resistor is connected across it?

A. 0.01 seconds
B. 0.02 seconds
C. 0.04 seconds
D. 0.06 seconds

4BE-3.14 How long does it take for an initial charge of 20-vdc to decrease to 0.37-vdc in a 0.01-microfarad capacitor when a 2 megohm resistor is connected across it?

A. 0.08 seconds
B. 0.6 seconds
C. 0.4 seconds
D. 0.2 seconds

4BE-3.15 How long does it take for an initial charge of 20-vdc to decrease to 0.13-vdc in a 0.01-microfarad capacitor when a 2 megohm resistor is connected across it?

A. 0.06 seconds
B. 0.08 seconds
C. 0.1 seconds
D. 1.2 seconds

4BE-3.16 How long does it take for an initial charge of 800-vdc to decrease to 294-vdc in a 450-microfarad capacitor when a 1 megohm resistor is connected across it?

A. 80 seconds
B. 294 seconds
C. 368 seconds
D. 450 seconds

4BE-3.17 How long does it take for an initial charge of 800-vdc to decrease to 108-vdc in a 450-microfarad capacitor when a 1 megohm resistor is connected across it?

A. 225 seconds
B. 294 seconds
C. 450 seconds
D. 900 seconds

4BE-3.18 How long does it take for an initial charge of 800-vdc to decrease to 39.9-vdc in a 450-microfarad capacitor when a 1 megohm resistor is connected across it?

A. 1350 seconds
B. 900 seconds
C. 450 seconds
D. 225 seconds

4BE-3.19 How long does it take for an initial charge of 800-vdc to decrease to 40.2-vdc in a 450-microfarad capacitor when a 1 megohm resistor is connected across it?

A. Approximately 225 seconds
B. Approximately 450 seconds
C. Approximately 900 seconds
D. Approximately 1350 seconds

4BE-3.20 How long does it take for an initial charge of 800-vdc to decrease to 14.8-vdc in a 450-microfarad capacitor when a 1 megohm resistor is connected across it?

A. Approximately 900 seconds
B. Approximately 1350 seconds
C. Approximately 1804 seconds
D. Approximately 2000 seconds

4BE-4.1 What is a Smith Chart?

A. A graph for calculating impedance along transmission lines
B. A graph for calculating great circle bearings
C. A graph for calculating antenna height
D. A graph for calculating radiation patterns

4BE-4.2 What type of coordinate system is used in a Smith Chart?

A. Voltage and current circles
B. Resistance and reactance circles
C. Voltage and current lines
D. Resistance and reactance lines

4BE-4.3 What type of calculations can be performed using a Smith Chart?

A. Beam headings and radiation patterns
B. Satellite azimuth and elevation bearings
C. Impedance and SWR values in transmission lines
D. Circuit gain calculations

4BE-4.4 What are the two families of circles which make up a Smith Chart?

A. Resistance and voltage
B. Reactance and voltage
C. Resistance and reactance
D. Voltage and impedance

4BE-4.5 What is the only straight line on a blank Smith Chart?

A. The reactance axis
B. The resistance axis
C. The voltage axis
D. The current axis

4BE-4.6 What is the process of normalizing with regard to a Smith Chart?

A. Reassigning resistance values with regard to the reactance axis
B. Reassigning reactance values with regard to the resistance axis
C. Reassigning resistance values with regard to the prime center
D. Reassigning prime center with regard to the reactance axis

4BE-4.7 What are the curved lines on a Smith Chart?

A. Portions of current circles
B. Portions of voltage circles
A. Portions of resistance circles
D. Portions of reactance circles

4BE-4.8 What is the third family of circles which are added to a Smith Chart during the process of solving problems?

A. Coaxial length circles
B. Antenna length circles
C. Standing wave ratio circles
D. Radiation pattern circles

4BE-4.9 What are the four basic steps for performing calculations on a Smith Chart?

A. (1) Normalize and plot impedance, draw constant SWR circle (2) Plot antenna gain (3) Determine attenuation or loss with second SWR circle (4) Read load
B. (1) Plot wavelength of transmission line (2) Plot antenna gain (3) Read transmission line loss (4) Calculate effective radiated power
C. (1) Normalize and plot impedance, draw constant SWR circle (2) Plot transmission line wavelength (3) Determine attenuation or loss with second SWR circle (4) Reverse normalization
D. (1) Normalize and plot impedance, draw constant SWR circle (2) Plot antenna frequency (3) Determine attenuation or loss with second SWR circle (4) Reverse normalization

4BE-4.10 How are the wavelength scales on a Smith Chart calibrated?

A. In portions of transmission line electrical frequency
B. In portions of transmission line electrical wavelength
C. In portions of antenna electrical wavelength
D. In portions of antenna electrical frequency

4BE-5.1 What is the impedance of a network comprised of a 0.1 microhenry inductor in series with 20 ohms resistance, at 30 MHz?

A. 20 + j19
B. 20 – j19
C. 19 + j20
D. 19 – j20

4BE-5.2 What is the impedance of a network comprised of a 0.1 microhenry inductor in series with 30 ohms resistance, at 5-MHz?

A. 30 – j3
B. 30 + j3
C. 3 + j30
D. 3 – j30

4BE-5.3 What is the impedance of a network comprised of a 10 microhenry inductor in series with 40 ohms resistance, at 500-MHz?

A. 40 + j31400
B. 40 − j31400
C. 31400 + j40
D. 31400 − j40

4BE-5.4 What is the impedance of a network comprised of a 1.0 millihenry inductor in series with 200 ohms resistance, at 30-kHz?

A. 200 − j188
B. 200 + j188
C. 188 + j200
D. 188 − j200

4BE-5.5 What is the impedance of a network comprised of a 10 millihenry inductor in series with 600 ohms resistance, at 10-kHz?

A. 628 + j600
B. 628 − j600
C. 600 + j628
D. 600 − j628

4BE-5.6 What is the impedance of a network comprised of a 100-picofarad capacitor in parallel with 4000 ohms resistance, at 500-kHz?

A. 1949 − j1551
B. 1949 + j1551
C. 1551 + j1949
D. 1551 − j1949

4BE-5.7 What is the impedance of a network comprised of a 0.001-microfarad capacitor in series with 400 ohms resistance, at 500-kHz?

A. 400 − j318
B. 318 − j400
C. 400 + j318
D. 318 + j400

4BE-5.8 What is the impedance of a network comprised of a 0.01-microfarad capacitor in parallel with 300 ohms resistance, at 50-kHz?

A. 150 − j159
B. 150 + j159
C. 159 + j150
D. 159 − j150

4BE-5.9 What is the impedance of a network comprised of a 0.1-microfarad capacitor in series with 40 ohms resistance, at 50 kHz?

A. 40 + j32
B. 40 − j32
C. 32 − j40
D. 32 + j40

4BE-5.10 What is the impedance of a network comprised of a 1.0-microfarad capacitor in parallel with 30 ohms resistance, at 5-MHz?

A. 0.000034 + j.032
B. 0.032 + j.000034
C. 0.000034 − j.032
D. 0.032 − j.000034

4BE-6.1 What is the impedance and phase angle of a network comprised of a 100 ohm reactance inductor in series with 100 ohms resistance?

A. Z = 121 ohms, Theta = 35 degrees
B. Z = 141 ohms, Theta = 45 degrees
C. Z = 161 ohms, Theta = 55 degrees
D. Z = 181 ohms, Theta = 65 degrees

4BE-6.2 What is the impedance and phase angle of a network comprised of a 100 ohm reactance inductor, 100 ohms capacitive reactance, and 100 ohms resistance, all in series?

A. Z = 100 ohms, Theta = 90 degrees
B. Z = 10 ohms, Theta = 0 degrees
C. Z = 100 ohms, Theta = 0 degrees
D. Z = 10 ohms, Theta = 100 degrees

4BE-6.3 What is the impedance and phase angle of a network comprised of a 100 ohm reactance capacitor in series with 100 ohms resistance?

A. Z= 121 ohms, Theta= −25 degrees
B. Z= 141 ohms, Theta= −45 degrees
C. Z= 161 ohms, Theta= −65 degrees
D. Z= 191 ohms, Theta= −85 degrees

4BE-6.4 What is the impedance and phase angle of a network comprised of a 100 ohm reactance capacitor in parallel with 100 ohms resistance?

A. Z= 31 ohms, Theta= −15 degrees
B. Z= 51 ohms, Theta= −25 degrees
C. Z= 71 ohms, Theta= −45 degrees
D. Z= 91 ohms, Theta= −65 degrees

4BE-6.5 What is the impedance and phase angle of a network comprised of a 300 ohm reactance inductor in series with 400 ohms resistance?

A. Z= 400 ohms, Theta= 27 degrees
B. Z= 500 ohms, Theta= 37 degrees
C. Z= 600 ohms, Theta= 47 degrees
D. Z= 700 ohms, Theta= 57 degrees

4BE-6.6 What is the impedance and phase angle of a network comprised of a 400 ohm reactance capacitor in series with 300 ohms resistance?

A. Z= 200 ohms, Theta= −22 degrees
B. Z= 300 ohms, Theta= −33 degrees
C. Z= 400 ohms, Theta= −43 degrees
D. Z= 500 ohms, Theta= −53 degrees

4BE-6.7 What is the impedance and phase angle of a network comprised of a 300 ohm reactance capacitor, 600 ohms reactance inductor, and 400 ohms resistance, all in series?

A. Z= 500 ohms, Theta= 37 degrees
B. Z= 400 ohms, Theta= 27 degrees
C. Z= 300 ohms, Theta= 17 degrees
D. Z= 200 ohms, Theta= 10 degrees

4BE-6.8 What is the impedance and phase angle of a network comprised of a 400 ohm reactance inductor in series with 300 ohms resistance?

A. Z= 400 ohms, Theta= 43 degrees
B. Z= 500 ohms, Theta= 53 degrees
C. Z= 600 ohms, Theta= 63 degrees
D. Z= 700 ohms, Theta= 73 degrees

4BE-6.9 What is the impedance and phase angle of a network comprised of a 100 ohm reactance inductor in parallel with 100 ohms resistance?

A. Z = 71 ohms, Theta = 45 degrees
B. Z = 81 ohms, Theta = 55 degrees
C. Z = 91 ohms, Theta = 65 degrees
D. Z = 100 ohms, Theta = 75 degrees

4BE-6.10 What is the impedance and phase angle of a network comprised of a 300 ohm reactance capacitor in series with 400 ohms resistance?

A. Z = 200 ohms, Theta = −10 degrees
B. Z = 300 ohms, Theta = −17 degrees
C. Z = 400 ohms, Theta = −27 degrees
D. Z = 500 ohms, Theta = −37 degrees

SUBELEMENT 4BF—Circuit Components (4 Questions)

4BF-1.1 What is the schematic symbol for an n-channel junction FET?

A.

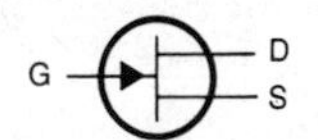

B.

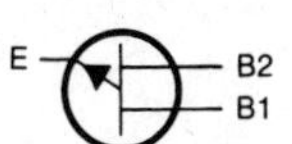

C.

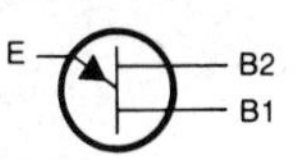

D.

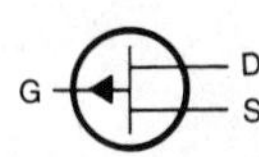

4BF-1.2 What is the schematic symbol for a p-channel junction FET?

A.

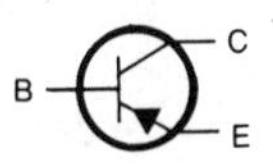

B.

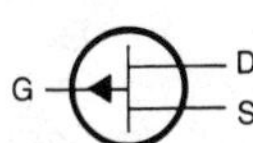

C.

D.

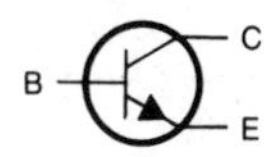

4BF-1.3 How does the input impedance of a field-effect transistor compare with that of a bipolar transistor?

A. One cannot compare input impedance without first knowing the supply voltage
B. An FET has low input impedance; a bipolar transistor has high input impedance
C. The input impedance of FETs and bipolar transistors is the same
D. An FET has high input impedance; a bipolar transistor has low input impedance

4BF-1.4 What are the two basic types of junction field-effect transistors?

A. N-channel and P-channel
B. High power and low power
C. MOSFET and GaAsFET
D. Silicon FET and germanium FET

4BF-1.5 What is the schematic symbol for an n-channel MOSFET?

A.

B.

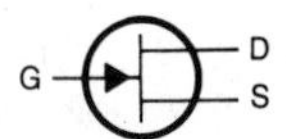

C.

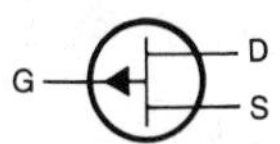

D.

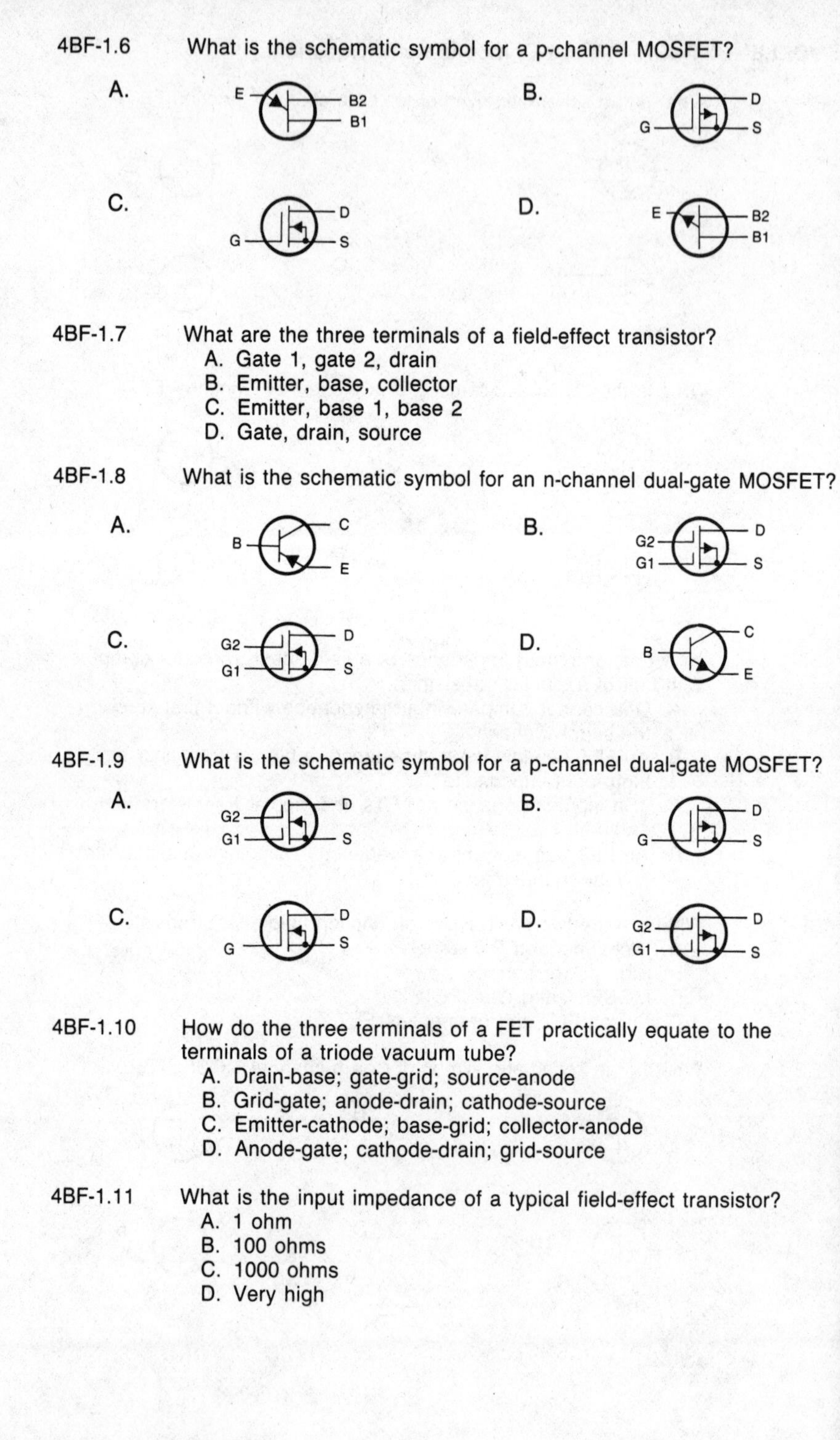

4BF-1.6 What is the schematic symbol for a p-channel MOSFET?

A. (E, B2, B1)

B. (G, D, S)

C. (G, D, S)

D. (E, B2, B1)

4BF-1.7 What are the three terminals of a field-effect transistor?

A. Gate 1, gate 2, drain
B. Emitter, base, collector
C. Emitter, base 1, base 2
D. Gate, drain, source

4BF-1.8 What is the schematic symbol for an n-channel dual-gate MOSFET?

A. (B, C, E)

B. (G2, G1, D, S)

C. (G2, G1, D, S)

D. (B, C, E)

4BF-1.9 What is the schematic symbol for a p-channel dual-gate MOSFET?

A. (G2, G1, D, S)

B. (G, D, S)

C. (G, D, S)

D. (G2, G1, D, S)

4BF-1.10 How do the three terminals of a FET practically equate to the terminals of a triode vacuum tube?

A. Drain-base; gate-grid; source-anode
B. Grid-gate; anode-drain; cathode-source
C. Emitter-cathode; base-grid; collector-anode
D. Anode-gate; cathode-drain; grid-source

4BF-1.11 What is the input impedance of a typical field-effect transistor?

A. 1 ohm
B. 100 ohms
C. 1000 ohms
D. Very high

4BF-1.12 What is a depletion-mode FET?

A. An FET that has a channel with no gate voltage applied; a current flows with zero gate voltage
B. An FET that has a channel that blocks current when the gate voltage is zero
C. An FET without a channel; no current flows with zero gate voltage
D. An FET without a channel to hinder current through the gate

4BF-1.13 What is an enhancement-mode FET?

A. An FET with a channel that blocks voltage through the gate
B. An FET with a channel that allows a current when the gate voltage is zero
C. An FET without a channel to hinder current through the gate
D. An FET without a channel; no current occurs with zero gate voltage

4BF-1.14 What is a CMOS IC?

A. A chip with only P-channel transistors
B. A chip with P-channel and N-channel transistors
C. A chip with only N-channel transistors
D. A chip with only bipolar transistors

4BF-1.15 Why are special precautions necessary in handling FET and CMOS devices?

A. They are susceptible to damage from static charges
B. They have fragile leads that may break off
C. They have micro-welded semiconductor junctions that are susceptible to breakage
D. They are light sensitive

4BF-1.16 Why do many MOSFET devices have built-in gate-protective Zener diodes?

A. The gate-protective Zener diode provides a voltage reference to provide the correct amount of reverse-bias gate voltage
B. The gate-protective Zener diode protects the substrate from excessive voltages
C. The gate-protective Zener diode keeps the gate voltage within specifications to prevent the device from overheating
D. The gate-protective Zener diode prevents the gate insulation from being punctured by small static charges or excessive voltages

4BF-2.1 What is the schematic symbol for an operational amplifier?

A. 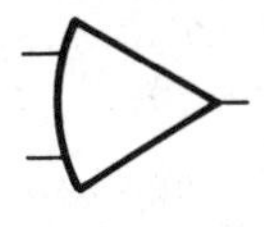

C. 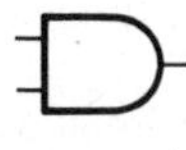

B. 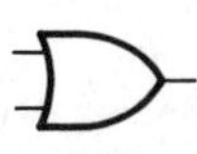

D. 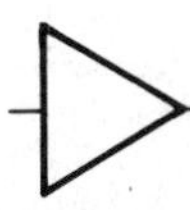

4BF-2.2 What is an operational amplifier?

A. A high-gain, direct-coupled differential amplifier whose characteristics are determined by components external to the amplifier unit
B. A high-gain, direct-coupled audio amplifier whose characteristics are determined by components external to the amplifier unit
C. An amplifier used to increase the average output of frequency modulated amateur signals to the legal limit
D. A program subroutine that calculates the gain of an RF amplifier

4BF-2.3 What would be the characteristics of the ideal op-amp?

A. Zero input impedance, infinite output impedance, infinite gain, flat frequency response
B. Infinite input impedance, zero output impedance, infinite gain, flat frequency response
C. Zero input impedance, zero output impedance, infinite gain, flat frequency response
D. Infinite input impedance, infinite output impedance, infinite gain, flat frequency response

4BF-2.4 What determines the gain of a closed-loop op-amp circuit?

A. The external feedback network
B. The collector-to-base capacitance of the PNP stage
C. The power supply voltage
D. The PNP collector load

4BF-2.5 What is an inverting op-amp circuit?

A. An operational amplifier circuit connected such that the input and output signals are 180 degrees out of phase
B. An operational amplifier circuit connected such that the input and output signal are in phase
C. An operational amplifier circuit connected such that the input and output are 90 degrees out of phase
D. An operational amplifier circuit connected such that the input impedance is held at zero, while the output impedance is high

4BF-2.6 What is a non-inverting op-amp circuit?

A. An operational amplifier circuit connected such that the input and output signals are 180 degrees out of phase
B. An operational amplifier circuit connected such that the input and output signal are in phase
C. An operational amplifier circuit connected such that the input and output are 90 degrees out of phase
D. An operational amplifier circuit connected such that the input impedance is held at zero while the output impedance is high

4BF-2.7 What is a differential op-amp circuit?

A. The output is held constant by the reference input when the variable input is different from the reference
B. One input signal is amplified and split into two different output signals
C. The output is the difference between the referenced input and the variable input
D. The logic state of the output is high only when the input logic states are different

4BF-2.8 What is a phase-locked loop circuit?

A. An electronic servo loop consisting of a ratio detector, reactance modulator, and voltage-controlled oscillator
B. An electronic circuit also known as a monostable multivibrator
C. An electronic circuit consisting of a precision push-pull amplifier with a differential input
D. An electronic servo loop consisting of a phase detector, a low-pass filter and voltage-controlled oscillator

4BF-2.9 What functions are performed by a phase-locked loop?

A. Wideband AF and RF power amplification
B. Comparison of two digital input signals, digital pulse counter
C. Photovoltaic conversion, optical coupling
D. Frequency synthesis, FM demodulation

4BF-2.10 How does a PLL function?

A. Output from a voltage controlled oscillator and frequency standard are compared; the sum of the two frequencies equals the error voltage that changes the phase detector frequency
B. Output from a voltage controlled oscillator and frequency standard are compared; the difference between the two frequencies equals the error voltage that changes the voltage controlled oscillator frequency
C. Output from a voltage controlled oscillator and frequency standard are compared; the sum of the two frequencies equals the error voltage that changes the voltage controlled oscillator frequency
D. Output from a voltage controlled oscillator and frequency standard are compared; the difference between the two frequencies equals the error voltage that changes the phase detector frequency

4BF-3.1 What is the recommended power supply voltage for 7400 series integrated circuits?

A. 12.00 volts at the input to the device
B. 50.00 volts at the input to the device
C. 5.00 volts at the input to the device
D. 13.60 volts at the input to the device

4BF-3.2 What type of digital-logic is implemented by the 7400 family of ICs?

A. Resistor-transistor logic
B. Transistor-transistor logic
C. Diode-transistor logic
D. Emitter-coupled logic

4BF-3.3 If TTL device inputs are left open, what logic state do the inputs assume?

A. A high logic state
B. A low logic state
C. The device becomes randomized and will not provide consistent high or low logic states
D. Open inputs on a TTL device are ignored

4BF-3.4 When operating with a plus 5-volt power supply, what level of input voltage is high in a TTL device?

A. 2.0 to 5.5 volts
B. 1.5 to 3.0 volts
C. 1.0 to 1.5 volts
D. −5.0 to −2.0 volts

4BF-3.5 When operating with a plus 5-volt power supply, what level of input voltage is low in a TTL device?

A. −2.0 to −5.5 volts
B. 2.0 to 5.5 volts
C. −0.6 to 0.8 volts
D. −0.8 to 0.4 volts

4BF-3.6 Why do circuits containing TTL devices have several bypass capacitors per printed circuit board?

A. To prevent RFI to receivers
B. To keep the switching noise within the circuit, thus eliminating RFI
C. To filter out switching harmonics
D. To prevent switching transients from appearing on the supply line

4BF-3.7 What is an AND gate?

A. A circuit that produces a logic "1" at its output only if all inputs are logic "1"
B. A circuit that produces a logic "0" at its output only if all inputs are logic "1"
C. A circuit that produces a logic "1" at its output if only one input is a logic "1"
D. A circuit that produces a logic "1" at its output if all inputs are logic "0"

4BF-3.8 What is an OR gate?

A. A circuit that produces a logic "1" at its output if any input is logic "1"
B. A circuit that produces a logic "0" at its output if any input is logic "1"
C. A circuit that produces a logic "0" at its output if all inputs are logic "1"
D. A circuit that produces a logic "1" at its output if all inputs are logic "0"

4BF-3.9 What is a NOT gate?

A. A circuit that produces a logic "O" at its output when the input is logic "1" and vice versa
B. A circuit that does not allow data transmission when its input is high
C. A circuit that allows data transmission only when its input is high
D. A circuit that produces a logic "1" at its output when the input is logic "1" and vice versa

4BF-3.10 What is a NOR gate?

A. A circuit that produces a logic "0" at its output only if all inputs are logic "0"
B. A circuit that produces a logic "1" at its output only if all inputs are logic "1"
C. A circuit that produces a logic "0" at its output if any or all inputs are logic "1"
D. A circuit that produces a logic "1" at its output if some but not all of its inputs are logic "1"

4BF-3.11 What is a NAND gate?

A. A circuit that produces a logic "0" at its output only when all inputs are logic "0"
B. A circuit that produces a logic "1" at its output only when all inputs are logic "1"
C. A circuit that produces a logic "0" at its output if some but not all of its inputs are logic "1"
D. A circuit that produces a logic "0" at its output only when all inputs are logic "1"

4BF-3.12 What type of digital IC is also known as a latch?

A. A decade counter
B. An OR gate
C. A flip-flop
D. An op-amp

4BF-3.13 How many states does a decade counter digital IC have?

A. 6
B. 10
C. 15
D. 20

4BF-3.14 What is the function of a decade counter digital IC?

A. Decode a decimal number for display on a seven-segment LED display
B. Produce one output pulse for every ten input pulses
C. Produce ten output pulses for every input pulse
D. Add two decimal numbers

4BF-3.15 What is a truth table?

A. A table of logic symbols that indicate the high logic states of an op-amp
B. A diagram showing logic states when the digital device's output is true
C. A list of input combinations and their corresponding outputs that characterize a digital device's function
D. A table of logic symbols that indicate the low logic states of an op-amp

4BF-4.1 What does the term CMOS mean?

A. Common mode oscillating system
B. Complementary mica-oxide silicon
C. Complementary metal-oxide semiconductor
D. Complementary metal-oxide substrate

4BF-4.2 What type of integrated circuit is the 4000 series?

A. TTL
B. ECL
C. RTL
D. CMOS

4BF-4.3 What is one major advantage of CMOS over other devices?

A. Small size
B. Low current consumption
C. Low cost
D. Ease of circuit design

4BF-4.4 Why do CMOS digital integrated circuits have high immunity to noise on the input signal or power supply?

A. Larger bypass capacitors are used in 4000-series circuit design
B. The input switching threshold is about two times the power supply voltage
C. The input switching threshold is about one-half the power supply voltage
D. Input signals are stronger

4BF-5.1 What is a vidicon?

A. A vacuum tube with a phosphorus layer, photoconductive mosaic, transparent conductive film and proton gun
B. A vacuum tube with a phosphorus layer, photoconductive mosaic, transparent conductive film and electron gun
C. A vacuum tube with a photoresistive layer, photoconductive mosaic, transparent conductive film and electron gun
D. A vacuum tube with a cadmium coated layer, photoconductive mosaic, transparent conductive film and electron gun

4BF-5.2 How is the electron beam deflected in a vidicon?

A. By varying the beam voltage
B. By varying the bias voltage on the beam forming grids inside the tube
C. By varying the beam current
D. By varying electromagnetic fields

4BF-5.3 What is the output impedance of a vidicon?

A. Very low
B. Varying between high and low, depending on the amount of light being exposed to the photoconductive layer
C. Very high
D. Varying between very low and medium, depending on the amount of beam current

4BF-5.4 What type of CRT deflection is better when high frequency waves are to be displayed on the screen?

A. Electromagnetic
B. Tubular
C. Radar
D. Electrostatic

BF-5.5 What is an ion trap?

A. A super-heated element used to remove air from a partially evacuated vacuum tube
B. A magnet installed in CRTs that prevents negative ions from burning a spot on the luminescent screen
C. A beam-forming electrode in cathode ray tubes that focuses charged particles being beamed to the screen
D. An electrode that cuts off the electron beam to prevent a CRT from displaying retrace lines

SUBELEMENT 4BG—Practical Circuits (4 Questions)

4BG-1.1 What is a flip-flop circuit?

A. A binary sequential logic element with one stable state
B. A binary sequential logic element with eight stable states
C. A binary sequential logic element with four stable states
D. A binary sequential logic element with two stable states

4BG-1.2 How many bits of information can be stored in a single flip-flop circuit?

A. 1
B. 2
C. 3
D. 4

4BG-1.3 What is a bistable multivibrator circuit?

A. An "AND" gate
B. An "OR" gate
C. A flip-flop
D. A clock

4BG-1.4 What is an astable multivibrator?

A. A circuit that alternates between two stable states
B. A circuit that alternates between a stable state and an unstable state
C. A circuit set to block either a 0 pulse or a 1 pulse and pass the other
D. A circuit that alternates between two unstable states

4BG-1.5 What is a monostable multivibrator?

A. A circuit that can be switched momentarily to the opposite binary state and then returns after a set time to its original state
B. A "clock" circuit that produces a continuous square wave oscillating between 1 and 0
C. A circuit designed to store one bit of data in either the 0 or the 1 configuration
D. A circuit that maintains a constant output voltage, regardless of variations in the input voltage

4BG-1.6 What is the schematic symbol for an AND gate?

A.

B.

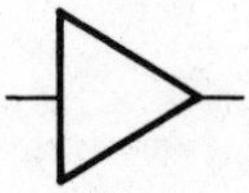

C.

D.

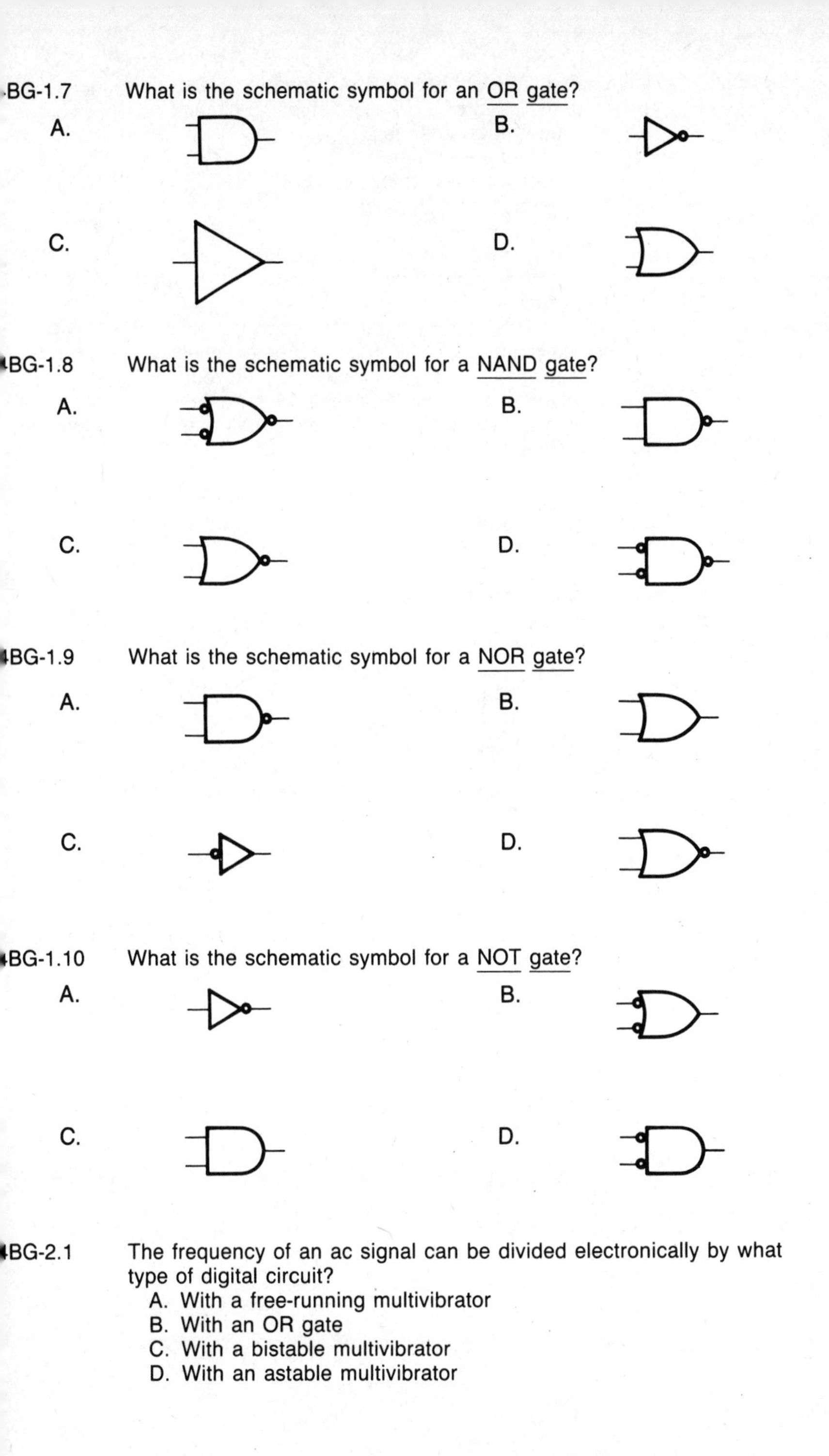

BG-1.7 What is the schematic symbol for an OR gate?

A. B.

C. D.

BG-1.8 What is the schematic symbol for a NAND gate?

A. B.

C. D.

BG-1.9 What is the schematic symbol for a NOR gate?

A. B.

C. D.

BG-1.10 What is the schematic symbol for a NOT gate?

A. B.

C. D.

BG-2.1 The frequency of an ac signal can be divided electronically by what type of digital circuit?

A. With a free-running multivibrator
B. With an OR gate
C. With a bistable multivibrator
D. With an astable multivibrator

4BG-2.2 How many output changes are obtained for every two trigger pulses applied to the input of a bistable T flip-flop circuit?

A. No output level changes
B. One output level change
C. Two output level changes
D. Four output level changes

4BG-2.3 What is a crystal-controlled marker generator?

A. A low-stability oscillator that "sweeps" through a band of frequencies
B. An oscillator often used in aircraft to determine the craft's location relative to the inner and outer markers at airports
C. A high-stability oscillator whose output frequency and amplitude can be varied over a wide range
D. A high-stability oscillator that generates a series of reference signals at known frequency intervals

4BG-2.4 What additional circuitry is required in a 100-kHz crystal-controlled marker generator to provide markers at 50- and 25-kHz?

A. An emitter-follower
B. Two frequency multipliers
C. One or more flip-flops
D. A voltage divider

4BG-2.5 What is the purpose of a prescaler circuit?

A. It converts the output of a JK flip-flop to that of an RS flip-flop
B. It multiplies an HF signal so a low-frequency counter can display the operating frequency
C. It prevents oscillation in a low frequency counter circuit
D. It divides an HF signal so a low-frequency counter can display the operating frequency

4BG-2.6 What does the accuracy of a frequency counter depend upon?

A. The internal crystal reference
B. A voltage-regulated power supply with an unvarying output
C. Accuracy of the AC input frequency to the power supply
D. Proper balancing of the power-supply diodes

4BG-2.7 Why is 1-MHz the standard crystal reference in many counter circuits?

A. A 1-MHz AT-cut crystal has excellent temperature stability
B. 1-MHz crystals are Y-cut
C. The series and parallel resonant points are the same at 1 MHz
D. 1-MHz crystals are X-cut

4BG-2.8 How many flip-flops are required to divide a signal frequency by 4?

A. 1
B. 2
C. 4
D. 8

BG-3.1 What are the advantages of using an op-amp over LC elements in an audio filter?

A. Op-amps are more rugged and can withstand more abuse than can LC elements
B. Op-amps are fixed at one frequency
C. Op-amps are available in more styles and types than are LC elements
D. Op-amps exhibit gain rather than insertion loss

BG-3.2 What determines the gain and frequency characteristics of an op-amp RC active filter?

A. Values of capacitances and resistances built into the op-amp
B. Values of capacitances and resistances external to the op-amp
C. Voltage and frequency of DC input to the op-amp power supply
D. Regulated DC voltage output from the op-amp power supply

BG-3.3 What are the principle uses of an op-amp RC active filter in amateur circuitry?

A. Op-amp circuits are used as high-pass filters to block RFI at the input to receivers
B. Op-amp circuits are used as low-pass filters between transmitters and transmission lines
C. Op-amp circuits are used as filters for smoothing power-supply output
D. Op-amp circuits are used as audio filters for receivers

BG-3.4 What type of capacitors should be used in an op-amp RC active filter circuit?

A. Electrolytic
B. Disc ceramic
C. Polystyrene
D. Paper dielectric

BG-3.5 How can unwanted ringing and audio instability be prevented in a multisection op-amp RC audio filter circuit?

A. Restrict both gain and Q
B. Restrict gain, but increase Q
C. Restrict Q, but increase gain
D. Increase both gain and Q

BG-3.6 Where should an op-amp RC active audio filter be placed in an amateur receiver?

A. In the IF strip, immediately before the detector
B. In the audio circuitry immediately before the speaker or phone jack
C. Between the balanced modulator and frequency multiplier
D. Within the AGC loop

BG-3.7 What is meant by the term offset voltage with regard to an op-amp?

A. The output voltage of the op-amp minus its input voltage
B. The difference between the output voltage of the op-amp and the input voltage required in the following stage
C. The potential between the amplifier-input terminals of the op-amp in a closed-loop condition
D. The potential between the amplifier-input terminals of the op-amp in an open-loop condition

4BG-3.8 What is the term for the temperature coefficient of offset voltage with respect to time in an op-amp?

A. Burst
B. Drift
C. Flutter
D. Thermal noise

4BG-3.9 What is the input impedance of a theoretically ideal op-amp?

A. 100 ohms
B. 1000 ohms
C. Very low
D. Very high

4BG-3.10 What is the output impedance of a theoretically ideal op-amp?

A. Very low
B. Very high
C. 100 ohms
D. 1000 ohms

4BG-4.1 What is meant by the term noise figure of a communications receiver?

A. The level of noise entering the receiver from the antenna
B. The relative strength of a received signal 3 kHz removed from the carrier frequency
C. The level of noise generated in the front end and succeeding stages of a receiver
D. The ability of a receiver to reject unwanted signals at frequencies close to the desired one

4BG-4.2 What two factors determine the sensitivity of a receiver?

A. Dynamic range and third-order intercept
B. Cost and availability
C. Intermodulation distortion and dynamic range
D. Bandwidth and noise figure

4BG-4.3 Which stage of a receiver primarily establishes its noise figure?

A. The audio stage
B. The IF strip
C. The RF stage
D. The local oscillator

4BG-4.4 What is the limiting condition for sensitivity in a communications receiver?

A. The noise floor of the receiver
B. The power-supply output ripple
C. The two-tone intermodulation distortion
D. The input impedance to the detector

4BG-4.5 What is the theoretical minimum noise floor of a receiver with a 400-Hertz bandwidth?

A. −141 dBm
B. −148 dBm
C. −174 dBm
D. −180 dBm

4BG-5.1 How can selectivity be achieved in the front end circuitry of a communications receiver?

A. By using an audio filter
B. By using a preselector
C. By using an additional RF amplifier stage
D. By using an additional IF amplifier stage

4BG-5.2 A receiver selectivity of 2.4-kHz in the if circuitry is optimum for what type of amateur signals?

A. CW
B. SSB voice
C. Double-sideband AM voice
D. FSK RTTY

4BG-5.3 What occurs during A1A reception if too narrow a filter bandwidth is used in the if stage of a receiver?

A. Undesired signals will reach the audio stage
B. Output-offset overshoot
C. Cross-modulation distortion
D. Filter ringing

4BG-5.4 What degree of selectivity is desirable in the if circuitry of an amateur emission F1B receiver?

A. 100 Hz
B. 300 Hz
C. 6000 Hz
D. 2400 Hz

4BG-5.5 A receiver selectivity of 10-kHz in the if circuitry is optimum for what type of amateur signals?

A. SSB voice
B. Double-sideband AM
C. CW
D. FSK RTTY

4BG-5.6 What degree of selectivity is desirable in the if circuitry of an emission J3E receiver?

A. 1 kHz
B. 2.4 kHz
C. 4.2 kHz
D. 4.8 kHz

4BG-5.7 What is an undesirable effect of using too wide a filter bandwidth in the if section of a receiver?

A. Output-offset overshoot
B. Undesired signals will reach the audio stage
C. Thermal-noise distortion
D. Filter ringing

4BG-5.8 How should the filter bandwidth of a receiver if section compare with the bandwidth of a received signal?

A. Filter bandwidth should be slightly greater
B. Filter bandwidth should be approximately half
C. Filter bandwidth should be approximately two times
D. Filter bandwidth should be approximately four times

4BG-5.9 What degree of selectivity is desirable in the if circuitry of an emission F3E receiver?

A. 1 kHz
B. 2.4 kHz
C. 4.2 kHz
D. 15 kHz

4BG-5.10 How can selectivity be acheived in the if circuitry of a communications receiver?

A. Incorporate a means of varying the supply voltage to the local oscillator circuitry
B. Replace the standard JFET mixer with a bipolar transistor followed by a capacitor of the proper value
C. Remove AGC action from the IF stage and confine it to the audio stage only
D. Incorporate a high-Q filter

4BG-6.1 What is meant by the dynamic range of a communications receiver

A. The number of kHz between the lowest and the highest frequency to which the receiver can be tuned
B. The maximum possible undistorted audio output of the receiver, referenced to one milliwatt
C. The ratio between the minimum discernable signal and the largest tolerable signal without causing audible distortion products
D. The difference between the lowest-frequency signal and the highest-frequency signal detectable without moving the tuning knob

4BG-6.2 What is the term for the ratio between the largest tolerable receive input signal and the minimum discernable signal?

A. Intermodulation distortion
B. Noise floor
C. Noise figure
D. Dynamic range

4BG-6.3 What is an acceptable optimum dynamic range obtainable with the present state of the art receiver?

A. An acceptable dynamic range for a present state-of-the-art receiver is 25 dB
B. An acceptable dynamic range for a present state-of-the-art receiver is 50 dB
C. An acceptable dynamic range for a present state-of-the-art receiver is 100 dB
D. An acceptable dynamic range for a present state-of-the-art receiver is 200 dB

BG-6.4 What type of problems are caused by poor dynamic range in a communications receiver?

A. Cross-modulation of the desired signal and desensitization from strong adjacent signals
B. Oscillator instability requiring frequent retuning, and loss of ability to recover the opposite sideband, should it be transmitted
C. Cross-modulation of the desired signal and insufficient audio power to operate the speaker
D. Oscillator instability and severe audio distortion of all but the strongest received signals

BG-6.5 The ability of a communications receiver to perform well in the presence of strong signals outside the amateur band of interest is indicated by what figure?

A. Noise figure
B. Blocking dynamic range
C. Signal-to-noise ratio
D. Audio output

BG-7.1 What voltage gain can be expected from the circuit in Figure 4BG-7 when R1 is 1000 ohms and Rf is 100 kilohms?

A. 0.1
B. 1
C. 10
D. 100

BG-7.2 What voltage gain can be expected from the circuit in Figure 4BG-7 when R1 is 1800 ohms and Rf is 68 kilohms?

A. 1
B. 0.03
C. 38
D. 76

BG-7.3 What voltage gain can be expected from the circuit in Figure 4BG-7 when R1 is 3300 ohms and Rf is 47 kilohms?

A. 28
B. 14
C. 7
D. 0.07

BG-7.4 What voltage gain can be expected from the circuit in Figure 4BG-7 when R1 is 10 ohms and Rf is 47 kilohms?

A. 0.00026
B. 9400
C. 4700
D. 2350

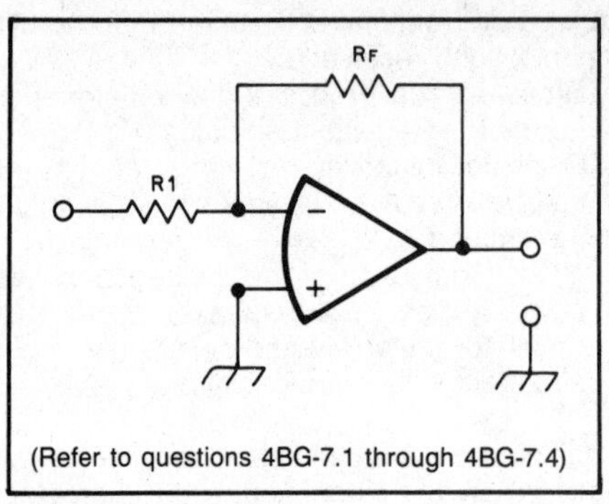

(Refer to questions 4BG-7.1 through 4BG-7.4)

FIGURE 4BG-7

4BG-7.5 What parameter must be selected when designing an audio filter using an op-amp?
- A. Bandpass characteristics
- B. Desired current gain
- C. Temperature coefficient
- D. Output-offset overshoot

4BG-8.1 What determines the input impedance in a FET common-source amplifier?
- A. The input impedance is essentially determined by the resistance between the drain and substrate
- B. The input impedance is essentially determined by the resistance between the source and drain
- C. The input impedance is essentially determined by the gate biasing network
- D. The input impedance is essentially determined by the resistance between the source and substrate

4BG-8.2 What determines the output impedance in a FET common-source amplifier?
- A. The output impedance is essentially determined by the drai resistor
- B. The output impedance is essentially determined by the inpu impedance of the FET
- C. The output impedance is essentially determined by the drai supply voltage
- D. The output impedance is essentially determined by the gate supply voltage

4BG-9.1 What frequency range will be tuned by the circuit in Figure 4BG- when L is 10-microhenrys, Cf is 156-picofarads, and Cv is 50-picofarads maximum and 2-picofarads minimum?
- A. 3508 through 4004 kHz
- B. 6998 through 7360 kHz
- C. 13.396 through 14.402 MHz
- D. 49.998 through 54.101 MHz

4BG-9.2 What frequency range will be tuned by the circuit in Figure 4BG-9 when L is 30-microhenrys, Cf is 200-picofarads, and Cv is 80-picofarads maximum and 10-picofarads minimum?

A. 1737 through 2005 kHz
B. 3507 through 4004 kHz
C. 7002 through 7354 kHz
D. 14.990 through 15.020 MHz

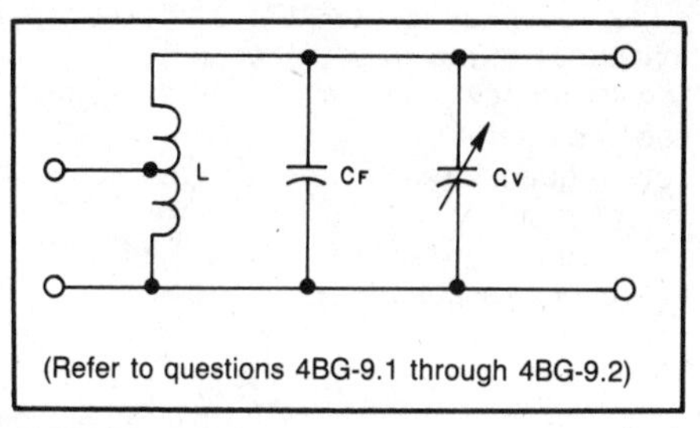

(Refer to questions 4BG-9.1 through 4BG-9.2)

FIGURE 4BG-9

4BG-9.3 What is the purpose of a bypass capacitor?

A. It increases the resonant frequency of the circuit
B. It removes direct current from the circuit by shunting DC to ground
C. It removes alternating current by providing a low impedance to ground
D. It acts as a voltage divider

4BG-9.4 What is the purpose of a coupling capacitor?

A. It blocks direct current and passes alternating current
B. It blocks alternating current and passes direct current
C. It increases the resonant frequency of the circuit
D. It decreases the resonant frequency of the circuit

4BG-9.5 How is the frequency response of a single-stage amplifier affected when the capacitance of the coupling capacitors is decreased?

A. Low-frequency response improves and high-frequency response is degraded
B. Low-frequency response and high-frequency response both improve
C. Low-frequency response is degraded
D. Frequency response stays the same

4BG-9.6 How is the frequency response of a single-stage amplifier affected when the capacitance of the by-pass capacitor is increased?

A. Low-frequency response improves
B. Low-frequency response is degraded and high-frequency response improves
C. Low-frequency response and high-frequency response are both degraded
D. Frequency response stays the same

4BG-9.7 How is the frequency response of a single-stage amplifier affected if the capacitance of the coupling capacitors is increased?

A. Low-frequency response improves
B. Low-frequency response is degraded and high-frequency response improves
C. Low-frequency response and high-frequency response are both degraded
D. Frequency response stays the same

4BG-9.8 How is the frequency response of a single-stage amplifier affected if the capacitance of the by-pass capacitor is decreased?

A. Low-frequency response improves and high-frequency response is degraded
B. Low-frequency response and high-frequency response both improve
C. Low-frequency response is degraded
D. Frequency response stays the same

UBELEMENT 4BH—Signals and Emissions (4 Questions)

BH-1.1 In a pulse-width modulation system, what parameter does the modulating signal vary?
A. Pulse duration
B. Pulse frequency
C. Pulse amplitude
D. Pulse intensity

BH-1.2 What is the type of modulation where the modulating signal varies the duration of the transmitted pulse?
A. Amplitude modulation
B. Frequency modulation
C. Pulse-width modulation
D. Pulse-height modulation

BH-1.3 In a pulse-position modulation system, what parameter does the modulating signal vary?
A. The number of pulses per second
B. Both the frequency and amplitude of the pulses
C. The duration of the pulses
D. The time at which each pulse occurs

BH-1.4 Why is the transmitter peak power in a pulse modulation system much greater than its average power?
A. The signal duty cycle is less than 100%
B. The signal reaches peak amplitude only when voice-modulated
C. The signal reaches peak amplitude only when voltage spikes are generated within the modulator
D. The signal reaches peak amplitude only when the pulses are also amplitude-modulated

BH-1.5 How is voice transmitted in a pulse modulation system?
A. A standard pulse is varied in either duration, position, frequency or code, and the amount of variation depends on the voice waveform at that instant
B. Voice is converted into a frequency-modulated RF signal, which varies the frequency at which pulses are transmitted
C. Voice is converted into an amplitude-modulated RF signal, which varies the frequency at which pulses are transmitted
D. A standard pulse is varied in either duration, position, amplitude or code, and the amount of variation depends on the voice waveform at that instant

BH-2.1 What is the duration of a 60 speed Baudot RTTY data pulse?
A. 11 milliseconds
B. 40 milliseconds
C. 31 milliseconds
D. 22 milliseconds

BH-2.2 What is the duration of a 60 speed Baudot RTTY start pulse?
A. 11 milliseconds
B. 22 milliseconds
C. 31 milliseconds
D. 40 milliseconds

4BH-2.3 What is the duration of a 60 speed Baudot stop pulse?

A. 11 milliseconds
B. 22 milliseconds
C. 31 milliseconds
D. 40 milliseconds

4BH-2.4 What is the meaning of the term baud rate?

A. The speed of transmission in kilobits per second
B. The speed of transmission in bytes per second
C. The speed of transmission in words per minute
D. The speed of transmission in number of discrete events per second

4BH-2.5 What is the baud rate for a 60 speed Baudot RTTY system?

A. Approximately 22
B. Approximately 45
C. Approximately 66
D. Approximately 90

4BH-3.1 What is amplitude compandored single sideband?

A. Reception of single sideband with a conventional CW receive
B. Reception of single sideband with a conventional FM receiver
C. Single sideband incorporating speech-compression at the transmitter and speech-expansion at the receiver
D. Single sideband incorporating speech expansion at the transmitter and speech compression at the receiver

4BH-3.2 What is meant by compandoring?

A. Compressing speech at the transmitter and expanding it at th receiver
B. Using an audio-frequency signal to produce pulse-length modulation
C. Combining amplitude and frequency modulation to produce a single-sideband signal
D. Detecting and demodulating a single-sideband signal by converting it to a pulse-modulated signal

4BH-3.3 What is the purpose of a pilot tone in an amplitude compandored single sideband system?

A. It permits rapid tuning of a mobile receiver
B. It replaces the suppressed carrier at the receiver
C. It permits rapid change of frequency to escape high-powered interference
D. It acts as a beacon to indicate the present propagation characteristic of the band

4BH-3.4 What is the approximate frequency of the pilot tone in an amplitud compandored single sideband system?

A. 1 kHz
B. 5 MHz
C. 455 kHz
D. 3 kHz

4BH-3.5 How many more voice transmissions can be packed into a given frequency band for amplitude compandored single sideband systems over conventional emission F3E systems?

A. 2
B. 4
C. 8
D. 16

4BH-4.1 What is the necessary bandwidth of a 170-Hertz shift, 60 speed Baudot emission F1B transmission?

A. 45 Hz
B. 250 Hz
C. 442 Hz
D. 600 Hz

4BH-4.2 What is the necessary bandwidth of a 170-Hertz shift, 60 speed Baudot emission J2B transmission?

A. 45 Hz
B. 250 Hz
C. 442 Hz
D. 600 Hz

4BH-4.3 What is the necessary bandwidth of a 170-Hertz shift, 100 speed Baudot emission F1B transmission?

A. 250 Hz
B. 278 Hz
C. 442 Hz
D. 600 Hz

4BH-4.4 What is the necessary bandwidth of a 170-Hertz shift, 100 speed Baudot emission J2B transmission?

A. 250 Hz
B. 278 Hz
C. 442 Hz
D. 600 Hz

4BH-4.5 What is the necessary bandwidth of a 200-WPM international Morse code emission A1A transmission?

A. Approximately 200 Hz
B. Approximately 400 Hz
C. Approximately 800 Hz
D. Approximately 1600 Hz

4BH-4.6 What is the necessary bandwidth of a 200-WPM international Morse code emission J2A transmission?

A. Approximately 200 Hz
B. Approximately 400 Hz
C. Approximately 800 Hz
D. Approximately 1600 Hz

4BH-4.7 What is the necessary bandwidth of a 1000-Hertz shift, 1200-baud ASCII emission F1D transmission?

A. 1000 Hz
B. 1200 Hz
C. 440 Hz
D. 2400 Hz

*This code speed is unrealistic—A1A and J2A signals are intended for reception by ear.

4BH-4.8 What is the necessary bandwidth of a 200-WPM international Mors code emission J2D transmission?

A. Approximately 200 Hz
B. Approximately 400 Hz
C. Approximately 800 Hz
D. Approximately 1600 Hz

4BH-4.9 What is the necessary bandwidth of a 4800-hertz frequency shift, 9600-baud ASCII emission F1D transmission?

A. 15.26 kHz
B. 9.6 kHz
C. 4.8 kHz
D. 5.76 kHz

4BH-4.10 What is the necessary bandwidth of a 4800-hertz frequency shift, 9600-baud ASCII emission J2D transmission?

A. 15.26 kHz
B. 9.6 kHz
C. 4.8 kHz
D. 5.76 kHz

4BH-4.11 What is the necessary bandwidth of a 5-WPM international Morse code emission A1A transmission?

A. Approximately 5 Hz
B. Approximately 10 Hz
C. Approximately 20 Hz
D. Approximately 40 Hz

4BH-4.12 What is the necessary bandwidth of a 5-WPM international Morse code emission J2A transmission?

A. Approximately 5 Hz
B. Approximately 10 Hz
C. Approximately 20 Hz
D. Approximately 40 Hz

4BH-4.13 What is the necessary bandwidth of a 170-Hertz shift, 110-baud ASCII emission F1B transmission?

A. 304 Hz
B. 314 Hz
C. 608 Hz
D. 628 Hz

4BH-4.14 What is the necessary bandwidth of a 170-Hertz shift, 110-baud ASCII emission J2B transmission?

A. 304 Hz
B. 314 Hz
C. 608 Hz
D. 628 Hz

4BH-4.15 What is the necessary bandwidth of a 170-Hertz shift, 75-WPM Baudot emission F1B transmission?

A. 0 Hz
B. 261 Hz
C. 591 Hz
D. 1000 Hz

4BH-4.16 What is the necessary bandwidth of a 170-Hertz shift, 75-WPM Baudot emission J2B transmission?

A. 0 Hz
B. 261 Hz
C. 591 Hz
D. 1000 Hz

4BH-4.17 What is the necessary bandwidth of a 13-WPM international Morse code emission A1A transmission?

A. Approximately 13 Hz
B. Approximately 26 Hz
C. Approximately 52 Hz
D. Approximately 104 Hz

4BH-4.18 What is the necessary bandwidth of a 13-WPM international Morse code emission J2A transmission?

A. Approximately 13 Hz
B. Approximately 26 Hz
C. Approximately 52 Hz
D. Approximately 104 Hz

4BH-4.19 What is the necessary bandwidth of a 170-Hertz shift, 300-baud ASCII emission F1D transmission?

A. 0 Hz
B. 0.3 kHz
C. 0.5 kHz
D. 1.0 kHz

4BH-4.20 What is the necessary bandwidth for a 170-Hertz shift, 300-baud ASCII emission J2D transmission?

A. 0 Hz
B. 0.3 kHz
C. 0.5 kHz
D. 1.0 kHz

4BH-5.1 What is the term for the amplitude of the maximum positive excursion of a signal as viewed on an oscilloscope?

A. Peak-to-peak voltage
B. Inverse peak negative voltage
C. RMS voltage
D. Peak positive voltage

4BH-5.2 What is the term for the amplitude of the maximum negative excursion of a signal as viewed on an oscilloscope?

A. Peak-to-peak voltage
B. Inverse peak positive voltage
C. RMS voltage
D. Peak negative voltage

4BH-6.1 What is the easiest voltage amplitude dimension to measure by viewing a pure sine wave signal on an oscilloscope?

A. Peak-to-peak voltage
B. RMS voltage
C. Average voltage
D. DC voltage

4BH-6.2 What is the relationship between the peak-peak voltage and the peak voltage amplitude in a symmetrical wave form?

A. 1:1
B. 2:1
C. 3:1
D. 4:1

4BH-6.3 What input-amplitude parameter is valuable in evaluating the signal-handling capability of a Class A amplifier?

A. Peak voltage
B. Average voltage
C. RMS voltage
D. Resting voltage

SUBELEMENT 4BI—Antennas and Feed lines (4 Questions)

4BI-1.1 What factors determine the receiving antenna gain required at an amateur station in earth operation?

A. Height, transmitter power and antennas of satellite
B. Length of transmission line and impedance match between receiver and transmission line
C. Preamplifier location on transmission line and presence or absence of RF amplifier stages
D. Height of earth antenna and satellite orbit

4BI-1.2 What factors determine the EIRP required by an amateur station in earth operation?

A. Satellite antennas and height, satellite receiver sensitivity
B. Path loss, earth antenna gain, signal-to-noise ratio
C. Satellite transmitter power and orientation of ground receiving antenna
D. Elevation of satellite above horizon, signal-to-noise ratio, satellite transmitter power

4BI-1.3 What factors determine the EIRP required by an amateur station in telecommand operation?

A. Path loss, earth antenna gain, signal-to-noise ratio
B. Satellite antennas and height, satellite receiver sensitivity
C. Satellite transmitter power and orientation of ground receiving antenna
D. Elevation of satellite above horizon, signal-to-noise ratio, satellite transmitter power

4BI-1.4 How does the gain of a parabolic dish type antenna change when the operating frequency is doubled?

A. Gain does not change
B. Gain is multiplied by 0.707
C. Gain increases 6 dB
D. Gain increases 3 dB

4BI-1.5 What happens to the beamwidth of an antenna as the gain is increased?

A. The beamwidth increases geometrically as the gain is increased
B. The beamwidth increases arithmetically as the gain is increased
C. The beamwidth is essentially unaffected by the gain of the antenna
D. The beamwidth decreases as the gain is increased

4BI-1.6 How is circular polarization produced using linearly-polarized antennas?

A. Stack two yagis, fed 90 degrees out of phase, to form an array with the respective elements in parallel planes
B. Stack two yagis, fed in phase, to form an array with the respective elements in parallel planes
C. Arrange two yagis perpendicular to each other, fed 90 degrees out of phase
D. Arrange two yagis perpendicular to each other, fed in phase

4BI-1.7 What is the beamwidth of a symmetrical pattern antenna with a gain of 20 dB as compared to an isotropic radiator?

A. 10.1 degrees
B. 20.3 degrees
C. 45.0 degrees
D. 60.9 degrees

4BI-1.8 What is the beamwidth of a symmetrical pattern antenna with a gain of 30 dB as compared to an isotropic radiator?

A. 3.2 degrees
B. 6.4 degrees
C. 37 degrees
D. 60.4 degrees

4BI-1.9 What is the beamwidth of a symmetrical pattern antenna with a gain of 15 dB as compared to an isotropic radiator?

A. 72 degrees
B. 52 degrees
C. 36.1 degrees
D. 3.61 degrees

4BI-1.10 What is the beamwidth of a symmetrical pattern antenna with a gain of 12 dB as compared to an isotropic radiator?

A. 34.8 degrees
B. 45.0 degrees
C. 58.0 degrees
D. 51.0 degrees

4BI-2.1 What is an isotropic radiator?

A. A hypothetical, omnidirectional antenna
B. In the northern hemisphere, an antenna whose directive pattern is constant in southern directions
C. An antenna high enough in the air that its directive pattern is substantially unaffected by the ground beneath it
D. An antenna whose directive pattern is substantially unaffected by the spacing of the elements

4BI-2.2 What is the antenna pattern for an isotropic radiator?

A. A figure-8
B. A unidirectional cardioid
C. A parabola
D. A sphere

4BI-2.3 What is the directivity of an isotropic radiator?

A. No directivity at all
B. Highly focused directivity; theoretically, an infinitely small beamwidth
C. 2
D. 4

4BI-2.4 When is it useful to refer to an isotropic radiator?

A. When comparing the gains of directional antennas
B. When testing a transmission line for standing wave ratio
C. When (in the northern hemisphere) directing the transmission in a southerly direction
D. When using a dummy load to tune a transmitter

4BI-2.5 What standard radiator provides a reference for comparative antenna measurements?

A. Quarter-wave vertical
B. Yagi
C. Bobtail curtain
D. Isotropic radiator

4BI-2.6 What type of directivity pattern does an isotropic radiator have?

A. A figure-8
B. A unidirectional cardioid
C. A parabola
D. A sphere

4BI-2.7 How much gain does a 1/2 wavelength dipole have over an isotropic radiator?

A. About 1.5 dB
B. About 2.1 dB
C. About 3.0 dB
D. About 6.0 dB

4BI-2.8 What purpose does an isotropic radiator serve?

A. It is used to compare signal strengths (at a distant point) of different transmitters
B. It is used as a standard of comparison for antenna gain measurements
C. It is used as a dummy load for tuning transmitters
D. It is used to measure the standing-wave-ratio on a transmission line

4BI-2.9 How much gain does an antenna have over a 1/2 wavelength dipole when it has 6 dB gain over an isotropic radiator?

A. About 3.9 dB
B. About 6.0 dB
C. About 8.1 dB
D. About 10.0 dB

4BI-2.10 How much gain does an antenna have over a 1/2 wavelength dipole when it has 12 dB gain over an isotropic radiator?

A. About 6.1 dB
B. About 9.9 dB
C. About 12.0 dB
D. About 14.1 dB

4BI-3.1 What is the radiation pattern of two 1/4 wavelength vertical antennas spaced 1/2 wavelength fed 180 degrees out of phase?

A. Unidirectional cardioid
B. Omnidirectional
C. Figure-8 broadside to the antennas
D. Figure-8 end-fire in line with the antennas

4BI-3.2 What is the radiation pattern of two 1/4 wavelength vertical antennas spaced 1/4 wavelength fed 90 degrees out of phase?

A. Unidirectional cardioid
B. Figure-8 end-fire
C. Figure-8 broadside
D. Omnidirectional

4BI-3.3 What is the radiation pattern of two 1/4 wavelength vertical antennas spaced 1/2 wavelength fed in phase?

A. Omnidirectional
B. Cardioid unidirectional
C. Figure-8 broadside to the antennas
D. Figure-8 end-fire in line with the antennas

4BI-3.4 How far apart should two 1/4 wavelength vertical antennas be spaced in order to produce a figure-8 pattern that is broadside to the plane of the verticals when fed in phase?

A. 1/8 wavelength
B. 1/4 wavelength
C. 1/2 wavelength
D. 1 wavelength

4BI-3.5 How many 1/2 wavelengths apart should two 1/4 wavelength vertica antennas be spaced to produce a figure-8 pattern that is in line wit the vertical antennas when they are fed 180 degrees out of phase?

A. one half wavelength apart
B. two half wavelengths apart
C. three half wavelengths apart
D. four half wavelengths apart

4BI-3.6 What is the radiation pattern for two 1/4 wavelength spaced 1/4 wavelength vertical antennas fed 180 degrees out of phase?

A. Omnidirectional
B. Cardioid unidirectional
C. Figure-8 broadside to the antennas
D. Figure-8 end-fire in line with the antennas

4BI-3.7 What is the radiation pattern for two 1/8 wavelength spaced 1/4 wavelength vertical antennas fed 180 degrees out of phase?

A. Omnidirectional
B. Cardioid unidirectional
C. Figure-8 broadside to the antennas
D. Figure-8 end-fire in line with the antennas

4BI-3.8 What is the radiation pattern for two 1/8 wavelength spaced 1/4 wavelength vertical antennas fed in phase?

A. Omnidirectional
B. Cardioid unidirectional
C. Figure-8 broadside to the antennas
D. Figure-8 end-fire in line with the antennas

4BI-3.9 What is the radiation pattern for two 1/4 wavelength spaced 1/4 wavelength vertical antennas fed in phase?

A. Substantially unidirectional
B. Elliptical
C. Cardioid unidirectional
D. Figure-8 end-fire in line with the antennas

4BI-3.10 What is the radiation pattern for two 1/4 wavelength spaced 1/4 wavelength vertical antennas fed 90 degrees out of phase?

A. Omnidirectional
B. Cardioid unidirectional
C. Elliptical
D. Figure-8 broadside to the antennas

4BI-4.1 What is a nonresonant rhombic antenna?

A. A unidirectional antenna terminated in a resistance equal to its characteristic impedance
B. An open-ended bidirectional antenna
C. An antenna resonant at approximately double the frequency of the intended band of operation
D. A horizontal triangular antenna consisting of two adjacent sides and the long diagonal of a resonant rhombic antenna

4BI-4.2 What is a resonant rhombic antenna?

A. A unidirectional antenna, each of whose sides is equal to half a wavelength and which is terminated in a resistance equal to its characteristic impedance
B. A bidirectional antenna open at the end opposite that to which the transmission line is connected and with each side approximately equal to one wavelength
C. An antenna with an LC network at each vertex (other than that to which the transmission line is connected) tuned to resonate at the operating frequency
D. A high-frequency antenna, each of whose sides contains traps for changing the resonance to match the band in use

4BI-4.3 What is the effect of a terminating resistor on a rhombic antenna?

A. It reflects the standing waves on the antenna elements back to the transmitter
B. It changes the radiation pattern from essentially bidirectional to essentially unidirectional
C. It changes the radiation pattern from horizontal to vertical polarization
D. It decreases the ground loss

4BI-4.4 What should be the value of the terminating resistor on a rhombic antenna?

A. About 50 ohms
B. About 75 ohms
C. About 800 ohms
D. About 1800 ohms

4BI-4.5 What are the advantages of a nonresonant rhombic antenna?

A. Wide frequency range, high gain and high front-to-back ratio
B. High front-to-back ratio, compact size and high gain
C. Unidirectional radiation pattern, high gain and compact size
D. Bidirectional radiation pattern, high gain and wide frequency range

4BI-4.6 What are the disadvantages of a nonresonant rhombic antenna?

A. It requires a large area for proper installation and has a narrow bandwidth
B. It requires a large area for proper installation and has a low front-to-back ratio
C. It requires a large amount of aluminum tubing and has a low front-to-back ratio
D. It requires a large area and four sturdy supports for proper installation

4BI-4.7 What is the characteristic impedance at the input of a nonresonant rhombic antenna?

A. 50 to 55 ohms
B. 70 to 75 ohms
C. 300 to 350 ohms
D. 700 to 800 ohms

4BI-5.1 What is the delta matching system for matching an antenna to a feed line?

A. A method of matching a high-impedance transmission line to a lower-impedance antenna by connecting the line to the driven element in two places, spaced a fraction of a wavelength on each side of the center of the driven element
B. A method by which the antenna current is made to flow around the perimeter of an equilateral triangle after leaving the transmission line and before entering the driven element
C. A method using a three-conductor transmission line connected to the driven element at points located 1/4, 1/2, and 3/4 of the distance from one end of the driven element to the other
D. A method by which a coaxial transmission line feeds the driven element both at the center and at one end of that element, with residual inductive reactance tuned out by a variable capacitor inserted between the center conductor and the driven element

4BI-5.2 What is the gamma matching system for matching an antenna to a feed line?

A. A balanced feed system in which the transmission line is connected to the antenna in two places, spaced a fraction of a wavelength on each side of the center of the driven element
B. A balanced feed system in which residual capacitances are tuned out at the antenna itself by a remotely controlled variable inductor, producing resonance at the precise frequency in use
C. An unbalanced feed system in which the driven element is fed at one end and also at 1/8 wavelength from that end
D. An unbalanced feed system in which the driven element is fed both at the center of that element and a fraction of a wavelength to one side of the center

4BI-5.3 What is the universal stub system for matching an antenna to a feed line?

A. Connecting the transmission line to the antenna at points 1/3 and 2/3 the distance from one end of the driven element to the other
B. Connecting only the center conductor of a coaxial transmission line to the antenna and allowing the outer braid to form a stub perpendicular to the transmission line and close to but not touching the antenna
C. Connecting a short section of transmission line near the antenna and perpendicular to the transmission line
D. Connecting the coaxial transmission line at both the center of the driven element, and a fraction of a wavelength to one side of the center

4BI-5.4 What should be the maximum capacitance of the resonating capacitor in a gamma matching circuit on a 1/2 wavelength dipole antenna for the 20 meter band?

A. 70 pF
B. 140 pF
C. 200 pF
D. 0.2 pF

4BI-5.5 What should be the maximum capacitance of the resonating capacitor in a gamma matching circuit on a 1/2 wavelength dipole antenna for the 10 meter band?

A. 70 pF
B. 140 pF
C. 200 pF
D. 0.2 pF

4BI-6.1 How does a 1/4 wavelength transmission line appear to a generator when the line is shorted at the far end?

A. As a very high impedance
B. As a very low impedance
C. The same as the characteristic impedance of the transmission line
D. The same as the generator output impedance

4BI-6.2 How does a 1/4 wavelength transmission line appear to a generator when the line is open at the far end?

A. As a very high impedance
B. As a very low impedance
C. The same as the characteristic impedance of the line
D. The same as the input impedance to the final generator stage

4BI-6.3 How does a 1/8 wavelength transmission line appear to a generator when the line is shorted at the far end?

A. As a capacitive reactance
B. The same as the characteristic impedance of the line
C. As an inductive reactance
D. The same as the input impedance to the final generator stage

4BI-6.4 How does a 1/8 wavelength transmission line appear to a generator when the line is open at the far end?

A. The same as the characteristic impedance of the line
B. As an inductive reactance
C. As a capacitive reactance
D. The same as the input impedance of the final generator stage

4BI-6.5 How does a 1/2 wavelength transmission line appear to a generator when the line is shorted at the far end?

A. As a very high impedance
B. As a very low impedance
C. The same as the characteristic impedance of the line
D. The same as the output impedance of the generator

4BI-6.6 How does a 1/2 wavelength transmission line appear to a generator when the line is open at the far end?

A. As a very high impedance
B. As a very low impedance
C. The same as the characteristic impedance of the line
D. The same as the output impedance of the generator

4BI-6.7 How does a 3/8 wavelength transmission line appear to a generator when the line is shorted at the far end?

A. The same as the characteristic impedance of the line
B. As an inductive reactance
C. As a capacitive reactance
D. The same as the input impedance to the final generator stage

4BI-6.8 How does a 3/8 wavelength transmission line appear to a generator when the line is open at the far end?

A. As a capacitive reactance
B. The same as the characteristic impedance of the line
C. As an inductive reactance
D. The same as the input impedance to the final generator stage

Element 4B Answer Key

Subelement 4BA

Question	Answer
4BA-1.1	B
4BA-1.2	A
4BA-1.3	A
4BA-1.4	D
4BA-1.5	C
4BA-1.6	B
4BA-1.7	A
4BA-1.8	C
4BA-2.1	D
4BA-2.2	A
4BA-2.3	D
4BA-2.4	B
4BA-2.5	A
4BA-2.6	D
4BA-2.7	C
4BA-2.8	D
4BA-3.1	B
4BA-3.2	D
4BA-3.3	A
4BA-3.4	D
4BA-4.1	A
4BA-4.2	B
4BA-5.1	B
4BA-5.2	A
4BA-5.3	C
4BA-5.4	D
4BA-5.5	A
4BA-5.6	C
4BA-6.1	B
4BA-6.2	B
4BA-6.3	A
4BA-7.1	B
4BA-7.2	A
4BA-7.3	B
4BA-7.4	A
4BA-7.5	D
4BA-7.6	C
4BA-7.7	C
4BA-8.1	C
4BA-8.2	B
4BA-8.3	A
4BA-9.1	C
4BA-9.2	D
4BA-9.3	A
4BA-9.4	C
4BA-9.5	B
4BA-9.6	C
4BA-9.7	B
4BA-10.1	D
4BA-10.2	D
4BA-10.3	A
4BA-10.4	C
4BA-11.1	A
4BA-11.2	B
4BA-11.3	D
4BA-11.4	C
4BA-11.5	B
4BA-11.6	A
4BA-11.7	C
4BA-11.8	C
4BA-12.1	D
4BA-12.2	A
4BA-12.3	C
4BA-12.4	B
4BA-12.5	B
4BA-12.6	D
4BA-13.1	A
4BA-13.2	D
4BA-13.3	A
4BA-13.4	D
4BA-13.5	D
4BA-13.6	A
4BA-14.1	C
4BA-14.2	B
4BA-14.3	A
4BA-14.4	D
4BA-14.5	A
4BA-14.6	B
4BA-15.1	B
4BA-15.2	B
4BA-15.3	B
4BA-15.4	B
4BA-16.1	A
4BA-16.2	B
4BA-16.3	D
4BA-16.4	B
4BA-17.1	A
4BA-17.2	B
4BA-17.3	B

4BA-17.4	A
4BA-17.5	C
4BA-17.6	B
4BA-17.7	A
4BA-18.1	D
4BA-18.2	A
4BA-18.3	C
4BA-18.4	A
4BA-19.1	D
4BA-19.2	A
4BA-20.1	D
4BA-20.2	B
4BA-20.3	A
4BA-20.4	D
4BA-21.1	A
4BA-21.2	A
4BA-21.3	A
4BA-21.4	A
4BA-21.5	A
4BA-21.6	A
4BA-21.7	A
4BA-22.1	A
4BA-22.2	D
4BA-23.1	D
4BA-23.2	A
4BA-24.1	A
4BA-24.2	A
4BA-25.1	D
4BA-25.2	A
4BA-26.1	D
4BA-26.2	A
4BA-27.1	B
4BA-28.1	B
4BA-28.2	B
4BA-29.1	D
4BA-29.2	D

Subelement 4BB

4BB-1.1	D
4BB-1.2	A
4BB-1.3	B
4BB-1.4	D
4BB-1.5	B
4BB-1.6	B
4BB-1.7	B
4BB-1.8	C
4BB-1.9	A
4BB-1.10	C
4BB-2.1	A
4BB-2.2	C
4BB-2.3	C
4BB-2.4	B
4BB-2.5	A
4BB-2.6	D
4BB-2.7	D
4BB-2.8	C
4BB-2.9	B
4BB-2.10	C

Subelement 4BC

4BC-1.1	D
4BC-1.2	B
4BC-1.3	A
4BC-1.4	D
4BC-1.5	B
4BC-2.1	B
4BC-2.2	C
4BC-3.1	A
4BC-3.2	C
4BC-3.3	C

Subelement 4BD

4BD-1.1	A
4BD-1.2	B
4BD-1.3	C
4BD-1.4	D
4BD-1.5	A
4BD-2.1	B
4BD-2.2	C
4BD-2.3	D
4BD-2.4	D
4BD-3.1	A
4BD-3.2	A
4BD-3.3	C
4BD-3.4	D
4BD-3.5	B
4BD-3.6	B
4BD-3.7	D
4BD-3.8	C
4BD-3.9	D
4BD-3.10	C
4BD-4.1	A
4BD-4.2	B
4BD-4.3	C
4BD-4.4	D
4BD-4.5	A
4BD-4.6	B
4BD-4.7	C

4BD-4.8 D
4BD-4.9 B
4BD-4.10 C

Subelement 4BE

4BE-1.1 B
4BE-1.2 A
4BE-1.3 D
4BE-1.4 C
4BE-1.5 D
4BE-1.6 A
4BE-1.7 B
4BE-1.8 C
4BE-1.9 A
4BE-1.10 D
4BE-2.1 D
4BE-2.2 C
4BE-2.3 B
4BE-2.4 A
4BE-2.5 D
4BE-2.6 D
4BE-2.7 B
4BE-2.8 A
4BE-2.9 C
4BE-2.10 D
4BE-3.1 C
4BE-3.2 A
4BE-3.3 B
4BE-3.4 B
4BE-3.5 C
4BE-3.6 C
4BE-3.7 B
4BE-3.8 D
4BE-3.9 A
4BE-3.10 A
4BE-3.11 B
4BE-3.12 A
4BE-3.13 D
4BE-3.14 A
4BE-3.15 C
4BE-3.16 D
4BE-3.17 D
4BE-3.18 A
4BE-3.19 D
4BE-3.20 C
4BE-4.1 A
4BE-4.2 B
4BE-4.3 C
4BE-4.4 C
4BE-4.5 B
4BE-4.6 C
4BE-4.7 D
4BE-4.8 C
4BE-4.9 C
4BE-4.10 B
4BE-5.1 A
4AE-5.2 B
4BE-5.3 A
4BE-5.4 B
4BE-5.5 C
4BE-5.6 D
4BE-5.7 A
4BE-5.8 D
4BE-5.9 B
4BE-5.10 C
4BE-6.1 B
4BE-6.2 C
4BE-6.3 B
4BE-6.4 C
4BE-6.5 B
4BE-6.6 D
4BE-6.7 A
4BE-6.8 B
4BE-6.9 A
4BE-6.10 D

Subelement 4BF

4BF-1.1 A
4BF-1.2 B
4BF-1.3 D
4BF-1.4 A
4BF-1.5 A
4BF-1.6 B
4BF-1.7 D
4BF-1.8 C
4BF-1.9 D
4BF-1.10 B
4BF-1.11 D
4BF-1.12 A
4BF-1.13 D
4BF-1.14 B
4BF-1.15 A
4BF-1.16 D
4BF-2.1 A
4BF-2.2 A
4BF-2.3 B
4BF-2.4 A
4BF-2.5 A

4BF-2.6	B
4BF-2.7	C
4BF-2.8	D
4BF-2.9	D
4BF-2.10	B
4BF-3.1	C
4BF-3.2	B
4BF-3.3	A
4BF-3.4	A
4BF-3.5	C
4BF-3.6	D
4BF-3.7	A
4BF-3.8	A
4BF-3.9	A
4BF-3.10	C
4BF-3.11	D
4BF-3.12	C
4BF-3.13	B
4BF-3.14	B
4BF-3.15	C
4BF-4.1	C
4BF-4.2	D
4BF-4.3	B
4BF-4.4	C
4BF-5.1	C
4BF-5.2	D
4BF-5.3	C
4BF-5.4	D
4BF-5.5	B

Subelement 4BG

4BG-1.1	D
4BG-1.2	A
4BG-1.3	C
4BG-1.4	D
4BG-1.5	A
4BG-1.6	A
4BG-1.7	D
4BG-1.8	B
4BG-1.9	D
4BG-1.10	A
4BG-2.1	C
4BG-2.2	C
4BG-2.3	D
4BG-2.4	C
4BG-2.5	D
4BG-2.6	A
4BG-2.7	A
4BG-2.8	B
4BG-3.1	D
4BG-3.2	B
4BG-3.3	D
4BG-3.4	C
4BG-3.5	A
4BG-3.6	D
4BG-3.7	C
4BG-3.8	B
4BG-3.9	D
4BG-3.10	A
4BG-4.1	C
4BG-4.2	D
4BG-4.3	C
4BG-4.4	A
4BG-4.5	B
4BG-5.1	B
4BG-5.2	B
4BG-5.3	D
4BG-5.4	B
4BG-5.5	B
4BG-5.6	B
4BG-5.7	B
4BG-5.8	A
4BG-5.9	D
4BG-5.10	D
4BG-6.1	C
4BG-6.2	D
4BG-6.3	C
4BG-6.4	A
4BG-6.5	B
4BG-7.1	D
4BG-7.2	C
4BG-7.3	B
4BG-7.4	C
4BG-7.5	A
4BG-8.1	C
4BG-8.2	A
4BG-9.1	A
4BG-9.2	A
4BG-9.3	C
4BG-9.4	A
4BG-9.5	C
4BG-9.6	A
4BG-9.7	A
4BG-9.8	C

Subelement 4BH

4BH-1.1	A
4BH-1.2	C

4BH-1.3	D
4BH-1.4	A
4BH-1.5	D
4BH-2.1	D
4BH-2.2	B
4BH-2.3	C
4BH-2.4	D
4BH-2.5	B
4BH-3.1	C
4BH-3.2	A
4BH-3.3	A
4BH-3.4	D
4BH-3.5	B
4BH-4.1	B
4BH-4.2	B
4BH-4.3	B
4BH-4.4	B
4BH-4.5	C
4BH-4.6	C
4BH-4.7	D
4BH-4.8	C
4BH-4.9	A
4BH-4.10	A
4BH-4.11	C
4BH-4.12	C
4BH-4.13	B
4BH-4.14	B
4BH-4.15	B
4BH-4.16	B
4BH-4.17	C
4BH-4.18	C
4BH-4.19	C
4BH-4.20	C
4BH-5.1	D
4BH-5.2	D
4BH-6.1	A
4BH-6.2	B
4BH-6.3	A

Subelement 4BI

4BI-1.1	A
4BI-1.2	A
4BI-1.3	B
4BI-1.4	C
4BI-1.5	D
4BI-1.6	C
4BI-1.7	B
4BI-1.8	B
4BI-1.9	C
4BI-1.10	D
4BI-2.1	A
4BI-2.2	D
4BI-2.3	A
4BI-2.4	A
4BI-2.5	D
4BI-2.6	D
4BI-2.7	B
4BI-2.8	B
4BI-2.9	A
4BI-2.10	B
4BI-3.1	D
4BI-3.2	A
4BI-3.3	C
4BI-3.4	C
4BI-3.5	A
4BI-3.6	D
4BI-3.7	D
4BI-3.8	A
4BI-3.9	B
4BI-3.10	B
4BI-4.1	A
4BI-4.2	B
4BI-4.3	B
4BI-4.4	C
4BI-4.5	A
4BI-4.6	D
4BI-4.7	D
4BI-5.1	A
4BI-5.2	D
4BI-5.3	C
4BI-5.4	B
4BI-5.5	A
4BI-6.1	A
4BI-6.2	B
4BI-6.3	C
4BI-6.4	C
4BI-6.5	B
4BI-6.6	A
4BI-6.7	C
4BI-6.8	C

Appendix A

U S Customary—Metric Conversion Factors

International System of Units (SI) — Metric Units

Prefix	Symbol		Multiplication Factor
exa	E	10^{18} =	1,000,000,000,000,000,000
peta	P	10^{15} =	1,000,000,000,000,000
tera	T	10^{12} =	1,000,000,000,000
giga	G	10^{9} =	1,000,000,000
mega	M	10^{6} =	1,000,000
kilo	k	10^{3} =	1,000
hecto	h	10^{2} =	100
deca	da	10^{1} =	10
(unit)		10^{0} =	1
deci	d	10^{-1} =	0.1
centi	c	10^{-2} =	0.01
milli	m	10^{-3} =	0.001
micro	μ	10^{-6} =	0.000001
nano	n	10^{-9} =	0.000000001
pico	p	10^{-12} =	0.000000000001
femto	f	10^{-15} =	0.000000000000001
atto	a	10^{-18} =	0.000000000000000001

Linear

1 meter (m) = 100 centimeters (cm) = 1000 millimeters (mm)

Area

1 m^2 = 1 × 10^4 cm^2 = 1 × 10^6 mm^2

Volume

1 m^3 = 1 × 10^6 cm^3 = 1 × 10^9 mm^3
1 liter (l) = 1000 cm^3 = 1 × 10^6 mm^3

Mass

1 kilogram (kg) = 1000 grams (g)
(Approximately the mass of 1 liter of water)

1 metric ton (or tonne) = 1000 kg

U S Customary Units

Linear Units

12 inches (in) = 1 foot (ft)
36 inches = 3 feet = 1 yard (yd)
1 rod = 5½ yards = 16½ feet
1 statute mile = 1760 yards = 5280 feet
1 nautical mile = 6076.11549 feet

Area

1 ft^2 = 144 in^2
1 yd^2 = 9 ft^2 = 1296 in^2
1 rod^2 = 30¼ yd^2
1 acre = 4840 yd^2 = 43,560 ft^2
1 acre = 160 rod^2
1 $mile^2$ = 640 acres

Volume

1 ft^3 = 1728 in^3
1 yd^3 = 27 ft^3

Liquid Volume Measure

1 fluid ounce (fl oz) = 8 fluidrams = 1.804 in^3
1 pint (pt) = 16 fl oz
1 quart (qt) = 2 pt = 32 fl oz = 57¾ in^3
1 gallon (gal) = 4 qt = 231 in^3
1 barrel = 31½ gal

Dry Volume Measure

1 quart (qt) = 2 pints (pt) = 67.2 in^3
1 peck = 8 qt

1 bushel = 4 pecks = 2150.42 in^3

Avoirdupois Weight

1 dram (dr) = 27.343 grains (gr) or (gr a)
1 ounce (oz) = 437.5 gr
1 pound (lb) = 16 oz = 7000 gr
1 short ton = 2000 lb, 1 long ton = 2240 lb

Troy Weight

1 grain troy (gr t) = 1 grain avoirdupois
1 pennyweight (dwt) or (pwt) = 24 gr t
1 ounce troy (oz t) = 480 grains
1 lb t = 12 oz t = 5760 grains

Apothecaries' Weight

1 grain apothecaries' (gr ap) = 1 gr t = 1 gr a
1 dram ap (dr ap) = 60 gr
1 oz ap = 1 oz t = 8 dr ap = 480 gr
1 lb ap = 1 lb t = 12 oz ap = 5760 gr

Temperature

°F = 9/5°C + 32
°C = 5/9 (°F − 32)
K = °C + 273
°C = K − 273

Multiply ⟶

Metric Unit = Conversion Factor × U.S. Customary Unit

⟵ Divide

Metric Unit ÷ Conversion Factor = U.S. Customary Unit

Metric Unit =	Conversion Factor ×	U.S. Unit
(Length)		
mm	25.4	inch
cm	2.54	inch
cm	30.48	foot
m	0.3048	foot
m	0.9144	yard
km	1.609	mile
km	1.852	nautical mile
(Area)		
mm^2	645.16	$inch^2$
cm^2	6.4516	in^2
cm^2	929.03	ft^2
m^2	0.0929	ft^2
cm^2	8361.3	yd^2
m^2	0.83613	yd^2
m^2	4047	acre
km^2	2.59	mi^2
(Mass)	(Avoirdupois Weight)	
grams	0.0648	grains
g	28.349	oz
g	453.59	lb
kg	0.45359	lb
tonne	0.907	short ton
tonne	1.016	long ton

Metric Unit =	Conversion Factor ×	U.S. Unit
(Volume)		
mm^3	16387.064	in^3
cm^3	16.387	in^3
m^3	0.028316	ft^3
m^3	0.764555	yd^3
ml	16.387	in^3
ml	29.57	fl oz
ml	473	pint
ml	946.333	quart
l	28.32	ft^3
l	0.9463	quart
l	3.785	gallon
l	1.101	dry quart
l	8.809	peck
l	35.238	bushel
(Mass)	(Troy Weight)	
g	31.103	oz t
g	373.248	lb t
(Mass)	(Apothecaries' Weight)	
g	3.387	dr ap
g	31.103	oz ap
g	373.248	lb ap

Standard Resistance Values

Numbers in **bold** type are ±10% values. Others are 5% values.

Ohms										*Megohms*				
1.0	3.6	**12**	43	**150**	510	**1800**	6200	**22000**	75000	0.24	0.62	1.6	4.3	11.0
1.1	**3.9**	13	**47**	160	**560**	2000	**6800**	24000	**82000**	**0.27**	**0.68**	**1.8**	**4.7**	**12.0**
1.2	4.3	**15**	51	**180**	620	**2200**	7500	**27000**	91000	0.30	0.75	2.0	5.1	13.0
1.3	**4.7**	16	**56**	200	**680**	2400	**8200**	30000	**100000**	**0.33**	**0.82**	**2.2**	**5.6**	**15.0**
1.5	5.1	**18**	62	**220**	750	**2700**	9100	**33000**	110000	0.36	0.91	2.4	6.2	16.0
1.6	**5.6**	20	**68**	240	**820**	3000	**10000**	36000	**120000**	**0.39**	**1.0**	**2.7**	**6.8**	**18.0**
1.8	6.2	**22**	75	**270**	910	**3300**	11000	**39000**	130000	0.43	1.1	3.0	7.5	20.0
2.0	**6.8**	24	**82**	300	**1000**	3600	**12000**	43000	**150000**	**0.47**	**1.2**	**3.3**	**8.2**	**22.0**
2.2	7.5	**27**	91	**330**	1100	**3900**	13000	**47000**	160000	0.51	1.3	3.6	9.1	
2.4	**8.2**	30	**100**	360	**1200**	4300	**15000**	51000	**180000**	**0.56**	**1.5**	**3.9**	**10.0**	
2.7	9.1	**33**	110	**390**	1300	**4700**	16000	**56000**	200000					
3.0	**10.0**	36	**120**	430	**1500**	5100	**18000**	62000	**220000**					
3.3	11.0	**39**	130	**470**	1600	**5600**	20000	**68000**						

Resistor Color Code

Color	*Sig. Figure*	*Decimal Multiplier*	*Tolerance (%)*
Black	0	1	
Brown	1	10	
Red	2	100	
Orange	3	1,000	
Yellow	4	10,000	
Green	5	100,000	
Blue	6	1,000,000	

Color	*Sig. Figure*	*Decimal Multiplier*	*Tolerance (%)*
Violet	7	10,000,000	
Gray	8	100,000,000	
White	9	1,000,000,000	
Gold	—	0.1	5
Silver	—	0.01	10
No color	—		20

Schematic Symbols Used in Circuit Diagrams

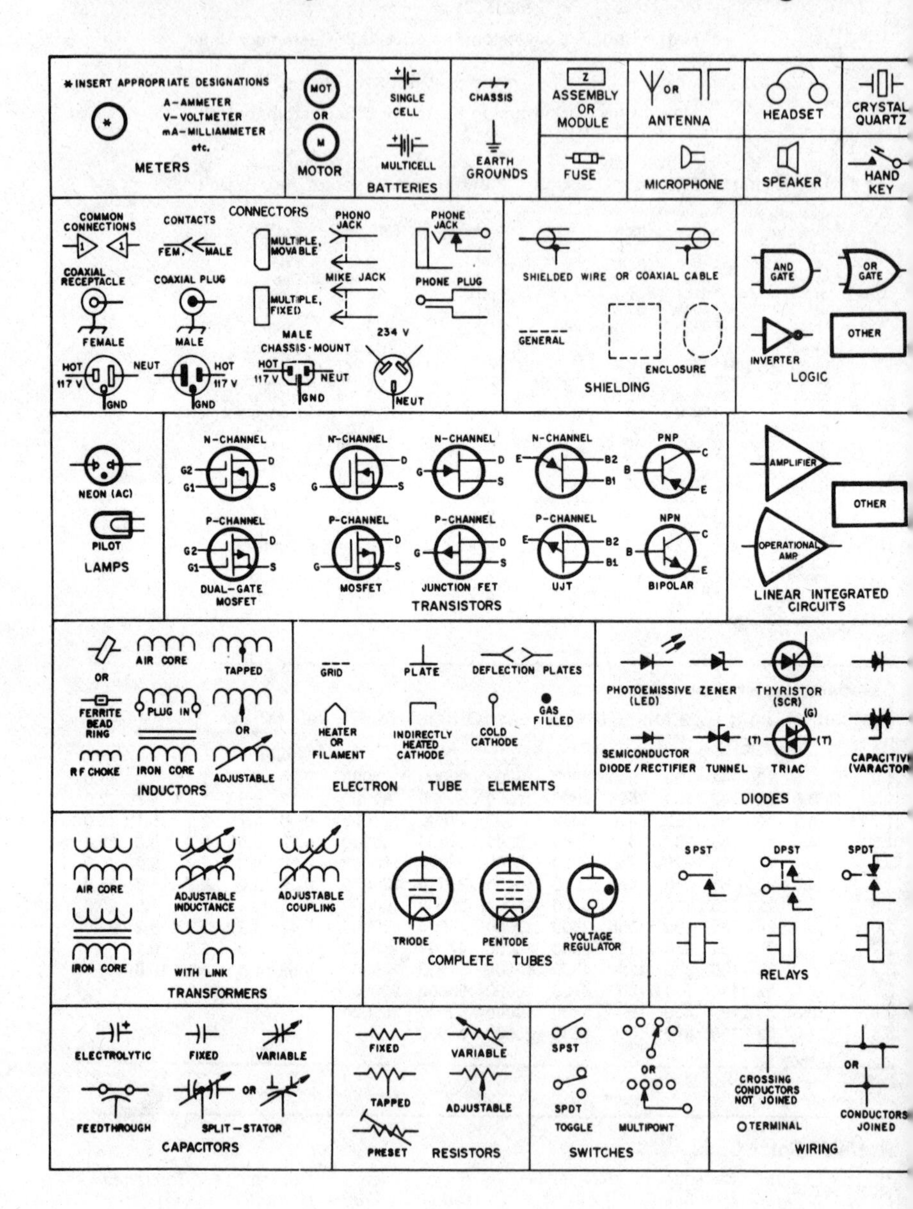

Standard Capacitance Values

pF	*pF*
0.3	470
5	500
6	510
6.8	560
7.5	600
8	680
10	750
12	800
15	820
18	910
20	1000
22	1000
24	1200
25	1200
27	1300
30	1500
33	1500
39	1600
47	1800
50	2000
51	2200
56	2500
68	2700
75	3000
82	3300
91	3900
100	4000
120	4300
130	4700
150	4700
180	5000
200	5000
220	5600
240	6800
250	7500
270	8200
300	10000
330	10000
350	20000
360	30000
390	40000
400	50000
470	

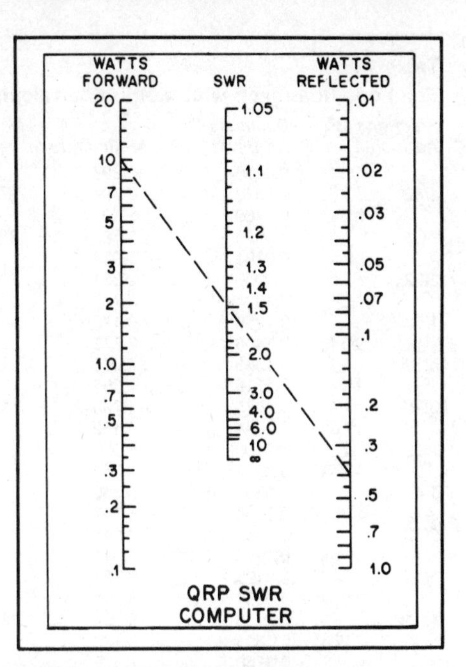

Nomograph of SWR versus forward and reflected power for levels up to 20 watts. Dashed line shows an SWR of 1.5:1 for 10 W forward and 0.4 W reflected.

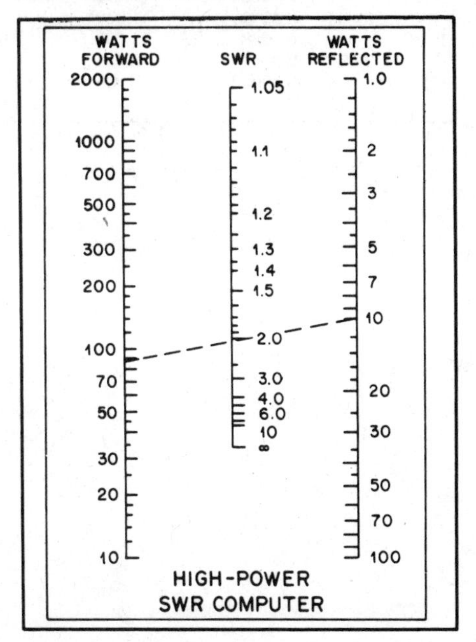

Nomograph of SWR versus forward and reflected power for levels up to 2000 watts. Dashed line shows an SWR of 2:1 for 90 W forward and 10 W reflected.

Table 1

Fractions of an Inch with Metric Equivalents

Fractions Of An Inch		*Decimals Of An Inch*	*Millimeters*	*Fractions Of An Inch*		*Decimals Of An Inch*	*Millimeters*
	1/64	0.0156	0.397		33/64	0.5156	13.097
1/32		0.0313	0.794	17/32		0.5313	13.494
	3/64	0.0469	1.191		35/64	0.5469	13.891
		0.0625	1.588	9/16		0.5625	14.288
	5/64	0.0781	1.984		37/64	0.5781	14.684
3/32		0.0938	2.381	19/32		0.5938	15.081
	7/64	0.1094	2.778		39/64	0.6094	15.478
1/8		0.1250	3.175	5/8		0.6250	15.875
	9/64	0.1406	3.572		41/64	0.6406	16.272
5/32		0.1563	3.969	21/32		0.6563	16.669
	11/64	0.1719	4.366		43/64	0.6719	17.066
3/16		0.1875	4.763	11/16		0.6875	17.463
	13/64	0.2031	5.159		45/64	0.7031	17.859
7/32		0.2188	5.556	23/32		0.7188	18.256
	15/64	0.2344	5.953		47/64	0.7344	18.653
1/4		0.2500	6.350	3/4		0.7500	19.050
	17/64	0.2656	6.747		49/64	0.7656	19.447
9/32		0.2813	7.144	25/32		0.7813	19.844
	19/64	0.2969	7.541		51/64	0.7969	20.241
5/16		0.3125	7.938	13/16		0.8125	20.638
	21/64	0.3281	8.334		53/64	0.8281	21.034
11/32		0.3438	8.731	27/32		0.8438	21.431
	23/64	0.3594	9.128		55/64	0.8594	21.828
3/8		0.3750	9.525	7/8		0.8750	22.225
	25/64	0.3906	9.922		57/64	0.8906	22.622
13/32		0.4063	10.319	29/32		0.9063	23.019
	27/64	0.4219	10.716		59/64	0.9219	23.416
7/16		0.4375	11.113	15/16		0.9375	23.813
	29/64	0.4531	11.509		61/64	0.9531	24.209
15/32		0.4688	11.906	31/32		0.9688	24.606
	31/64	0.4844	12.303		63/64	0.9844	25.003
1/2		0.50000	12.700	1		1.0000	25.400

Appendix B

Equations Used in this Book

$$E = hf \quad \text{(Eq. 5-1)}$$

$$R = \frac{\rho \ell}{A} \quad \text{(Eq. 5-2)}$$

$$\sigma = \frac{1}{\rho} \quad \text{(Eq. 5-3)}$$

$$G = \frac{1}{R} \quad \text{(Eq. 5-4)}$$

$$\tau = RC \quad \text{(Eq. 5-5)}$$

$$R_T \text{ (series)} = R1 + R2 + R3 + \ldots + Rn \quad \text{(Eq. 5-6)}$$

$$R_T \text{ (parallel)} = \frac{1}{\frac{1}{R1} + \frac{1}{R2} + \frac{1}{R3} + \ldots + \frac{1}{Rn}} \quad \text{(Eq. 5-7)}$$

$$C_T \text{ (series)} = \frac{1}{\frac{1}{C1} + \frac{1}{C2} + \frac{1}{C3} + \ldots + \frac{1}{Cn}} \quad \text{(Eq. 5-8)}$$

$$C_T \text{ (parallel)} = C1 + C2 + C3 + \ldots + Cn \quad \text{(Eq. 5-9)}$$

$$V_{(t)} = E\left(1 - e^{\frac{-t}{\tau}}\right) \quad \text{(Eq. 5-10)}$$

$$v_{(t)} = E\left(e^{\frac{-t}{\tau}}\right) \quad \text{(Eq. 5-11)}$$

$$I_{(t)} = \frac{E}{R}\left(1 - e^{\frac{-t}{\tau}}\right) \quad \text{(Eq. 5-12)}$$

$$\tau = \frac{L}{R} \quad \text{(Eq. 5-13)}$$

$$I_{(t)} = \frac{E}{R}\left(e^{\frac{-t}{\tau}}\right) \quad \text{(Eq. 5-14)}$$

$$Z = \frac{E}{I} = \frac{E_R + E_X}{I} \quad \left(\begin{array}{c}\text{Series}\\\text{Circuit}\end{array}\right) \qquad \text{(Eq. 5-15)}$$

$$Z = \frac{E}{I} = \frac{E}{I_R + I_X} \quad \left(\begin{array}{c}\text{Parallel}\\\text{Circuit}\end{array}\right) \qquad \text{(Eq. 5-15A)}$$

$$X_L = 2\pi fL \qquad \text{(Eq. 5-16)}$$

$$X_C = \frac{1}{2\pi fC} \qquad \text{(Eq. 5-17)}$$

$$Z = \sqrt{R^2 + X^2} \qquad \text{(Eq. 5-18)}$$

$$\tan\theta = \frac{\text{side opposite the angle}}{\text{side adjacent to the angle}} \qquad \text{(Eq. 5-19)}$$

$$I_R = \frac{E_R}{R} = E_R \times G \qquad \text{(Eq. 5-20)}$$

$$I_X = \frac{E_X}{X} = E_X \times B \qquad \text{(Eq. 5-21)}$$

$$X = Z \sin\theta \qquad \text{(Eq. 5-22)}$$

$$R = Z \cos\theta \qquad \text{(Eq. 5-23)}$$

$$V_{out} = \frac{Rf}{R1} V_{in} \qquad \text{(Eq. 6-1)}$$

$$V_{gain} = \frac{V_{out}}{V_{in}} \qquad \text{(Eq. 6-2)}$$

$$V_{gain} = \frac{\frac{Rf}{R1} V_{in}}{V_{in}} = \frac{Rf}{R1} \qquad \text{(Eq. 6-3)}$$

$$\text{Gain (dB)} = 20 \log(V_{gain}) \qquad \text{(Eq. 6-4)}$$

$$V_{out} = \frac{R1 + Rf}{R1} V_{in} \qquad \text{(Eq. 6-5)}$$

$$V_{gain} = \frac{V_{out}}{V_{in}} = \frac{\frac{R1 + Rf}{R1} V_{in}}{V_{in}} = \frac{R1 + Rf}{R1} \qquad \text{(Eq. 6-6)}$$

$$V_{out} = \frac{R1 + Rf}{R2 + R3} \; \frac{R3}{R1} V2 - \frac{Rf}{R1} V1 \qquad \text{(Eq. 6-7)}$$

$$\overline{A \cdot B} = \overline{A} + \overline{B} \qquad \text{(Eq. 6-8)}$$

$$\overline{A + B} = \overline{A} \cdot \overline{B} \qquad \text{(Eq. 6-9)}$$

$$T = 1.1\ RC \qquad \text{(Eq. 7-1)}$$

$$f = \frac{1.46}{(R1 + (2 \times R2))\ C} \qquad \text{(Eq. 7-2)}$$

$$V_{out} = \frac{Rf}{R1} V_{in} \qquad \text{(Eq. 7-3)}$$

$$V_{gain} = \frac{Rf}{R1} \qquad \text{(Eq. 7-4)}$$

$$V_{out} = \frac{R1 + Rf}{R1} V_{in} \qquad \text{(Eq. 7-5)}$$

$$V_{gain} = \frac{R1 + Rf}{R1} \qquad \text{(Eq. 7-6)}$$

$$V_{out} = \left(\frac{R1 + Rf}{R2 + R3}\right)\left(\frac{R3}{R1}\right)(V2) - \left(\frac{Rf}{R1}\right)(V1) \qquad \text{(Eq. 7-7)}$$

$$V_{out} = \frac{Rf}{R1}(V2 - V1) \qquad \text{(Eq. 7-8)}$$

$$R1 = \frac{Q}{2\ \pi\ f_o\ A_v\ C1} \qquad \text{(Eq. 7-9)}$$

$$R2 = \frac{Q}{(2Q^2 - A_v)\ (2\ \pi\ f_o C_1)} \qquad \text{(Eq. 7-10)}$$

$$R3 = \frac{2Q}{2\pi\ f_o\ C1} \qquad \text{(Eq. 7-11)}$$

$$R4 = R5 \approx 0.02 \times R3 \qquad \text{(Eq. 7-12)}$$

$$G_m = \frac{\Delta\ I_D}{\Delta\ E_{GS}} \qquad \text{(Eq. 7-13)}$$

$$A_v = G_m\ R_L \qquad \text{(Eq. 7-14)}$$

$$NF = \frac{\dfrac{S_{in}}{N_{in}}}{\dfrac{S_{out}}{N_{out}}} = \frac{S_{in}\ N_{out}}{S_{out}\ N_{in}} \qquad \text{(Eq. 7-15)}$$

$$P_n = k\,T_o\,B \qquad \text{(Eq. 7-16)}$$

$$\text{Blocking dynamic range} = \text{noise floor} - \text{blocking level} \qquad \text{(Eq. 7-17)}$$

$$\text{IMD DR} = \text{noise floor} - \text{IMD level} \qquad \text{(Eq. 7-18)}$$

$$f_o = \frac{1}{2\pi\sqrt{L\,C}} \qquad \text{(Eq. 7-19)}$$

$$BW = 2M + 2DK \qquad \text{(Eq. 7-20)}$$

$$X_C = \frac{1}{2\pi\,f\,C} \qquad \text{(Eq. 7-21)}$$

$$T_b = \frac{1}{B} \qquad \text{(Eq. 8-1)}$$

$$Bw = B \times K \qquad \text{(Eq. 8-2)}$$

$$\frac{1\text{ word}}{\text{min}} = \frac{50\text{ elements}}{\text{min}} = \frac{50\text{ elements}}{60\text{ s}} = 0.83\text{ bauds} \qquad \text{(Eq. 8-3)}$$

$$\text{WPM} \times 0.83 = \text{bauds} \qquad \text{(Eq. 8-4)}$$

$$\text{bauds} = \frac{\text{WPM}}{1.2} \qquad \text{(Eq. 8-5)}$$

$$Bw = \left(\frac{\text{WPM}}{1.2}\right) \times K \qquad \text{(Eq. 8-6)}$$

$$Bw = \text{WPM} \times (3 / 1.2) = \text{WPM} \times 2.5 \qquad \text{(Eq. 8-7)}$$

$$Bw = \text{WPM} \times (5 / 1.2) = \text{WPM} \times 4.2 \qquad \text{(Eq. 8-8)}$$

$$Bw = \text{WPM} \times 4 \qquad \text{(Eq. 8-9)}$$

$$Bw = (K \times \text{Shift}) + B \qquad \text{(Eq. 8-10)}$$

$$Bw = (1.2 \times 170\text{ Hz}) + 45 = 249\text{ Hz} \qquad \text{(Eq. 8-11)}$$

$$\text{peak-peak} = \text{peak positive} - \text{peak negative} \qquad \text{(Eq. 8-12)}$$

$$\text{peak-to-peak} = 2 \times \text{peak positive} = -2 \times \text{peak negative} \qquad \text{(Eq. 8-13)}$$

$$\text{dBd} = \text{dBi} - 2.14 \qquad \text{(Eq. 9-1}$$

$$\text{dBi} = \text{dBd} + 2.14 \qquad \text{(Eq. 9-2}$$

$$\text{dB} = 10 \log \left(\frac{\text{P2}}{\text{P1}}\right) \qquad \text{(Eq. 9-3}$$

$$\text{Gain Ratio} = \frac{41{,}253}{\theta_E \times \theta_H} \qquad \text{(Eq. 9-4}$$

$$\text{Gain Ratio} = \frac{25{,}000}{\theta_E \times \theta_H} \qquad \text{(Eq. 9-5}$$

$$\text{Gain Ratio} = \frac{41{,}253}{\theta^2} \qquad \text{(Eq. 9-6}$$

$$\text{Beamwidth} = \sqrt{\frac{41{,}253}{\text{Gain Ratio}}} = \frac{203}{\sqrt{\text{Gain Ratio}}} \qquad \text{(Eq. 9-7}$$

Index

ARRL MEMBERS

This proof of purchase may be used as a $0.50 credit on your next ARRL purchase or renewal, 1 credit per member. Validate by entering your membership number — the first 7 digits on your *QST* label — below:

EXTRA CLASS
LM

PROOF OF
PURCHASE